The Philosophy of Online Manipulation

Are we being manipulated online? If so, is being manipulated by online technologies and algorithmic systems notably different from human forms of manipulation? And what is under threat exactly when people are manipulated online?

This volume provides philosophical and conceptual depth to debates in digital ethics about online manipulation. The contributions explore the ramifications of our increasingly consequential interactions with online technologies such as online recommender systems, social media, user-friendly design, microtargeting, default-settings, gamification, and real-time profiling. The authors in this volume address four broad and interconnected themes:

- What is the conceptual nature of online manipulation? And how, methodologically, should the concept be defined?
- Does online manipulation threaten autonomy, freedom, and meaning in life, and if so, how?
- What are the epistemic, affective, and political harms and risks associated with online manipulation?
- What are legal and regulatory perspectives on online manipulation?

The Philosophy of Online Manipulation brings these various considerations together to offer philosophically robust answers to critical questions concerning our online interactions with one another and with autonomous systems. It will be of interest to researchers and advanced students working in moral philosophy, digital ethics, philosophy of technology, and the ethics of manipulation.

Fleur Jongepier is Assistant Professor of Digital Ethics at the Radboud University Nijmegen, The Netherlands. She is currently working on a research project on the impact of algorithms on our capacity for autonomy and the ways in which algorithms are said to know us "better than we know ourselves". She is interested in feminist ethics, self and identity, moral pedagogy and is actively engaged in public philosophy.

Michael Klenk is Assistant Professor of Philosophy at Delft University of Technology, The Netherlands. His work is at the intersection of metaethics, moral psychology, and the philosophy of technology. He held a Niels Stensen Fellowship to study manipulation and social media and has published journal articles, book chapters, and outreach pieces on the nature and ethics of manipulation, specifically in the context of technology. He is the editor of *Higher-Order Evidence and Moral Epistemology* (Routledge, 2020) and the co-editor of *Philosophy in the Age of Science? Inquiries into Philosophical Progress, Method, and Societal Relevance* (2020).

The sophisticated way in which data-driven technologies are able to manipulate our thinking and actions raises fundamental ethical questions about – among other things – freedom, legitimacy, and integrity in our networked society. By bringing together philosophical discussions on manipulation, human–machine interaction, and digital ethics, this volume provides an in-depth and much-needed analysis of the key concepts and questions underpinning these challenges.

Esther Keymolen, *Tilburg University, The Netherlands*

Routledge Research in Applied Ethics

Self-Defense, Necessity, and Punishment
A Philosophical Analysis
Uwe Steinhoff

Ethics and Error in Medicine
Edited by Fritz Allhoff and Sandra L. Borden

Care Ethics and the Refugee Crisis
Emotions, Contestations, and Agency
Marcia Morgan

Corporate Responsibility and Political Philosophy
Exploring the Social Liberal Corporation
Kristian Høyer Toft

The Ethics of War and the Force of Law
A Modern Just War Theory
Uwe Steinhoff

Sexual Ethics in a Secular Age
Is There Still a Virtue of Chastity?
Edited by Eric J. Silverman

The Ethics of Virtual and Augmented Reality
Building Worlds
Erick Jose Ramirez

The Philosophy of Online Manipulation
Edited by Fleur Jongepier and Michael Klenk

For more information about this series, please visit: www.routledge.com/Routledge-Research-in-Applied-Ethics/book-series/RRAES

The Philosophy of Online Manipulation

Edited by Fleur Jongepier
and Michael Klenk

Routledge
Taylor & Francis Group
NEW YORK AND LONDON

First published 2022
by Routledge
605 Third Avenue, New York, NY 10158

and by Routledge
4 Park Square, Milton Park, Abingdon, Oxon, OX14 4RN

Routledge is an imprint of the Taylor & Francis Group, an informa business

Library of Congress Cataloging-in-Publication Data
A catalog record for this book has been requested

ISBN: 978-1-032-03001-2 (hbk)
ISBN: 978-1-032-07114-5 (pbk)
ISBN: 978-1-003-20542-5 (ebk)

DOI: 10.4324/9781003205425

Typeset in Sabon
by Apex CoVantage, LLC

Contents

Acknowledgements

The idea for this volume was born pre-pandemic during Michael's Niels Stensen Fellowship on the ethics of social media, which focused on questions about manipulation. This being before the pandemic, an international workshop was planned, with the usual invitations, trans-continental flights, paper presentations, and a dinner in mind. Things turned out differently, and we had to change tack. With a physical workshop out of sight, and budget for the workshop freed up, a plan for an open access volume began to emerge. Fleur worked on related topics during her VENI project on algorithms and autonomy, and we decided to team up. With Fleur aboard, we set out to plan the volume open access and, in addition, to take a risk and set up an online workshop series dedicated to the discussion, sharing, and improving early drafts of the individual contributions of this volume and the book as a whole. Together with our contributors and the (regular) participants of the workshop series, we got a chance not only to discuss new philosophical approaches, distinctions, and conceptual tools but also to experiment with creating and engaging in an online research community. It has been an intense and illuminating experience, and we hope that our contributors and participants found it as rewarding and stimulating as we did. What could have been "just" another philosophy workshop, fun and illuminating for a few, turned into an accessible workshop series of ten sessions, attended by a considerable number of people, and ultimately led to this book, available for everyone. It was an opportunity afforded by the changed circumstances of the pandemic. It took more time to organise, and we missed many of the more free-flowing and rewarding interactions of in-person meetings. So, at least as online philosophy is concerned, we come down with a measured judgement: the online world offers great opportunities for making philosophy accessible and open to everyone across the globe, yet we were also powerfully reminded of the value of real-life interactions and hope to meet many of our contributors and participants in the offline world in the future.

We are immensely grateful to the Niels Stensen Fellowship and the Dutch Organisation for Scientific Research (NWO) that made the open access publication of this volume possible. We hereby also want to thank everyone who contributed and participated for their patience with digital breakdowns, but most of all for their enthusiasm, their curiosity, their highly constructive feedback, and the many inspiring conversations about online manipulation.

1 Introduction and overview of chapters

Fleur Jongepier and Michael Klenk

Nor from mine own weak merits will I draw
The smallest fear or doubt of her revolt,
For she had wifi and chose me.

– Revised passage from *Othello*, Act 3, Scene 3

1 Modern-day Iago

Shakespeare's Othello depicts a paradigmatic case of manipulation: Iago is jealous of Othello's relationship with Desdemona and forges a deceitful plan to tear them apart by making Othello believe – falsely – that Desdemona is cheating on him. Amongst other things, he places a handkerchief in the luggage of one of Othello's close confidants that Othello gave as a gift to Desdemona. Upon finding the handkerchief, Othello falls for Iago's trap and believes that he was betrayed by Desdemona. Iago's plan succeeds: a clear case of interpersonal manipulation.[1]

Interpersonal manipulation can also happen online. A modern-day Iago may have arranged for Othello to find misleading but suggestive messages on Desdemona's social media account to achieve the same effect. Or he may have harnessed more sophisticated technological means to manipulate messages exchanged between Othello and Desdemona through their voice assistant or smart fridge. And perhaps, there are new forms of interpersonal manipulation that an online modern-day Iago could realise, for example moderating and influencing what people see online and which content they are exposed to. Manipulation is as old as the history of mankind. And yet there are important reasons to be especially concerned about manipulation taking place online, in particular the scale and the nature of online manipulation. First, the scale: what is perhaps most striking about the online world is our increased interaction with algorithms and (autonomous) machines. One editor of this volume, for instance, has screen time warning pop-ups installed but happily clicks *Ignore warning for today* in order to continue scrolling on Twitter and Instagram. The other editor deleted emails from their phone but simply keeps logging back in through the browser. Worldwide, people spend about two and a half hours on social media *every day*

DOI: 10.4324/9781003205425-1

(people in the Philippines winning – or losing – the match with a whopping three hours and 53 minutes).[2] Netflix has 72.9 million users on average, and YouTube almost 200 million, with 80% of US parents of children of ages 11 and below indicating that their kids watch YouTube.[3] Almost 40% of the US population uses voice assistants.[4] Most important of all, when it comes to scale, is the breaking down of the online/offline or real-life/digital life boundary, given that our lives are becoming increasingly immersed with (online) technologies.

Of course, the pervasiveness of technology in our daily lives and how technologically blended our lives have become is, as such, no reason to think manipulation must be everywhere, too. It is, however, a reason to be especially alert in light of the tremendous influence that technology seems to have on us. The modern-day Iago is not the CEO of Google or Alibaba per se; Iago may also be hiding in our smartwatch, our Wifi-controlled lights, our robot vacuum cleaner, our care and sex robots, our children's smart dolls, and our pets' remotely controlled food machine. So yes, manipulation has always been around, and we've known billboards and dubious salesmen for a long time. Right now, however, looking at how our interactions are shaped online, we appear to be dealing with salesmen on steroids and billboards that follow us around and that change depending on who's looking at them.

A second reason to be especially concerned with modern-day Iagos concerns the nature of online manipulation. Iago is a bad and cunning person, but at least we can understand, conceptually, his cunningness to some extent and have some sense, morally, how to evaluate his actions when his evil ways are brought to the surface. Human manipulation can be just as awful – perhaps even more awful – than technologically mediated manipulation, but we typically know, who manipulated us, and which moral–emotional responses would be (very roughly) appropriate.

All of this is very unclear when it comes to being manipulated by YouTube videos, voice assistants, personalised Google search results, CandyCrush, political parties-using-Facebook, and so on. It is often unclear that we are manipulated. Online manipulation is rarely "brought to the surface." Whereas in Othello there is Emilia who, in the end, uncovers Iago's manipulation, there are not many online equivalents of Emilia in the digital age. The question of "who" manipulated us is even harder to answer, if that question makes sense at all. And rather than disappointment or anger that many of us experience in light of human manipulation, the typical moral–emotional response when one is subject to online manipulation is either confusion, a feeling of powerlessness, or simply indifference or fatigue ("ah, another scandal"). The type of agency and intentionality (not) exhibited by algorithms and more advanced online machines is complex and unclear, making societal–philosophical questions about their manipulative potential all the more acute. This volume aims to address these and other questions about the conceptual and moral nature of online manipulation. Here, we will discuss the aim of this volume in some more detail and provide an overview of the chapters.

2 This volume

Behind the recent public and academic "techlash" seems to be the growing concern that the influence exerted on us by algorithmic systems or more advanced technological machines like robots can be distinctly *manipulative* and for that reason especially problematic. At the same time, the debate about online manipulation rests on philosophically vexed and, to some extent, underexplored territory. Philosophical attention to manipulation is luckily on the rise (see, for instance, Coons and Weber 2014), and scholars have begun to explore how manipulation differs (or not) from coercion, persuasion, nudging, and other forms of influence, as well as whether manipulation necessarily constitutes a moral wrong of some kind, and, if so, why. The existing literature is still relatively scarce, however, and when it comes to the literature on online manipulation, it is often simply stipulated or suggested, rather than argued for, that a certain technology or technological development is manipulative, and it is sometimes just assumed that *because* its manipulative it must therefore be morally problematic. However, manipulation might well in some cases be morally unproblematic or indeed desirable, so the inference from "manipulative" to "immoral" is not always evident. Also, various technologies, actions, or developments might turn out to be morally problematic not because they are manipulative but because they are coercive (say). Finally, it is not always clear whether some technological tool or online design would be immoral rather than merely (very) annoying for internet users.

All in all, many fundamental questions about both the nature of online manipulation and its normative status deserve more systematic attention. For instance, must online manipulation (always) involve "intentions" of some sort, and is such a thing as manipulation by a non-human agent possible? Is online manipulation necessarily opaque, or can one be manipulated online "out in the open"? As for questions in the normative domain, is online manipulation always morally wrong, and if so, why? Can online manipulation also be morally acceptable or even a morally good thing to do? Does being manipulated online threaten autonomy, and if so, what do we take autonomy to be?

This edited volume aims to fill a critical gap in current discussions regarding the conceptual nature and moral status of online manipulation. We aim to provide theoretical and normative depth and nuance to debates in digital ethics about the manipulative influence of algorithms and autonomous systems. Thereby, we aim not only to enrich "applied" debates about online manipulation by bringing in contemporary developments from the philosophical debate regarding manipulation but importantly to also enrich and sharpen the philosophical debate by putting existing theories to the test by applying them to online cases and contexts. Finally, we hope to make a methodological contribution by offering a type of applied philosophy that is solidly anchored in philosophical theory whilst strictly in the service of contributing to contemporary societal questions and challenges.

3 Overview of chapters

This volume is the first to explicitly address the philosophy of online manipulation. It contains 20 previously unpublished chapters and brings together leading international philosophers and several promising scholars at earlier stages in their careers. We sought to illuminate the questions surrounding online manipulation specifically from a perspective informed by moral and political philosophy. The chapters in this volume fall under the following four parts:

Part I: Conceptual and methodological questions
Part II: Threats to autonomy, freedom, and meaning in life
Part III: Epistemic, affective, and political harms and risks
Part IV: Legal and regulatory perspectives

Any ordering of contemporary contributions to novel philosophical and societal developments is bound to be artificial to some extent, and this volume is no different in that regard, as most authors cover more than one, and sometimes all, of the aforementioned broader themes. Still, it is possible to observe differences in emphasis and focus. For instance, contributions falling under the first heading are primarily concerned with the conceptual question of what manipulation is, how we should go about defining the notion, and how (if at all) online manipulation is different from offline manipulation. Chapters falling under the second heading are principally concerned with the moral dimension of manipulation, addressing the question of what, if anything, would make online manipulation immoral, and what exactly is at stake or threatened when a person is manipulated online, with a specific focus on threats to autonomy, freedom, and meaning in life. Contributors clustered under the third header consider possible threats to knowledge, control of our emotions, and political legitimacy. Finally, a separate heading is reserved for contributions that zoom in on a specific technology (such as real-time profiling) and then go on to ask how, for that technology, regulation is currently arranged and how it might be improved.

3.1 Part I: Conceptual and methodological questions

In the opening chapter, titled "Online manipulation: charting the field," we – the editors – present an overview of what we consider to be some of the core questions surrounding the nature and normative dimension of offline and online manipulation. Our aim is not to settle these questions once and for all but to provide an overview of the theoretical landscape so that the reader is in a better position to locate and appreciate what is at stake in the other chapters that follow. We touch upon some methodological and conceptual preliminaries and then give a brief overview of so-called outcome- and process-based accounts of manipulation, noting their advantages and disadvantages. In the second part of the chapter, we consider what we call "aggravating factors" that help explain the distinct problems raised by manipulative online technologies, such as personalisation and opacity.

In the opening chapter, we mention quite a number of philosophical controversies and nuances regarding the conceptual nature of manipulation. Indeed, many discussions about manipulation, online or offline, involve asking the question "Are these kinds of influence *actually* instances of manipulation?" However, in chapter 3, "How philosophy might contribute to the practical ethics of online manipulation," Anne Barnhill argues that asking that question might not be the most productive way for philosophy to contribute to the debate and that we should be careful not to get bogged down in philosophical definitions and demarcation issues. Instead, she suggests that when online influence is called "manipulative," we should try to figure out what kinds of *concerns* are being registered by calling it manipulative and then query whether influence of that particular form is problematic and why.

In Chapter 4, Massimiliano L. Cappuccio, Constantine Sandis, and Austin Wyatt turn to the very distinction between online and offline manipulation in their chapter "Online manipulation and agential risk." They ask how manipulation enabled by AI-based technology that mediates our interactions online (such as recommender systems on social media) differs from other forms of manipulation. The authors draw on developments in communication science to suggest that different technologies enable different "communication paradigms" which, in turn, engender different forms of manipulation. They then turn to what they refer to as "engagement-maximization-based online manipulation" and argue that this is best thought of as an emergent phenomenon, not traceable to the explicit or implicit intentions of any individual agent but more akin to collective action.

The next two chapters address the very possibility of speaking sensibly about online manipulation or manipulation by machines. In Chapter 5, titled "Manipulative machines," Jessica Pepp, Rachel Sterken, Matthew McKeever, and Eliot Michaelson ask how the contemporary concept of manipulation could capture current and future instances of manipulation by machines. They provide a clear overview of the different theoretical positions one could take and introduce helpful insights from the conceptual engineering literature. They suggest that one might use the concept of manipulation as if machines could manipulate us, even if they don't literally do so. And they present an ameliorative approach which involves asking which *purpose* is served by having a certain concept and also allowing to change our concept of manipulation in order to make better sense of, and make room for, genuine machine manipulation.

In Chapter 6 "Manipulation, injustice, and technology," Michael Klenk defends a specific proposal about manipulation by technology. Understanding technology quite broadly, he shows that it has considerable effects on us independently of whether it is (artificially-) intelligent, autonomous, or embodied. He argues that being manipulated should be understood differently than manipulating. On his account, a manipulated mental state is one that is explained in the relevant way by an injustice. Drawing on considerations about epistemic injustice and the affordances of technology, he argues that technology can contribute to injustices that explain our mental states in relevant ways. Therefore, we can be manipulated by technology, independently of whether technology has, for example, intention.

3.2 Part II: Threats to autonomy, freedom, and meaning in life

When it comes to making online choices, an oft-heard concern is that these choices are manipulated and therefore not autonomous. In Chapter 7 "Commercial online choice architecture: when roads are paved with bad intentions," Thomas Nys and Bart Engelen turn to the question of what exactly is manipulative about commercial online choice architectures (COCAs) and in what way they threaten personal autonomy. They argue that considering the intentions of the manipulator is key, both conceptually and normatively speaking. They end their chapter by pointing out that even in cases where the intentions of internet users and COCA designers happen to align, there is still cause for concern as the latter are typically completely indifferent towards the aims of the former.

Fleur Jongepier and Jan Willem Wieland pick up the thread relating to indifference in Chapter 8 "Microtargeting people as a mere means." In this chapter, Jongepier and Wieland focus on political microtargeting and propose that what is wrong about employing such techniques is that they involve treating people as a mere means, which they argue involves genuinely caring about people's consent to be used in certain ways. They go on to explain what "caring about consent" comes down to in digital contexts and argue that political microtargeting typically, though not necessarily, involves treating people as a mere means due to a lack of care about people's consent to be used as a means towards the microtargeter's ends.

Next, Marianna Capasso argues in Chapter 9 "Manipulation as digital invasion: a neo-republican approach" that neo-republicanism can provide conceptual and normative tools to analyse and address the problem of manipulation in relation to digital nudges. The neo-republican approach offers a promising account of the connection between digital choice architecture and human freedom given its emphasis on social and political relations as well as collective and shared responsibility. Capasso individuates specific criteria to assess when digital nudges can amount to dominating manipulative interferences or "invasions." She argues that the main worry about digital nudges is not (just) the fact that they are typically not transparent but that it involves alien control and a lack of democratic means of empowerment, communication, and contestation.

In Chapter 10 "Gamification, manipulation, and domination," Moti Gorin remains within the Republican framework and focuses specifically on gamification, that is, the attempt to turn an activity into a game, to make it fun, engaging, and motivating. One of the examples of online gamification discussed by Gorin is Twitter, whose system of likes, retweets, and so on can be seen as introducing the so-called game reasons into human discourse, where such reasons would not ordinarily exist. Gamification turns out to be manipulation on Gorin's account because it is a kind of influence that makes people do something for game reasons rather than any other reasons that may ordinarily exist. Based on this analysis, Gorin presents an analysis of the wrong-making features of manipulation inspired by Republican worries about domination and offers an account of domination which he calls "interactive domination" that differs from the structural domination articulated by republican theories.

W. Jared Parmer likewise focuses on gamification in Chapter 11 "Manipulative design through gamification." Parmer focuses on gamification as it offers a useful starting point for understanding manipulative design more generally. Gamification is the implementation of inducements to 'striving play' for the sake of purposes beyond those typically found in games, such as to learn a skill or to develop certain habits. According to Parmer, gamification becomes manipulative when it involves deception, on the part of the manipulator, about her purposes. Parmer points out that one of the dangers about manipulative design is that it stands in the way of making our lives more meaningful because it can make it harder to work out and act on what we care about..

The relation of manipulation and meaning in life brought out by Parmer nicely connects with Chapter 12, "Technological manipulation and threats to meaning in life," by Sven Nyholm. Nyholm first offers a helpful overview of the different positions that one may take on the question of whether technology can manipulate humans. He then turns to the more general question regarding the relation of manipulation and meaning in life and provides an overview of different constituents or contributors to a meaningful life. Nyholm then argues that technological manipulation threatens some or all of these factors, thus endangering the opportunities of those interacting with the technology to enjoy meaning in life. Nyholm's chapter contributes to a better understanding of the normative dimension of manipulation as it suggests that it is a type of influence the effects of which are particularly harmful.

Geoff Keeling and Christopher Burr then consider the question of what distinguishes morally permissible from morally impermissible behavioural influencing strategies by software agents. They argue that morally impermissible instances of behavioural influence by software agents undermine the "mental integrity" of human users. In other words, such strategies diminish people's capacity for authentic decision-making. Such strategies, they argue, are morally permissible only if behavioural influence by software agents affords due respect to the mental integrity of the user.

3.3 Part III: Epistemic, affective, and political harms and risks

Within the focus on normative and evaluative aspects of manipulation, we then shift perspective to consider which epistemic, affective, and political harms and risks may be associated with manipulation.

In Chapter 14 "Is there a duty to disclose epistemic risk?" Hanna Kiri Gunn focuses on personalisation of online platforms and in particular on the epistemic risks involved. In many online spaces, she argues, we risk undermining our ability to be in reasonable control of our epistemic capacities, for instance through the personalisation of search engine results or being exploited by bots to spread fake news or emotional persuasion. Gunn argues that internet users are placed at risk of social-epistemic harm without their informed consent and that there is a moral duty to disclose the social-epistemic risks of using online services to prospective users. She closes the chapter by zooming in on moral responsibility and the many hands problem.

Lukas Schwengerer is likewise concerned with the epistemic dimension in Chapter 15 "Promoting vices: designing the web for manipulation." He is primarily concerned with normative and evaluative questions surrounding the problem with manipulation, which he approaches through a discussion of user-friendly design. Schwengerer takes an innovative virtue epistemic perspective to suggest that user-friendly design promotes an "overly trusting attitude" towards the information provided by the website. Schwengerer argues that artefacts like websites can warrant trust to a given degree. Trusting them beyond that degree "destroys the virtue of intellectual carefulness." When we lack that virtue, we are easier targets for manipulation because we might more readily and less critically believe, feel, or desire what the website's creator wants us to believe, feel, or desire. The virtue epistemic perspective makes it easy to see why that would be bad, and it is interesting in the context of our volume for making explicit the link between epistemic vices and potential for manipulation.

Next up are two chapters that deal, in different ways, with the link between online manipulation and emotions. Nathan Wildman, Natascha Rietdijk, and Alfred Archer focus on "Affective online manipulation" or the online influence on people's affective states. They begin by considering four key questions to distinguish different types of manipulation, such as whether it is active or passive, done intentionally or unintentionally, based on a top-down or bottom-up mechanism, and finally whether the aim is primarily to influence affective states or, ultimately, behaviour. Their next step is to consider why any of this would constitute manipulation.

They consider three prominent accounts and suggest, in a pragmatic vein, that each of them can account for the manipulativeness of online affective influence, albeit in different ways. The authors argue that in extreme cases, online affective manipulation constitutes a distinct type of injustice, namely "affective powerlessness," in which someone (or something) wields a large amount of power over the emotional states of the user, rendering the user affectively powerless.

The focus on the affective component of manipulation is continued in Chapter 17, "Manipulation and the affective realm of social media," by Alexander Fischer. He focuses on both the nature of manipulation and its moral evaluation. Fischer argues that manipulation manifests itself in changing the victim's evaluation of a given end as pleasurable or displeasurable. Hence, unlike coercion, which may force a given end upon the victim, manipulation merely moderates the attractiveness of an end and thus its likelihood to be chosen. In the second part of the chapter, Fischer turns to social media, and he gives several examples and cases to illustrate how social media impacts our affective states, thus making it a powerful tool for manipulation.

In Chapter 18, "Social media, emergent manipulation, and political legitimacy," Adam Pham, Alan Rubel, and Clinton Castro begin by observing that political advertising and disinformation campaigns on social media can have a significant effect on democratic politics. Pham, Rubel, and Castro point out that often the moral concerns with these activities are reduced to the effects they have on individuals, such as the fact that their autonomy is undermined. The authors instead suggest, by introducing and analysing the concept of "emergent manipulation," that the presence of manipulation in electoral politics threatens the legitimacy of the elections themselves, and thus that the wrongness of such activities is to be found at the group level.

3.4 Part IV: Legal and regulatory perspectives

Kalle Grill's chapter, "Regulating online defaults," concerns the normative aspects of manipulation, which he explores through a discussion of online defaults and how they may be regulated. A default option is an option from which one can only opt out by taking an action. Grill shows how online defaults – which have become inevitable features of online environments – can distract, misinform, harm, and eventually manipulate people. Grill's second main contribution is to consider principles for the regulation of defaults, including that they should be set to favour non-consumption, that data collection is minimised, and "that information provided by default is true, or at least not demonstrably false or against expert consensus."

In the final chapter of the volume, Jiahong Chen and Lucas Miotto discuss the morality of real-time profiling, that is, the collection of information about an individual's present status to generate a profile in an attempt to influence the individual's actions in the immediate future based on that profile. Zooming in on real-time profiling, they argue, allows us to see what is morally problematic with manipulation more generally. The authors argue that real-time profiling is morally wrong because it involves "psychological hijacking" and because, by making the user more vulnerable, it makes them more likely to be wronged in other ways too. The authors then turn to regulatory measures and discuss the implications for consumer protection law and data protection law and their limitations, arguing that a more targeted regulatory approach is needed to effectively address the unique challenges of real-time profiling.

4 General observations and concluding remarks

The contributions in this volume span across a wide spectrum, not just in terms of how conceptually or normatively oriented they are but also in terms of the technologies the authors focus on and the methodologies they (explicitly or implicitly) use. It should not be surprising this volume as a whole would not give us the true theory of online manipulation and why it is or isn't morally problematic. More than anything else, the chapters taken together give the reader a clear view of the state of the art when it comes to the philosophy of online manipulation. This view is bound to be kaleidoscopic because it includes philosophers who are very much concerned with getting the philosophical definition of "manipulation" right before moving on to the "online" adjective (whereas others get right to it); philosophers who are very much concerned with threats to individual persons (and others much more with threats to the collective, social, or political order); and so on. In other words, this volume will not give the reader "the" approach to studying online manipulation. However, it will, we hope, give a rich, kaleidoscopic view of many of the concepts, methodologies, moral concerns, and applications that are at stake in this debate that has only started to unfold.

When we consider all the chapters taken together, a few observations can be made. First, it is interesting to see how many chapters in this volume do not just "employ" concepts and theories from the philosophical debate about (offline) manipulation but really – as we, as editors, hoped – also challenge and test these theories by applying them to the online sphere. Second, it is interesting to see that many (though not all) contributors in the volume do not have a detailed and settled position on what they take manipulation to be, what exactly sets it apart from persuasion or coercion, whether it is necessarily opaque or intentional, why it's wrong, and so on. This can be indicative of the fact that both the philosophical debate about manipulation and the debate about online manipulation are still very much in development and there is as yet no clear "map" on which to position oneself. Also, it might be indicative of an (implicit) pragmatic methodological approach (to be discussed in the next chapter), namely that it is possible to have illuminating discussions about various aspects of online manipulation without necessarily providing a fine-grained definition of manipulation first.

Third, across all chapters the terms "online" and "technology" really stand in for a wide range of phenomena. We have discussions that understand "online" or "tech" in terms of highly general design approaches such as user-friendly design, default-settings, or gamification which are applicable to all technological designs. Others discuss more specific affordances of recent online and algorithmic technology such as social media, real-time profiling, and augmented many-to-many communication. Each individual contribution makes clear what the relevant factors are that may be seen as aggravating the problem of manipulation.

The fourth and final observation. In the original call for chapters for this volume, we were operating with a distinct "conceptual" and a "normative" part for the prospective book. As it turns out, this two-part ordering of the book did not make much sense in the end. Even though a couple of authors are clearly more concerned with either the conceptual side of online manipulation or with the normative side, by far most of the contributors really have an equal interest in both. In other words, we could say that to answer normative questions about why certain forms of online manipulation would be problematic in some way, one inevitably needs to enter some theoretical terrain (if only briefly). The converse is also true: to make progress on the question of what online manipulation is, conceptually speaking, it is hard if not impossible to say something about the instances in which it is (or appears to be) morally problematic. This, on its turn, may tell us something about whether or not "online manipulation" is a so-called thick concept (which is something discussed in the subsequent chapter).

It is important to point out some of the limits of this volume. Though this book, with its many chapters and diverse approaches, is very comprehensive, many other questions remain to be addressed and answered. For instance, this volume is heavy on the (moral and theoretical) theory and relatively light on the "what now?" question. Two chapters explicitly address regulation and policy issues, and many other authors also briefly discuss what the practical consequences of their account might be. Still, the emphasis is more on *understanding* online manipulation and applying new and existing philosophical resources to do so. Second, even though some authors make use of material from other disciplines (law, social sciences, and so on), this is not an interdisciplinary volume on online manipulation. It is a philosophy-based book on online manipulation, which has the aim of making certain developments in philosophical debates relevant to (as well as testing them against) developments and technologies in the online world. Despite it not being an interdisciplinary volume on the subject, we of course do very much hope that it will – by bringing in a lot of (sometimes neglected) philosophy – be of use to scholars from other disciplines working on online manipulation and related topics. Taken together, if there were going to be a second volume or follow-up to this book, it would take an interdisciplinary approach from the get-go, and it would be heavier on the "what now?" side.

We hope that this volume will help us and others to continue the discussion and motivate and inspire further work on this societally acute and philosophically intriguing topic.

Notes

1. Both authors contributed equally to this chapter. Fleur Jongepier's work was supported by a VENI grant. Michael Klenk's work was supported by the European Research Council under the Horizon 2020 programme under grant agreement 788321.
2. https://techjury.net/blog/time-spent-on-social-media/
3. https://techjury.net/blog/netflix-statistics/
4. www.emarketer.com/content/voice-assistant-and-smart-speaker-users-2020

Part I

Conceptual and methodological questions

2 Online manipulation

Charting the field

Fleur Jongepier and Michael Klenk

1 Introduction

When we introduced the main research questions and the contributions of this volume in the previous chapter, we touched upon two broad and fundamental topics. First, what is manipulation? Second, is online manipulation simply "regular" manipulation gone online or a new phenomenon? In this chapter, we tackle both questions and chart the overall terrain of online manipulation, critically considering existing and new possible answers to these questions. Our aim is to provide a conceptual map to the reader and allow them to locate the contributions in this volume on it.[1]

2 Three preliminary questions

In this section, we introduce and discuss three important preliminary questions concerning the study of manipulation. First, what is a good method to do study (online) manipulation and how can we gauge its success? We start with this question because it concerns fairly general points about philosophical methodology that are important to studying online manipulation. It has been pointed out that "manipulation" refers to a number of different phenomena, not all of which overlap in their interesting features (Cave 2007), which puts pressure on the question of how we should go about analysing manipulation, if such a thing can even be done. The subsequent two questions involve asking whether "manipulation" is a *thick* concept (2.2) and whether manipulation is necessarily intentional (2.3).

Though our discussion is critical – mentioning problems and worries where applicable – our aim in this chapter is not to argue for any particular answer to any of these questions. Rather, we want to chart the field and bring to the surface not just which positions are out there but also which challenges or worries one is likely to face when adopting them.

2.1 Method

How should we go about the study of manipulation? More specifically, is conceptual analysis a promising method for the study of manipulation?

DOI: 10.4324/9781003205425-3

Very roughly, conceptual analysis seeks to decompose a concept into its constituent parts.[2] A common and influential interpretation of that method has been to provide an explicit intension that is measured against the intuitive extension (the set of all things to which the concept applies) of a given concept (Queloz 2021, 23). This would lead to specifications of the necessary and sufficient conditions for the correct employment of a concept. We would have mastered a concept at the point where we can say whether the concept applies in any situation, and the criterion for application (e.g. "x is a G") is the concept's intension (Queloz 2021, 25).

Most of the existing philosophical work on manipulation proceeds by conceptual analysis and, therefore, it is worthwhile to enquire about its pedigree (cf. Coons and Weber 2014, 6).[3] The method of cases exemplifies this strand of conceptual analysis, whereby a proposed set of necessary and sufficient conditions is tested by considering (hypothetical) cases to see if the proposed conditions correctly qualify something as manipulation.

There are several reasons to be sceptical about conceptual analysis. Some of these reasons are perfectly general in that they pertain to the viability of conceptual analysis across the board. Conceptual analysis as understood here relies on assumptions about the nature of concepts that come from the classical theory of concepts. According to the classical theory of concepts, a concept like manipulation has a definitional structure that is composed of simpler concepts that express necessary and sufficient conditions for falling under the concept or qualifying as manipulation. The truth of the classical theory of concepts is presumed once we embark on conceptual analysis as interpreted here. But if the classical theory of concepts suffers from problems, then conceptual analysis – as understood here – would also be a method of doubtful pedigree. Existing worries about the classical theory of concepts that carryover to the study of manipulation for instance includes the worry regarding the very existence of conceptual essences that conceptual analysis aims to reveal.

A second challenge about conceptual analysis and studying manipulation more generally comes from experimental philosophy. There are serious questions about the reliability of our intuitions that arguably are the core "data" for conceptual analysis. In particular, there is a question about the legitimacy of claims to universality derived from the conceptual analyses pondered in philosophy. Conceptual analysis is supposed to uncover *the* meaning of a concept by drawing on "our" intuitions as evidence (cf. Climenhaga 2018). But who is the "we" here? Intuitions may differ temporarily and geographically, and the analyses on offer may reflect the highly idiosyncratic intuitions of philosophers from WEIRD – Western, educated, industrialized, rich, and democratic – societies (cf. Henrich, Heine, and Norenzayan 2010) and thus have limited scope. Experimental philosophy, and psychological research on manipulation more specifically, may alleviate some of these worries by systematically eliciting a more diverse set of intuitions (Knobe and Nichols 2008).[4] At the same time, however, such experimental approaches need to answer questions about the method's validity and reliability, especially if manipulation turns out to be a technical concept that requires some expertise to grasp (cf. Pölzler 2020).[5] For example, it is not clear to what extent we can rely on survey studies that prompt the intuitions of laypeople about manipulation to make inferences about the nature and value of manipulation.

There is also a challenge more specific to the study of manipulation as pointed out by Coons and Weber (2014). They wonder whether the concept of manipulation – quite independently of general worries about concepts – lacks core features that unify all cases of manipulation. Some scholars go as far as suggesting that manipulation lacks core cases because it is "too varied" (Baron 2003, 37) and some thus proclaim the attempt at a conceptual analysis is a "fruitless endeavour" (Kligman and Culver 1992, 175). We do maintain that there are core cases of manipulation (such as the case of Othello discussed in the Introduction), but we remain open as to whether all of them share a set of necessary and sufficient conditions. The concept of manipulation may exhibit what Alston (1967, 220) calls "combinatorial vagueness", which is present in cases where

> [W]e have a variety of conditions, all of which have something to do with the application of the term, yet are not able to make any sharp discriminations between those combinations of conditions which are, and those which are not, sufficient and/or necessary for application.
>
> (cited in Ackerman 1995, 337)

This is a relevant suggestion because there are several conditions often associated with manipulation (which we discuss in more detail here), and yet it is unclear how many or which of them are strictly necessary and sufficient for manipulation (cf. Coons and Weber 2014, 7).[6]

The overall worry here is that a concept like manipulation may simply evade analysis (even if the classical theory of concepts is true), just like the concepts "disability" in law or "species" in science, or indeed concepts like "love" or "consciousness". Concepts that allow for borderline cases may evade successful discovery of necessary conditions. The attempt to boil them down to their highest common factor by conceptual analysis may be the wrong approach to take. There will be counterexamples to almost any interesting intension, as any feature that is not strictly a necessary condition will eventually fall prey to counterexamples. This may leave us, at best, with an analysis that is too thin to be interesting and informative (cf. Queloz 2021, 25).

Arguably, the study of manipulation does not stand or fall with the propensity of the concept "manipulation" to bend to complete analysis in terms of necessary and sufficient conditions. Manipulation, though perhaps vague, varied, and beset with borderline cases, may yet be unified by Wittgensteinian family resemblance, that is, not a set of shared properties but a resemblance to paradigm cases of manipulation. Borderline cases would be those where the resemblance is not clear or not strong enough (cf. Coons and Weber 2014, 6). Assuming that there are some paradigm cases, and many grey areas, we can still usefully study manipulation.[7] For instance, it would be interesting to say just what the paradigms have in common and how they unify the other cases of manipulation. And even if there are no paradigms at all, there may be a focal core of the concept that we can study. To illustrate, with respect to the complex concept of an "epiphany", Sophie Grace Chappell helpfully describes the notion of a focal-case concept in the following passage, here replaced by the notion of "manipulation" (Chappell 2019, 97):

> There are clear and central cases of [manipulation]. . . . But there are also less clear and less central cases, which we might still want to call [cases of manipulation]; or there again, might not. Nothing much turns on where *exactly* we draw the boundaries around the proper use of the term "[manipulation]". The central territory of the concept is not threatened by minor demarcation disputes about its borders. There are certainly grey areas, and they certainly have their interest. There are equally certainly *non*-grey areas: for instance, the black ones and the white ones. . . . True, there are no non-stipulative necessary and sufficient conditions for something's being an [instance of manipulation]. . . . There are no non-stipulative necessary and sufficient conditions for something's being a mountain, either, and the category of the mountainous typically fades out around its edges into literally small-scale phenomena. That does not stop the geologist from studying mountains, nor the alpinist from climbing them.

The view that manipulation might exhibit some kind of vagueness, admit borderline cases, and lack a clear conceptual core would also have a noteworthy moral implication, for it may well make moral evaluations of manipulation more difficult. If there is no necessary condition common to all cases of manipulation, there cannot be the *same* moral reason against all cases of manipulation because there is no necessary element shared by all manipulative acts. More sceptical approaches about finding any unity, however, may also be positive, as many authors in this volume illustrate, as it may also help our understanding as to why some but not all cases of manipulation are morally problematic. Manipulation may be anything that resembles doing x, y, or z and so we might investigate the moral status of x, y, and z and find differing verdicts (sulking to get your way is always bad, but comforting your friend is ok, though both are, arguably, manipulative). Sometimes, it can be more useful to get a better view on the overall ballpark, as it were, even if the ballpark has a few items that shouldn't be there than having the clearest view on one item in it.

Of course, various methodological approaches will ideally be operative in any concept-heavy debate such as the (online) manipulation debate; and indeed, this volume is itself an illustration of methodological diversity. The central aim and conclusion following from this brief methodological discussion therefore is not that scholars try and work towards methodological consensus and agree on a shared and unified methodology within the online manipulation debate. Rather it is to make implicit methodologies *explicit* so as to learn from their differences and respective strengths and weaknesses and to find ways for various methodological approaches to be complementary and fruitfully run parallel, even if they are methodologically at odds.

2.2 Thickness

We turn now to the question of whether the concept picked out by the word "manipulation" is a thick or moralized concept and of whether and how manipulation depends on intentions (2.3).

Thick normative or thick evaluative concepts have both a significant degree of descriptive content and are normatively loaded. The concept "kindness", for example, may denote descriptive qualities like being self-less, helpful, and caring towards other people. At the same time, characterizing someone as kind typically involves expressing a pro-attitude towards the person or their behaviour and thus an evaluative statement as well. Being kind is being caring towards others, *and that is a good thing*. If "manipulation" is a thick concept, then it also has a significant degree of descriptive content as well as being normatively loaded. It would not only be a particular type of influence (assuming that this is what manipulation is) but it would be a particularly *good* or *bad* – and not normatively *neutral* – type of influence.

Another way to express the thought that manipulation is a thick concept is to say that manipulation may have a normative or evaluative status as a conceptual matter. Ackerman (1995), for instance, notes that several of the features commonly associated with manipulation – such as deceptiveness or using others for one's own benefit – are prima facie immoral. Grasping the concept would thus involve grasping a particular normative or evaluative status. Just like grasping kindness is to understand that being kind is *good*, grasping manipulation may be to understand that it is *bad*. If that were the case, then any analysis of manipulation would have to involve an account of its descriptive content as well as an account of its normative or evaluative content. The analysis need not involve two separate steps, of course. For example, if manipulation is analysed as deceptive influence, then it has both a descriptive component (influence of some sort) and a normative component already built into the concept of deception.[8] Importantly, whether or not manipulation is a thick or moralized concept is largely independent of the question of whether it can sometimes be permissible to manipulate. Manipulation may be, say, morally bad as a conceptual matter. But one might still maintain from a moral philosophical standpoint that it is merely a pro tanto wrong, which can be outweighed by other factors (e.g. beneficial consequences) (Baron 2003).[9]

Numerous considerations challenge the idea that a moral dimension is part of the concept of manipulation. As a first approximation of that point, consider that the word "manipulation" – and thus presumably the concept expressed by the word – is also appropriately used in ways that are clearly morally neutral. We speak of manipulating inanimate objects like sticks and stones, and we manipulate research subjects in experiments that are cleared by the research ethics board. Thus, there are instances of manipulation that do not appear to carry a specific normative or evaluative judgement with them. To adopt the "thick" reading thus involves explaining why and how manipulation within scientific studies is morally wrong.

The word "manipulation" may express different concepts, and defenders of the thickness of manipulation may claim that there is a distinct thick concept of interpersonal manipulation after all. Accordingly, how we use the word manipulation and the corresponding concepts in cases that do not relate to interpersonal interaction may be beside the point. Still, we can find examples where manipulation of persons is referred to in a morally neutral or even laudatory way. Allan Wood (2014) gives the example of a politician who silences a heckler at a political rally through skillful manipulation, rather than resorting to brute force by calling for security to remove the heckler. We might applaud and praise the politician for this action. It also seems that artists who seek to create a certain effect in us, or politicians who aim for structural reforms, may sometimes do so by means of manipulation and still be applauded for it. Especially the case of the artist may prompt us to consider that manipulation may sometimes be not even pro tanto wrong. Hence, there might be examples of manipulation being appropriately used in a normatively and evaluatively neutral or even positive way. This speaks against the thickness of manipulation, unless we can reasonably maintain that there is at least some pro tanto wrongness associated with manipulation in all of these cases, or if one works out why contrary to appearances this is morally wrong overall and not only in a pro tanto way.[10]

There is also a more general consideration that speaks against the thickness of manipulation. Allan Wood (2014) points out how we use manipulation in the course of moral explanation. For example, when someone enquires what exactly is morally problematic about (aspects of) social media, some may offer as an explanation that social media can be manipulative. Such an explanation would seem entirely reasonable and informative. However, if manipulation is a bad or immoral type of influence as a conceptual matter, then that explanation would lose some of its force. There will be descriptive information conceived in virtue of the descriptive content of the concept of information, but it will not be illuminating in a normative sense because the negative evaluation would be a matter of conceptual course. An additional normative explanation would be superfluous.[11]

A final methodological possibility to consider, which draws on the pragmatic methodology discussed in the previous section, is to grant what might be the main worry with respect to manipulation as a thick concept, namely that sometimes manipulation is harmless or even good and so to allow that manipulation isn't *always* bad or immoral and certainly not *necessarily* so, but then to go on and say: but it is, most of the time. The strategy is thus to agree that, *strictly speaking*, manipulation isn't a thick concept but then to suggest perhaps we should not talk so strictly. This pragmatic approach won't satisfy all philosophers, perhaps, but it might help the online manipulation debate move forward.

2.3 *Intentionality*

A third preliminary question concerns the relation of manipulation and intentions. Manipulation is almost always portrayed as requiring intentionality on the part of the manipulator. Marcia Baron gives the insightful example of the following apology, which seems strange: "I am so sorry that I manipulated you [treated you manipulatively]. I didn't mean to; I didn't realize I was manipulating you, and I never would have acted as I did had I known" (Baron 2014, 102). The reason this apology is strange, Baron suggests, is because it suggests manipulation can be unintentional. Instead, she and many others claim that manipulators must be capable of having or forming intentions and acting intentionally. Thus, at least on the standard conception of agency, manipulators must be *agents*. We can call this the *general intentionality requirement* for manipulation.

The general intentionality requirement is extensionally plausible because typical cases of manipulation indeed involve agents who perform manipulative actions. Moreover, it looks like manipulation may always be a reason for blame or praise. Since the latter is often thought to be applicable only in cases where we deal with subjects capable of forming intentions, these normative practices related to manipulation support the general intentionality requirement for manipulation.

The general intentionality requirement is particularly interesting in the context of this volume because we will be looking at the relation of manipulation and online technology. If technology, whatever it is, cannot be intentional, then any contribution that technology makes to a manipulative act may seem at best purely instrumental to a real (i.e., human) agent. A manipulative act, perpetuated by an individual or group agent, may turn out to be more effective, more consequential, or, as we dub it in Section 5, "aggravated" in some sense because of the use of technological artefacts. For example, real-time profiling on the web may allow manipulators to wield more powerful influences. We could on such a view allow that technology has a meaningful influence on the agent's choice and behaviour (Klenk 2020), which may change our normative assessment of the situation and the warranted political or legal repercussions (e.g. by partly excusing manipulators). But given the general intentionality requirement, the contribution of non-intentional technology should not change our assessment of the situation as manipulative.

Apart from broader intentionality requirements, some theorists further advocate a *specific intentionality requirement* for manipulation. Rather than requiring intentionality in the aforementioned, general sense, manipulative action may also require intentions with a particular content (Noggle 1996; Gorin 2014b). These intentions, in turn, could be either topical in the sense that they must be intentions to manipulate or more generally about something else. Perhaps, a manipulator must *intend to x*, where *x* is whatever set of necessary conditions manipulation might have. Several scholars have suggested that there is such a specific intentionality requirement for manipulation. Robert Noggle, for example, argues that manipulators *intend* to have their victims violate some norm that regulates belief, desire, or emotion (1996). Others think this condition is too strong, and that the act of manipulation requires only the fact that people "could have done otherwise", not that they had the explicit intention to violate specific norms. Some theorists, like Kate Manne (2014), go very light on the intentions and instead argue that there can even be something like unwitting manipulation.

The discussion of the intentionality requirements for manipulation nicely leads to a discussion of the conditions of manipulation which we outline in the next section that deals with the demarcating factors of (online) manipulation. Both the general and specific intentionality requirements may be bona fide necessary conditions for manipulation. Still, we do not discuss them as demarcating factors for two reasons. First, as will become apparent, the search for a plausible account of manipulation is often the search for conditions that distinguish manipulation from coercion and persuasion. The intentionality requirement would presumably cut across this discussion. That is, whatever we say about intentionality requirements for manipulation will presumably apply to coercion as well.[12]

3 Manipulation and the search for demarcating factors

In this section, we will introduce and review recent analyses of manipulation. We propose to understand recent work on manipulation as the search for *descriptive demarcating factors* that distinguish manipulative from other types of interpersonal influence.

3.1 The demarcation problem for manipulation

The "demarcation problem" for manipulation is the problem of giving an account of manipulation that demarcates it from neighbouring forms of social influence such as persuasion and coercion (cf. Klenk 2021b). The demarcation problem thus prompts us to say *how* manipulative influence can be described as a distinct form of social influence, with particular wrong-making features.

Like coercion and persuasion, manipulation is a kind of interpersonal or social *influence* (Coons and Weber 2014, 8). It is widely held to be characteristically distinct from coercion and persuasion in kind or degree (cf. Faden and Beauchamp 1986). But how, precisely?

We should first note the close proximity of manipulation and the typical effects of coercion in terms of autonomy loss and blame-related practices. Wood (2014), for instance, suggests that manipulation is a type of influence on a continuum with coercion, with the latter being more heavy-handed than manipulation. Manipulation "influences choice without quite removing it" (Wood 2014, 26). Similarly, Baron (2003, 42) suggests that manipulation may become so strong so as to be indistinguishable from coercion at some point.[13] Greenspan (2003) suggests that manipulation "seems to have a foot in both the usual categories of intentional interference's in another agent's autonomy, coercion and deception" but is unlike both. Unlike being coerced, being manipulated supposedly never entails being a fully passive victim or instrument. Some autonomy is retained. Likewise, Alm (2015, 256) suggests that manipulatees have "whatever type of control is needed for responsibility". Hence, the manipulated person still *does* something *voluntarily* (Coons and Weber 2014, 8). All the same, manipulation is often, though perhaps not always, seen as antithetical to autonomy (we discuss autonomy violation as demarcating factor later) and some suggest that manipulation implies autonomy loss (Susser, Roessler, and Nissenbaum 2019a, 2019b).[14]

This debate about the autonomy- and blame-related *effects* of manipulation is partly informed by the debate about incompatibilism, and many philosophers (Kane 1996; Pereboom 2001) suggest that an agent in "manipulation cases" is not free, though he or she acts on her own volition (Sripada 2012).[15] The examples of manipulation discussed in that debate are usually much crasser (think of neurological, deterministic interference with people's choices) than the ordinary cases of manipulation that we are concerned with in this volume. The debate is illustrative nonetheless as it suggests how manipulation, understood as detrimental to freedom (of choice), need not undermine volition or autonomy.

Nudging helpfully illustrates the proximity of coercion and manipulation. Nudges influence choices without removing them. Their apparent non-coercive influence is why some consider nudges as morally unproblematic (Thaler & Sunstein, 2008). At the same time, it appears to others that nudges are still a problematic way to influence people, partly because they seem to structure choices in worrisome ways, some of which may lower or hamper autonomy (e.g. Levy 2019). Some types of nudging may thus appear as a paradigmatic way to manipulate people: arguably, they do not remove autonomy, but they may hamper it by structuring choices and thus guiding people's decisions (cf. Sunstein 2016a). Both manipulation and nudging are typically described as being forms of non-coercive influence (Schmidt and Engelen 2020).

However, one must say more than that, because the negative definition "influence that is not coercive" is not very illuminating. One reason is that this strategy relies on a clear account of the notions of coercion and persuasion to begin with. However, neither coercion nor persuasion is very well understood as a type of influence. There is obviously a tremendous amount of work on coercion, but a lot of it concentrates on characterizing coerced actions and, in particular, their effect on blameworthiness and accountability.[16] Another reason is that we need to find a set of conditions that individually or jointly applies to manipulation but not to coercion or persuasion to solve the demarcation problem for manipulation.

To illustrate the problem, consider the view that, like coercion, manipulation removes, nullifies, or threatens autonomy and, unlike coercion, it operates covertly (e.g. Susser, Roessler, and Nissenbaum 2019b). The suggestion that it operates covertly may, on this view, demarcate manipulation from coercion. But if that is denied, and we will discuss this in the following, then we lose our handle on the distinction between coercion and manipulation.

We can now turn to our search for demarcating factors. We present three families of views that tackle the demarcation problem for manipulation. On what we will call *outcome* views, manipulation requires a particular outcome. On *process* views, manipulation requires a particular process of influence. On *norm* views, manipulation requires the violation of particular norms.

3.2 *Outcome views*

On outcome views of manipulation, manipulation always, or at least typically, directly, or indirectly, leads to actions or behaviours with particular features. We will discuss just two types of outcome views, according to which manipulation leads to harm or the violation of self-interest or to a loss of autonomy.

3.2.1 *Self-interest and harm*

Manipulative influences typically go against the interest of the person being manipulated. That is, they lead to outcomes that are directly or indirectly unbeneficial or outright harmful for the person being influenced. Direct outcomes of manipulative influence may include beliefs, emotions, or desires formed by the manipulated person. It is often in one's self-interest to form true beliefs, and to have appropriate emotions, and worthy desires. Manipulation may directly frustrate these. Indirectly, your (false) beliefs or (inappropriate) emotions may lead you to do things that frustrate your self-interest. You may vote for the wrong candidate, buy the product you do not need at a price that is much too high, or stay at the slot machine for hours on end. The frustration of self-interest is thus often linked to harm, and manipulation may be said to involve harm to the manipulatee. Many paradigmatic cases of manipulation feature frustrations of self-interest on the part of manipulated persons. For example, our introductory case of Othello suffers great harm as a result of the manipulation (he ends up killing his beloved Desdemona). Typical cases of manipulation online, such as voter manipulation or endless doomscrolling, go against self-interest, too.

However, frustration of self-interest and harm is unlikely to be a necessary feature of manipulation. Nudging, at least in some forms, seems to be manipulative. Nudging, at least in some forms, seems to be manipulative, and many such nudges are meant to serve the self-interest of the nudged persons.[17] In many paradigmatic cases it is actually meant to *serve* people's self-interest. Similarly, romantic love is not against self-interest, and yet it is sometimes considered to involve manipulation, especially at the early stages. For example, you may manipulate by presenting yourself better than you actually are or by flattering the other person. When we understand these manipulative manoeuvres as integral parts of romantic relationships and also consider the latter to be unproblematic or even fun, then it becomes problematic to accept the necessity manipulation as always going against self-interest.

Both nudging and romantic love may thus be counterexamples to the view that the frustration of self-interest is a necessary ingredient of manipulation. There is room to argue that these counterexamples are inconclusive. For example, the very act of influence may directly and instantaneously be against self-interest (e.g., there may have been better ways to influence one in a nudging or love relationship) while the situation may be all things considered good for the manipulatee. Moreover, one might have concerns with the method of cases that the aforementioned strategy of showing that manipulation can improve self-interest, relies on.

There are two more general problems with the self-interest proposal, though. First, we are looking for a demarcating factor to distinguish manipulation as a non-coercive type of influence. Coercion typically frustrates self-interest, as least in the minimal sense that a different type of influence may often be better for the person being influenced. So, frustration of self-interest and harm will also be an ingredient of coercion. It does not help us to demarcate manipulation from coercion. At best, such a theory of manipulation would be incomplete.

Second, manipulation is unlikely to be exhaustively characterized by any end state (i.e., the direct or indirect result of the influence) and should at least include features of the process through which manipulation occurs. The reason is that end states like having one's self-interest frustrated may be arrived at in a multitude of ways, and not all of them will be manipulative.[18] This is a more general formulation of the demarcation to coercion. Many types of influences or events may frustrate people's self-interest. If we want to single out manipulation, we have to find further demarcating factors.

3.2.2 Autonomy

Manipulation as autonomy undermining is an account that shares with the self-interest account a focus on the (direct or indirect) end result of manipulation. It is not always clear whether proposals that link manipulation and autonomy are attempts to spell out its wrong-making features or attempts to give an account of manipulation in descriptive terms (insofar as autonomy can be understood descriptively). As noted earlier, there is surely a close relation between manipulation and autonomy. But how plausible is it that the undermining of autonomy is a necessary feature of manipulative interaction? Paradigmatic examples of manipulation indeed often seem to deprive agents of their autonomy. But, again, that need not imply that the undermining of autonomy is a necessary criterion of manipulation. Manipulation need not interfere with autonomy (Blumenthal-Barby 2012) and may even enhance it (Buss 2005; Gorin 2014b; Klenk and Hancock 2019); for example, when manipulative influence allows you to reach your goals and bring your desires and urges in line with your higher-order volition and desires, as in Harry Frankfurt's classic account.

Notice that autonomy may be lost by means other than manipulation, and so the loss of autonomy is not sufficient for manipulation. The counterexamples to the conceptual link between autonomy and manipulation in the literature suggest that it is not necessary, either. But even if it would be necessary for manipulation, we would need to say more about the nature of manipulation to distinguish it from coercion, in the context of the demarcation problem. As noted earlier, the *outcome* of coercion is *also* less autonomy. So, the autonomy view does not, without further explanation (e.g., distinguishing different types of autonomy), seem sufficient to demarcate manipulation. Of course, whether we can bolster the account by identifying further factors in addition to autonomy loss that, together, demarcate manipulation from coercion remains to be seen.

3.3 Process views

Process views of manipulation interpret manipulation in terms of characteristic processes or modes of influence that lead to a given behaviour or action.

3.3.1 Covert influence

Covert or hidden influence has often been suggested as a defining feature of manipulation in the sense of being a typical (e.g., Baron 2003; Rudinow 1978) or even a necessary condition for manipulation (Susser, Roessler, and

Nissenbaum 2019a; Handelman 2009). This is a plausible proposal because both coercion and rational persuasion take place "out in the open." Persuaders need to get their interlocutors to "see" their reasons for acting, and so do coercers. Manipulators, in contrast, seem to operate undercover. Iago, for example, also deceives Othello and his plan would not succeed if Othello would know what is going on. It seems very plausible that, in order to succeed, manipulation must be hidden in the sense that the intentions of the manipulator, the process of influence, or direct or indirect target outcome remain hidden from the manipulatee.[19] And indeed also in the online manipulation debate, the view that manipulation must be hidden is popular (cf. Susser et al. 2019a).

However, it can be argued that covert or hidden influence is not a necessary condition for manipulation. Again, there are several counterexamples (Gorin 2014b; Krstić and Saville 2019; Barnhill 2014; Klenk 2021b). For example, manipulative guilt trips can be obvious and still be very effective. We can be lucidly aware that we're being manipulated into feeling guilt, even as we feel guilt and act on it (Barnhill 2014, 58).

Counterexamples to the covertness view purport to depict manipulative influences that are not hidden from the manipulatee. With respect to online manipulation, one might wonder whether, say, the manipulation as conducted by Netflix's auto-play or Facebook's newsfeed is going on *non-transparently*. Do internet users in the twenty-first century, post-Cambridge Analytica not know they are being manipulated after all?

Relatedly, it would seem perfectly appropriate to complain about manipulative influence. For example, you may be surveilled and be annoyed by the obviously manipulative attempt of some marketeer to get you to buy a product that you do not need. We've all been irritated by advertisements of things we bought the day before and by targeted ads for camping gear that appear on our screens just after we watched the odd outdoor documentary on Netflix. If manipulation were hidden by definition, our frustration at these ostensibly manipulative influences would betray a conceptual mistake. After all, given that we were aware of the influence, it cannot be classified as a manipulative influence (if the covertness view is true). Clearly, that is not the case.

Again, there is room to resist this conclusion. One can challenge the counterexamples. For instance, it may be argued that what seems like overt manipulative influence that takes place out in the open is actually coercion. Guilt trips, pressuring tactics, and perhaps highly advanced and emotionally salient targeting online may thus fall under coercive and not manipulative influences. This is an attractive answer if we hold on to the view that manipulation and coercion are only gradually distinct. Then there are several ways to refine the thesis. Proponents of the covertness view could, for example, distinguish between covertness being a feature of the influence (i.e., it is actually hidden from the manipulatee) or merely an intended feature of the influence (i.e., the manipulatory intends for the influence to be covert) that need not be actualized. Netflix and Facebook, the twenty-first century notwithstanding, are still trying to manipulate precisely because they are trying to keep their cunning influence hidden. The task for the defender of the covertness thesis is thus to show exactly what would remain (truly) hidden in manipulative influence. Alternatively, the covertness thesis advocate might want to distinguish between different types of knowing or being aware of being manipulated: one might know, in some "cognitive" sense that one is being manipulated by Facebook (or one's first date), but one still fails to know, in a different sense (whilst being wholly engaged online or enthralled by a date) that one is being manipulated. We will discuss covertness and transparency in some more detail in the following.

3.3.2 Bypassing rationality

Another possible demarcating factor of manipulation is the bypassing of reason (e.g., Noggle 1996; Scanlon 1998). The intuitive idea is that manipulation is a type of influence that does not (adequately) engage the victim's rational capacities (e.g., Sunstein 2016b).[20] It is important to be clear in spelling out what it takes to "bypass reason" and, according to Gorin (2014a), one can understand such accounts in several ways.

One way is to interpret manipulation as actively interfering with rational capacities in the sense that one generates psychological states that are "incompatible with the proper functioning of the person's rational capacities" (Gorin 2014a, 53). Alternatively, one may understand manipulative influence as bypassing rationality in the sense that one impedes the rational capacities of one's victim from functioning, where their functioning can be understood "narrowly" in terms of functioning given the information, beliefs, and preferences available to the agent or "broadly" in terms of functioning given whatever reasons there objectively are (Gorin 2014a, 54–57).

The bypassing-reason view explains well many paradigmatic cases of manipulation. Charming, using olfactory and visual influences, using someone's emotional outbursts, or playing on their jealousy (as in our introductory example of Othello) all seem like paradigmatic cases of manipulation that also seem to bypass the rational capacities of the victim, at least on some interpretations of "bypassing rationality" explicated earlier. For example, charming tactics may impede the proper functioning of your rational capacities by preventing them from picking up the reasons against giving in to your suitor. Many of the phenomena that give rise to a worry about online manipulation such as increasing polarization also seem to drive on emotional and often irrational tendencies of users, for example, a bias in favour of one's in-group.[21]

We can immediately see how the bypassing reason account would help to address the demarcation problem. It is an account that focuses on the process of influence, rather than the end result. And we noted earlier, how persuasion and coercion require that victims recognize and act on reasons to succeed. Hence, the bypassing-reason criterion is a promising one to resolve the demarcation problem.

However, Gorin (2014a) and several others have documented at length how manipulation can sometimes proceed precisely by exploiting rational facilities (Klenk 2021a; Barnhill 2016). For example, consider a politician, convinced of the rationality of their voters, who finds that voters are very much concerned with saving the environment. The politician proceeds to give good arguments for the protection of the environment, and she is voted into office. The politician herself, however, does not care about the environment herself at all (Gorin 2014b, 91). This seems to be a case of manipulation: she uses voters purely instrumentally. However, it is false in this case that the manipulator aims to make the manipulatees fall short of the ideals that govern their emotions or beliefs, respectively. For example, it is reasonable to accept good arguments for a true conclusion, if anything is.

Moreover, the idea that manipulative acts proceed through some specific pathways – in this case, the process of bypassing rationality – is questionable because "the processing route" or "origin" of an idea or mental state is unlikely to be always unequivocally bad. Certain beliefs or emotions may well have resulted from bypassed rationality (e.g., the result of being madly in love or deeply angry), but that doesn't mean these states are necessarily suspect – quite the contrary (cf. Jongepier 2017). Also, Barnhill (2016) makes a convincing case that the bypassing of rationality cannot convincingly be held to consist in using emotional, non-rational influences because the former are also sometimes bona fide ways to engage with the world. More generally, philosophers have long pointed out that emotional ways of responding to the world are *rational* responses; for example, reacting with a negative emotion towards an injustice.

This doesn't mean that accounts according to which something counts as manipulation in case it (minimally) involves bypassing the rationality of persons are doomed to fail. It's still plausible – if we take the case of propaganda, for instance – that debilitating people's capacity to think clearly and instead to dig their heels in emotional responses such as fear is worrisome. The point, rather, is that bypassing accounts need to explain why and when some bypassed states or emotional ways of responding to the world are bona fide processes and what separates those from the *mala fide* types.

3.4 Norm views

We have reviewed the most promising *outcome-* and *process*-oriented accounts of manipulation and seen their advantages and disadvantages. A different and increasingly influential type of account are *norm-based views of manipulation*. According to norm-based views, manipulation is associated with behaviour or action that violates norms (Scanlon 1998; Barnhill 2014; Noggle 1996, 2018a; Gorin 2014a, 2014b, 2018; Klenk 2020, 2021b; Sunstein 2016a). There are considerable differences as to how the norm violation that constitutes manipulation is understood. For example, Noggle's influential account of manipulation suggests that manipulation is constituted by the attempt to make someone else (the manipulatee) violate a norm, whereas others like Gorin and Klenk suggest that manipulation is constituted by the manipulator violating a norm of proper influence.

Norm-based accounts are promising and influential in the philosophical literature, but they have not received much uptake in the digital ethics literature yet. The unifying thought behind norm-based accounts of manipulation is that we can explicate the concept of manipulation in terms of epistemic, moral, or practical norms that manipulation violates.

The difference between outcome- and process-oriented views, on the one hand, and norm-based views, on the other hand, is subtle. After all, the fact that an action violates a norm may also be a particular *outcome* of a given interaction, just like some types of *processes* may constitute norm violations. What seems to set norm-based views apart is that the norm violation is constitutive of manipulation, rather than a (common or necessary) side effect.

Norm-based views may seem suspect insofar as they would seem to foreclose the debate about the thickness of manipulation. After all, it would seem that an account of manipulation in terms of a norm-violating social influence would imply that manipulation carries with it a normative or evaluative judgement as a conceptual matter. But that conclusion would be premature. First, insofar as we can give a descriptive account of norms (e.g., in terms of social expectations) we need not conclude that a norm-based account of manipulation implies the thickness of manipulation. Moreover, manipulation may turn out to be morally problematic in all cases without that fact being a constituent part of the concept. As mentioned earlier, these two things should be kept apart. Finally, the question very much depends on the details of the norm-based view under consideration. For example, Noggle's view suggests only that manipulative influence is the attempt to get someone else fall short of certain norms. And while there may be pro tanto norms against attempting such a thing, Noggle does not define manipulation in terms of the attempt to violate that norm. This may be a consequential difference to norm-based views like that of Gorin and Klenk, who analyse manipulation as falling short of certain interactional norms. In either case, however, the thickness of the concept need not be associated with a moral one, as manipulation may also be constituted by a violation of epistemic or practical norms, rather than moral ones.[22]

On Noggle's influential view, manipulation involves a violation of norms that pertain to the outcomes of an interaction, such as the behaviour or action exhibited by the victim of the manipulative influence. According to Noggle (1996, 44):

> There are certain norms or ideals that govern beliefs, desires, and emotions. Manipulative action is the attempt to get someone's beliefs, desires, or emotions to violate these norms, to fall short of these ideals.

For example, Iago intended for his actions to make Othello believe a falsehood (namely, that Desdemona was cheating on him), and thus he intentionally made Othello violate the norm that legislates believing truths.[23] What norms matter, on Noggle's account? The relevant norms or ideals are the ones that the manipulator envisions for the manipulatee. This retains a parallel with deception (where it matters what the deceiver takes to be the truth, from which he deviates), and it avoids the potential problem of committing to and identifying objective norms that govern belief, desires, or emotions. Most proponents of norm views follow Noggle in classifying manipulation as an "intentionally characterised" action (Noggle 1996) and specify it quite broadly in terms of attempting one's victim to violate some belief, desire, or emotion-related norm (see, for example, Barnhill 2014 and Gorin 2014a). In effect, the breadth of the different norms for emotions, beliefs, and desire that we recognize gives the account tremendous breadth and explanatory power. Thus, a norm-based account avoids the mistake of trying to shoehorn manipulation into the mold of necessary violation of some allegedly more basic outcome or process.

However, the norm-based view has problems with counterexamples, too. For instance, pressuring or charming tactics cannot be explained by the view even though they seem like bona fide cases of manipulation (Noggle 2018b). For example, consider emotional blackmail or related pressuring tactics. It would seem pressuring others provides them with good reasons to act. In light of the threat or the pressure to conform to someone else's demands it may make good sense to believe, desire, or feel just as the manipulator wants. In many cases of pressuring, the pressuring itself creates good practical reasons to yield to the threat. Indeed, the reason-generating nature of pressure is what the perpetrator relies on when they utter their threat. There is thus in that sense no violation of a norm. Indeed, it would seem that the manipulator in these cases relies on the manipulatee to be responsive to the reasons he or she provides in the form of pressure or, more generally, a threat (cf. Klenk 2021a). Insofar as using your emotional power over your significant other, (peer) pressuring your colleague into accepting the undesirable task, or seducing your online date is manipulative, the norm-based view cannot explain it. Since such cases appear to be bona fide cases of manipulation that we should want to explain, that is a problem for norm-based views.

Naturally, these counterexamples may be challenged. Perhaps, the norm-based view and its focus on norm violations could be coupled with additional conditions to account for these cases, such as a violation of self-interest (cf. Barnhill 2014). However, a deeper concern with the view is that it gives undue attention to the intentions of the manipulator as they concern the manipulatee. Noggle, for instance, suggests that manipulation is constituted by the attempt to make someone else fall short of norms that govern belief, emotion, or desire. Why make the demanding assumption that manipulators aim to have their victims violate a norm, rather than merely assuming that they influence their victims in a way that constitutes or results in a norm-violation?

A variant of the norm-based view that seeks to address this concern is the view that manipulation is negligent influence (Klenk 2020, 2021a). The negligence account is motivated by two problems. First, the aforementioned counterexamples to existing norm-based views of manipulation and the desire to account for these examples as manipulative influence. Second, the observation that these examples can be accounted for on normative terms only at the expense of introducing a proliferation in the type and scope of norms that manipulation violates as a constitutive matter. For example, Noggle's view could account for pressure cases by suggesting that manipulation is constituted by the violation of interactional norms that, amongst other things, imply that pressuring is prohibited. In effect, rather than just considering norms as they supposedly apply to the manipulatee, norm-based accounts would also have to invoke norms as they apply to the manipulator.

The core proposal of the negligence account is to suggest that the latter suffice to satisfactorily account for manipulative influence. Manipulators uniformly seem negligent regarding their chosen means of influence. However they influence their victims, their choice of influence is arguably not best explained by its "reason-revealingness" (to wit, its propensity to reveal reasons to the influenced person) but by its effectiveness in getting people to do what the manipulator wants. This kind of negligence is proposed as the common factor that unifies all cases of manipulation (Klenk 2021a). Marcia Baron suggests a similar line of thought when she writes a manipulator has "the aim of getting the other person do what one wants, together with recklessness in the way that one goes about reaching that goal" (Baron 2014, 103).

The negligence account would amount to a significant shift in thinking about manipulation. Manipulation would not be demarcated from coercion by what it does or adds to it but by what it lacks. Unlike coercion and persuasion, manipulators do not primarily care for reasons (they sometimes might, when it serves their purpose, but it is not an integral part of their endeavour). Gorin (2014b, in this volume) suggests a view along these lines when he analyses manipulation disjunctively as a violation of at least one of four types of norms, amongst them norms that demand being motivated by someone else's reasonable ends. Like the negligence account, Gorin's view also shifts the domain of norms whose violation constitutes manipulation to norms that apply to the manipulator. The open question is how to spell out those norms in detail and how many different types of norms are violated by manipulation as a constitutive matter.

In any case, the advantage of a negligence-type of account would be that the distinction to coercion could clearly be maintained because coercers *do* care about reasons but manipulators do not (cf. Schelling 1997). After all, coercers rely on their victims being able to appreciate that they are given good reasons (e.g., a threat to life) to comply with what the coercer wants them to do. A lunatic who cares not about reasons can be harmed, but not coerced.

A problem about the negligence account is that it may complicate matters too much when thinking about manipulation and thus be too far removed from ordinary discourse about manipulation (see Coons and Weber 2014 for related discussion). Also, depending on how the negligence relation is spelled out (to wit, the precise sense in which a manipulator fails to acknowledge or care for reasons), there is a question about whether or not norms or duties of care determine domains where manipulation can occur or whether we should better characterize negligent influences in domains without norms of care as benign forms of manipulation (or not as manipulation at all). Finally, and this will connect to the next section, we can ask whether manipulators need to have the capacity to be governed by an absence of negligence or a presence of norms of care to qualify as manipulators in the first place.

4 Intermediary conclusions

We can draw the following intermediary conclusions. First, we should be careful about the intentionality required for manipulation because it may concern the capacity for intention (what we called the general intentionality requirement) or the specific intention to manipulate or do something associated with manipulation (what we called the specific intentionality requirement).

A second major point is that manipulation is a type of influence that is distinct (in kind or degree) from coercion, and manipulated people still do something *voluntarily*. From this observation, we developed the demarcation challenge which is the challenge to define manipulation in contrast to coercion. Coercion notably has normative implications for (moral) responsibility, and it will be important to determine to what extent manipulation exculpates.

Finally, our discussion brings to the surface an important methodological assumption in the philosophical manipulation debate that is transported easily to the digital ethics debate, namely the anti-pluralist assumption that *one* of the accounts of what manipulation is must be right – not a combination of two or more views. The anti-pluralist assumption makes sense. After all, it's strange to think that in some cases what makes it a case of manipulation is that it involves negligence and in others it's because it involves bypassing rationality. Letting go of the anti-pluralist assumption would thus come at substantial explanatory costs of explaining how manipulation can be so multifaceted and still say it's manipulation across all cases. But it may not be impossible, especially if one were to adopt a "focal case concept" of manipulation. It also depends a great deal on *why* one wants a definition (or better understanding of) manipulation. Is it for getting a better understanding of digital manipulation? Is it for getting a better view on the harms for internet users or the wrongs of digital manipulators? Is it to develop new policy or legal regulations? Depending on the aims, accepting (a degree of) pluralism or conceptual messiness can range from being highly problematic to potentially productive.

The take-home message for this sub-section about theories of manipulation is thus, above anything else, the need to be explicit first of all about one's preferred theory of manipulation, second about one's methodology, and finally about one's aims.

5 Aggravating factors

Having discussed the relevant philosophical terrain and the rich variety of positions to be taken when it comes to defining manipulation and why it's bad or wrong (if it is), it is now time to look at the "techy" side of things. Which technologies can be considered manipulative or used in manipulative ways by corporations? Which aspects of the existing technologies make them effective manipulative tools (if tools they are)? Which technological advancements are especially worrying from a moral point of view?

These questions are the domain of the field of digital ethics, though they are not only questions in the field of digital ethics. The tech side is a vast territory and is, importantly, interdisciplinary territory. The aforementioned questions have also been addressed – often earlier, in fact – by legal scholars, computer scientists and communication scholars, and many others working on (digital) technologies for whom addressing questions about the manipulative and morally problematic nature of these technologies have been inevitable.

When it comes to studying the manipulative and immoral potential of new technologies, there are different approaches one might take. A common approach taken in the wider digital ethics literature is the "ethics of (insert technology)" approach. There are papers covering, for instance, the ethics of recommender systems, the ethics of algorithms, the ethics of automation, self-driving cars, social robots, voice assistances, and so on. The "ethics of" approach is valuable because each new technology or technological implementation will come with its own technical and moral characteristics. Recommender systems and self-driving cars, for instance, are entirely different, each giving rise to different conceptual and moral questions. It's important not to throw everything on one big pile, since doing so feeds into the already all-too-common slogans that "digital technology" as such is manipulating us and undermining our freedom (cf. Harari 2018).

While the "ethics of x" approach is valuable as well as necessary, it's also important that it is not the only approach on offer within the wider discipline of digital ethics. This is because, despite obvious and deep differences between various new technologies, there will also be important similarities in terms of what makes them especially manipulative and/or morally problematic. It's possible to attend to these shared features without having to make sweeping statements about digital technology in general undermining our freedom *tout court*.

These shared features are what we will refer to as "aggravating factors". An aggravating factor is a factor that sometimes or typically either (a) makes manipulation more effective, its effects worse or morally wrong, or (b) makes it harder for individuals to avoid or contest manipulative practices and technologies. In the following, we discuss what we regard as four noteworthy aggravating factors: personalization, opacity, flow, and lack of user control.

5.1 Personalization

Not just our Google searches and the ads we see online but also the health trackers we wear, the TVs we watch, and (future) fridges we use are increasingly personalized, in short, adapted to who we are. The terms "personalized" and "targeted" are often used interchangeably, though a distinction between them can be made. Personalization is typically understood as the way in which (e.g., machine learning) algorithms are designed such that they can deliver something that is in line with the user's preferences, personality, and so on. Targeting can be understood as the active steps, for example, a marketer can take to send specific ads to specific groups. In short, content is personalized (usually to individuals), whereas people are targeted (usually to groups).

In terms of aggravating factors for online manipulation, the main focus is thus on personalization. A first thing to note is that there's nothing wrong about personalization as such, quite the contrary. After all, it's quite nice to enter a record shop and receive personalized advice on the latest Jeff Tweedy or Mavis Staples album you absolutely need to listen to, and it's nice (if sometimes painful) to get tailored love advice from a close friend. Likewise, it can be great to receive personalized recommendations from platforms like Spotify or Netflix, just as it can, in principle, be convenient to be recommended products you might need or like.

However, personalization inside and outside of online contexts also offers opportunities not just for welcome advice but also unwelcome influence. The reason for thinking personalization is a serious aggravating factor when it comes to manipulation is recognizable also outside of discussions about digital influence. The better someone knows us, the greater impact their advices, statements, and warnings have on us because they can tailor their advice to who we are. The existence of the well-researched phenomenon of gaslighting – a manipulative strategy "aimed at getting another not to take herself seriously as an interlocutor" (Abramson 2014) – illustrates this clearly. Gaslighting can be as manipulative as it is precisely because the gaslighter knows the gaslightee all too well, her vulnerabilities in particular.

Having a lot of knowledge about someone isn't the same as "personalization". However, when such knowledge is put towards certain ends and becomes part of the particular things one says or does to someone, it can become – and in most social contexts, inevitably ends up being – personalized. Answering a person's question about how to get to x by giving them the answer straight is not personalization; telling your friend to get to x via y because you know there's a large flea market going on that they would enjoy (or hate), is. As is apparent, we haven't thereby yet said anything about such personalized advice being problematic or not.

As for online personalization, Susser et al. likewise mention targeting (which they seem to equivocate with personalization) as an exacerbating condition of manipulative technologies, writing that "the more targeted manipulation is the more we ought to worry about it". Or as Alexander Nix said in 2016, when he was still Cambridge Analytica's CEO, by building a psychographic model of "every single adult in the US" and thus by knowing "the personality of the people you're targeting, you can nuance your messaging to resonate more effectively with key audience groups", for instance on "specific issues such as the Second Amendment" (Concordia 2016).

Needless to say, there can also be personalized instances of online manipulation that aren't worrisome and in fact may be welcomed. Various forms of digital healthcare and mental self-care tools can be considered here. There are apps, for instance, that have virtual chat bots that adapt to the often-personal input given to them. There are many things to worry about when it comes to personalized mental health apps, such as privacy, data sale, hacking, undiagnosed conditions, less visits to GPs, and so on. *In principle*, though, online personalization might be desirable and thus not worrisome at all, even if in practice it (almost) always turns out to be.

The phrase of content, ads, or technologies being "adapted to who we are" should of course be taken with a considerable grain of salt. After all, what matters from a commercial or effectiveness perspective is first and foremost the digital profile that is constructed based on online traces a person leaves behind, not who the person really is. That being said, finding ever closer connections to people's "offline selves" – especially given that the online and offline worlds cannot be properly distinguished anymore – is of course also a way of being able to bring personalization to a higher level and influence people more effectively.

Though personalization is a serious aggravating factor when it comes to what makes technologies manipulative, we should also avoid thinking of personalization as something that is necessary to what makes certain online practices or techniques manipulative. It's also important to bear in mind the impact of impersonal or "sweeping" forms of online manipulation. Again, it's helpful to consider the offline context here. Take propaganda for instance, which is known to have a potentially enormous impact on people's beliefs, values, and actions, but it is not a personalized type of influence, historically it has often been quite the contrary (see Stanley 2015). By steering on feelings of anger or fear, propaganda is typically a broad-scale, sweeping type of influence that intends to resonate with something that large groups of people might fall for. Similarly, online disinformation might manipulate large crowds of people without necessarily doing so in a personalized fashion.

Finally, we need to be aware that it's often also precisely the data mining corporations and political consultant firms who stress the significant impact of personalized influence. In Nix's lecture from which the previous quote was taken, he was outright bragging about the impact of psychographic profiling, mentioning that today "we need not guess" anymore about what solution may or may not work because we now know "exactly which messages are going to appeal to which audiences". This makes good *corporate* sense, but contemporary science tells a much more nuanced story. Scholars keep pointing out that measuring the efficacy of profiling techniques is difficult and that the impact is sometimes said to be questionable (cf. Zarouali et al. 2020). This is not to say personalized online influence is entirely ineffective.

In short, we need to tell a nuanced story: personalization can be a genuine aggravating factor, and thus a serious cause for concern, even if it isn't always necessary to manipulate people online and even if it isn't the "magical marionette technique" that some make it out to be.

5.2 Opacity

Not knowing about someone's manipulative strategies – its being *opaque* or *non-transparent* to someone – generally makes one a lot more prone to being manipulated. Just as with magic: if you see another's trick, the trick won't fool you or not quite in the same way. The experienced online or offline manipulator will therefore generally try to make it the case that you don't see the trick, that you don't realize attempts are being made to steer you in a particular direction.

As mentioned earlier, there is a lot of philosophical discussion about whether or not opacity is a necessary condition for manipulation, and naturally this dispute extends into the domain of online manipulation. Some think it is necessary (Susser et al.) while others don't. It may be worthwhile to adjudicate whether or not it is necessary, but it may equally be more fruitful to agree on the existing common ground: not knowing that one is being manipulated is an aggravating factor to actually being (successfully) manipulated, regardless of whether there might also be ways of being manipulated in broad, digital daylight.

Also, a question that is perhaps worth more attention than it is currently getting is the question of what transparency and opacity in the digital domain mean exactly, given that it is a highly ambiguous concept. Depending on what we take transparency to mean, there's the further question of whether (online) transparency is even a worthwhile ideal to strive towards. Though important work has already been done with respect to both the conceptual and normative questions about transparency, many questions still remain to be answered, indeed formulated (Ananny & Crawford; Sandis & Sellen; Pasquale).

A recurring topic, also in this issue, is what type of communicability or explicitness by a corporation or government institution is sufficient for a type of influence to count as transparent or no longer opaque. Does, for instance, a hard-to-find page on an organization's website suffice as being "transparent" about potentially manipulative techniques such as micro-targeting? And isn't it transparent to us, post-Cambridge Analytica, that social media platforms attempt to manipulate us? These questions cannot be answered in a black-and-white fashion; instead, they require teasing apart the different meanings of transparency and opacity in different contexts.

Though the following is highly incomplete, a rudimentary list of different types of transparency may include the following:

Organizational Transparency: the type of explicit transparency that an organization gives about their digital strategies and means of influence. In this issue, Jared W. Palmer for instance gives the example of the gamifying language platform Duolingo, who made no secret about the fact that it generated profits for its owners by offering the translation services, which were done for free by its language-learning users, to businesses. Duolingo's founder mentioned this explicitly on Duolingo's own forums.

Active Outreach Transparency: this is the type of transparency an organization might give to its subscribers, share- and stakeholders and the broader public about their digital strategies and policies, which takes the form not just of a one-on possibly hard-to-find public message but as part of a continuing project. The messaging app Signal is a possible case in point, which regularly communicates about the technologies they (don't) use and their privacy policies and ethics on their own blog and Twitter, also clarifying how Signal differs from Facebook/WhatsApp and so on.[24]

Factive Transparency: in this type of transparency, an individual knows as a matter of fact (or as a matter of high likelihood) that a service or tool they are using, such as their smart fitness watch or voice assistant, is trying to steer them in certain directions and perhaps selling their data for commercial purposes. A test for factive transparency is simply the positive and explicit answer individuals would give when asked whether they think they are being manipulated by x on platform y through method z.

Engaged Opacity: in this case, an individual has the relevant knowledge just as in Factive Transparency except (1) their knowledge is not available for conscious awareness and (2) they are unaware in this way because they are (kept) engaged in their online behaviour or "in digital flow."

Needless to say, these types of transparency/opacity are hardly exhaustive and for each of these, many sub-types need to be distinguished. But a rudimentary list like this would already be helpful when claims are made about organizations (not) being transparent or something (not) being transparent to individuals. The distinction between factive and engaged transparency, for instance, allows us to recognize that a person might know (as a matter of factive transparency) that Facebook or their smartwatch is trying to steer them in certain ways whilst failing to know (as a matter of engaged opacity) that this is going on (because they're doomscrolling or trying to break personal running records). Making these distinctions also helps us in getting clear on what type of transparency is valuable and what organizations might need to do to "be transparent", as well as bringing out the fact that many corporations do their utmost to prevent people from attaining engaged transparency.

5.3 Flow

Engaged opacity brings out something that ought to be mentioned as a serious aggravating factor in its own right: online flow. Technology is usually, and understandably, designed for comfortable user experience – nothing is as frustrating as websites or gadgets not doing (immediately) what they should be doing. At the same time, being in online flow can prevent one from being aware of relevant knowledge, can hamper one's opportunities to reflect, can bypass one's rationality, and thus prevents one from gearing one's behaviour in directions that better fit one's larger or deeper desires or ideals. This aspect has been well researched for instance by (post)-phenomenologists of technology, who stress that the seamless phenomenological experience of the online world makes that people "forget" that they're not just running in the world but running with a smartwatch, that is, running with a tiny for-profit organization clutched to one's wrist (cf. Keymolen 2018). It is also a topic for philosophers working on how the digital world affects autonomy, authenticity, and weakness of will (e.g., Williams 2018) and numerous authors in this volume).

The topic of online flow – which, given the collapse between the online and offline worlds, usually just amounts to flow in the world – also merits attention because of how it paves the way for thinking about how disrupting flow might counteract existing manipulative forces. Some scholars have for instance begun to examine the potential of introducing "friction" in tech design (Terpstra et al. 2019). If a user's flow is disrupted, this might make it easier for people to stop and think about whether they really want to watch another video, scroll for another half hour, or insert data about one's menstruation cycle and symptoms in one's health watch.

5.4 Lack of user control

Another aggravating factor is the lack of control of the technologies that attempt to manipulate us. When it comes to being trapped in a filter bubble on YouTube or social media, there is typically little one can do to get out of one's bubble and enter another one. When it comes to recommender systems, again, there is little influence individuals have in changing the values, the settings, the input, and so on, of the technologies they use. In theory, though not in practice, it would be possible for users to select, say, more random news items or getting news from an "anti-bubble", for example, to receive news that is on the opposite end of what your political, social, or moral views are (or in any case what its algorithms believe your views are). Likewise, it is possible in theory, but not in practice, to actively tweak and improve what Spotify or Netflix think you like to listen to or watch, and the same goes for what smart homes recommend to their users. And, finally and most dramatically, it is possible in theory for users to refuse being microtargeted and tracked across the web and to have some control about the extent to which they want to give up on privacy or data traces in return for (free) services or alternatively to have the option to pay for them – but again, not possible in practice. In practice, internet users and owners of smartwatches and smart homes and what not are usually faced with a "take it or leave it" situation. If you want the robot vacuum cleaner, it comes with the corporation knowing not just the size of your rooms but also where your dinner table is and when you're (not) home. One can refuse of course, but in most cases, the service or product fails to work properly or fails to work at all. Lack of control and quasi-coercive circumstances or offers have a distinct way of making people susceptible to manipulation.

One problem about lack of user control is that of accuracy: a lack of user control also obstructs better accuracy of digital profiles. If users had more control about the technologies they engage with, the technologies would be better adapted to "who they really are" and what services or goods they are after (personalization and lack of control are thus importantly connected). But ironically, at the same time, lower accuracy due in part to a lack of user control also makes people *less* susceptible to manipulation. This is because manipulation tends to be more effective the better certain strategies are tailored to individuals' personalities and vulnerabilities. By not being able to change or adapt the digital profile or "digital persona" (Clarke 1994) that is made about us, we might also get out of some of the tech giant's digital clutches.

On the other hand, a lack of user control more often makes one more susceptible to being manipulated, especially if the need for using the technology is high (or quasi-coercive). This can be so because one is repeatedly being exposed to certain influences even if they do not fit one's digital profile, such as being confronted with political messages that do not necessarily fit one's political views, (perhaps because one has no formulated views as yet) or being constantly confronted with products one has no desire to buy (until one sees them often enough). Without being able to "influence the influence", individuals can slide into certain ways of thinking or behaving. The worst kinds of lack of user control are "dark patterns" (e.g., Gray et al.), such as when users are deliberately refrained from changing, meddling, or refusing certain options or settings (e.g., privacy- or profiling-friendly ones).

Also, it is conceivable – again in theory – for certain services such as social media and the way news is shown to users, to *require* of users to express their preferences, to ask them whether they prefer to be shown news in line with the profile they (the for-profit organization) has constructed of them or whether they prefer an anti-bubble, or alternating filter and anti-bubbles, and so on. Such algorithmic self-governance may help make individuals more robust against manipulation. Commercial corporations are, however, unlikely, depending on their moral compass (see the following) to implement degrees of algorithmic self-governance in their services and products, hence this discussion is mostly a purely idealistic one.

5.5 An organization's moral compass

The list of possible aggravating factors is only a small and non-exhaustive list of factors that can contribute to certain technologies being manipulative. We have here described a few that we believe are particularly acute, but there are many other possible factors that are likely to contribute, such as the *free use* (financially speaking) of technological services which can create the (implicit) thought that being surveilled or manipulated in return is acceptable. Another factor is the *human-likeness* of technologies or their possible anthropomorphic nature which is especially relevant in the context of robots' potential of being manipulative. Yet another is the possible *rogueness* of technologies, that is, when technologies such as self-driving cars or war-drones start doing things on their own account, deviating from human design and plans.

Also, apart from being only a start, the list of aforementioned possible aggravating factors is just that: *possible* aggravating factors. Digital technologies, when they have one or many of the said factors, aren't necessarily manipulative. In fact, most of the factors that can make certain technologies more likely to be manipulative are also the factors that make it that certain technologies can be put to virtuous ends. Care robots that have some human-like aspects (e.g., eyes) and which operate with great flow, and which are designed to be opaque to some degree (given that people in need of care, for example those suffering from dementia or autism benefit from a degree of opacity), are likely to be more effective, for instance. The aggravating factors, then, are not necessarily sure-fire signs that a certain technology is manipulative, or manipulative in a morally problematic way.

So when should we (not) be worried about opacity, flow, or lack of control? One important guide is the overall *moral compass* of (private or public) organizations (see, e.g., Van de Poel and Royakkers 2011; Vallor 2006; Leonelli 2016). Which values does a corporation or government institution implicitly and explicitly ascribe to? What is their business model and how do the organization's moral values relate to non-moral values such as profit maximization? Which values does it have at heart and which values does it actually carry out? Which risks and problems does it anticipate? How quickly and effectively does it react when such values (autonomy, privacy, human dignity, freedom of speech) are violated? How easy or difficult is it to get non-automated or *human* responses to requests or concerns? Depending on the answer to these questions, the aggravating factors can be worrisome to more or lesser degrees. We should be less worried about high flow and opacity when it comes to a non-profit start-up that builds privacy-friendly apps to improve women's knowledge of their menstruation cycle and moods compared to high flow and opacity when it comes to a corporation like Cambridge Analytica. Which is not to say we have no reason to be concerned even in the first case, as moral compasses of new and rapidly growing tech companies tend to change too.

Needless to say, what an organization's moral compass is, is a notoriously hard question to get an answer to. However, there are some handles to get clues including written statements on the organization's own website, the formulation and design of their Terms and Conditions, whether they have ethicists on board and/or how their ethics committee is chosen and which authority they are assigned, the way they respond to concerns or incidents, whether they engage in ethics washing, and so on.

It is the combination of an analysis of the possible aggravating factors of certain technologies in combination with a sense of an organization's moral compass that designs those technologies or puts them to use that we can get a picture of the level of concern about how likely, and just how impactful, manipulation will be.

6 Conclusion

In this chapter, we have charted the field of the contemporary debate concerning online manipulation. As for the method of studying (online) manipulation, we have discussed the classical conceptual analysis approach and mentioned its problems as well as novel alternative methodological approaches such as the "focal case concept" approach. We also mentioned that, when studying manipulation, one needs to decide and/or be explicit about (1) whether or not one thinks manipulation is a so-called thick or moralistic concept and (2) whether manipulation necessarily involves intentionality and if so, in what sense.

We then moved on to discuss the concept of manipulation and which features might help us distinguish it from coercion and persuasion. To this end, we distinguished outcome-based views (in terms of (3.2.1) self-interest and harm and (3.2.2) autonomy), process-based views (in terms of (3.3.1) covertness or (3.3.2) bypassing rationality), and norm-based views (including the negligence-based view).

In the second half of this chapter we mentioned numerous possible aggravating factors, that is, factors that make manipulation worse or that make it harder for people to get out of a manipulator's clutches. We focused in particular on (5.1) personalization, (5.2) opacity, (5.3) flow, and (5.4) lack of control. Finally, we mentioned that taking into account an organization's *moral compass* – in spite of often being a near-impossible endeavour – is key to knowing whether the said factors are indeed cause for concern.

It should be stressed at this point that "the field" we have chosen to chart has been only a small piece of a larger landscape. As we discussed in the Introduction to this volume, several important and intriguing aspects of (online) manipulation such as its legal, political, and psychological aspects cry out for further study, and they promise much intriguing insight.

Notes

1. We are grateful to Anne Barnhill, Thomas Nys, and Robert Noggle for very helpful written comments on an earlier version of this chapter. The audience at our online workshop series also provided helpful comments and suggestions on an early presentation of the material collected here. Both authors contributed equally to this chapter. Michael Klenk drafted initial versions of Sections 2, 3, and 4, and Fleur Jongepier drafted initial versions of Sections 5 and 6. Michael Klenk's work on this chapter was supported first by a Niels Stensen Fellowship and later by the European Research Council under the Horizon 2020 programme under grant agreement 788321. Fleur Jongepier's work on this chapter was supported by an NWO Veni grant (VI.Veni.191F.056).
2. We simplify the debate about the nature of analysis here for ease of exposition. See Beaney (2021) for further discussion.
3. Though see the discussion by Houk (2018) on alternative approaches.
4. See Feurer and Fischer (2021) and Klenk, Xun Liu, and Hancock (2021) for examples of the nascent experimental work on manipulation.
5. Especially considering the question of whether manipulation has – as a conceptual matter – a normative or evaluative component. See Hopster and Klenk (2020) for further discussion on the limits and benefits of using empirical methods in ethics.
6. The view that several conditions such as deception, autonomy loss, and harm are associated with manipulation is also supported by initial experimental research on non-philosopher's views about manipulation (cf. Klenk, Xun Liu, and Hancock 2021).
7. Of course, if there isn't even a paradigm, as suggested by Baron (2003, 37), then even this approach is put into doubt.
8. Thanks to Anne Barnhill for prompting us to clarify this point.
9. Compare the discussion of the thickness of manipulation in Wood (2014), who like us understands it as a question about the meaning of the concept, versus

the sense in which manipulation is a pro tanto wrong, as discussed by Baron (2014).

10. This almost sounds like a contradiction in terms and is impossible to pull off, but it becomes more feasible if one were to distinguish moral and non-moral forms of laudability. It is sometimes also said of certain populist politicians (clearly not all of them) that they are cunning in such a way that demands our respect, even if their cunningness is used for immoral purposes and so do not demand our *moral* respect.
11. In addition, Coons and Weber (2014) note that we may wonder about whether anything is truly right or wrong independently of some idiosyncratic perspective – as various sceptical challenges in philosophy and beyond demonstrate – but we do not wonder about the reality of manipulation. Proponents of a thick view on manipulation could maintain that a subjective evaluation is part of the concept, but it would be less plausible to suggest that manipulation is stance-independently moralized as a conceptual matter. Based on this sceptical view, there must be some descriptive account of the concept of manipulation independently of moralized considerations.
12. Also note that the process, outcome, and norm-based accounts of manipulation that we discuss in Section 3 may be presented in what we might call deontic or telic fashion. Deontic versions of these accounts portray the demarcating features as the object of an intention. For instance, a deontic covertness thesis would have the manipulator intend to covertly influence her victim. A telic or consequential version would do without intentions and merely require that there is an influence that leads to the manipulatee remaining oblivious about some important feature of the interaction. The distinction between what we call deontic and telic versions of different accounts of manipulation is not always made explicit, nor are decisions for or against a particular view defended. But it seems to be a reasonable and noteworthy distinction to draw. This is especially so given the focus of this volume on interactions mediated by and perhaps with machines that may lack intentionality.
13. Several scholars, like Baron (2003), suggest that manipulation merely limits options, rather than removing them, and that this may be a useful demarcating factor. See also Handelman (2009), who defends the view that manipulation is about presenting some specific choice as best to the agent.
14. See also the debate about incompatibilism, free will, and moral responsibility. The important manipulation cases are supposed to involve a victim performing the manipulator's course of action on its own volition, cf. Sripada (2012). See also Cave (2007) for a discussion of the charge that motive manipulation is morally bad, which seems to be similar to Fischer (2017); Fischer and Illies (2018).
15. The incompatibilism debate is interesting in this context. Incompatibilists argue that the "not fully free" intuition is sensitive to the agent in a manipulation case not being the ultimate source of his or her action. Compatibilists, in contrast, suggest that this intuition is sensitive to the fact that manipulation damages or impairs the agent's cognitive, evaluative, or affective capacities.
16. Garnett (2018) being an illuminating exception. Note also that in fields outside philosophy (e.g., communication studies) persuasion is used to describe tactics commonly associated with manipulation. Note also that it is not entirely clear that an analysis of patient behaviour – such as coerced or manipulated action – allows for inferences about agent behaviour – such as coercive or manipulative action. There may be benefits to dissociating analyses of manipulated from analyses of manipulating action and to offer accounts that are partly independent, for example, Klenk, in this volume.
17. Thanks to Anne Barnhill for prompting us to clarify this point.

18. Thanks to Thomas Nys for prompting us to clarify this point.
19. See also Cohen (2018).
20. A variant of that view might be the one suggested by Blumenfeld (1988), who suggests that manipulation bypasses character, which he understands as an amalgam of reasons, motives, and desires integrated in the manipulatee's character.
21. Some theorists have suggested that manipulation works not only by bypassing reasons but – more specifically – by exploiting vulnerabilities in the subject. Again, this may be correct as a causal statement about manipulation because manipulation may often happen to proceed in these ways. But the claim interpreted as a conceptual claim is more difficult to maintain. The primary problem with this is that vulnerabilities are likely relative to context. For example, the gustatory "bias" to prefer sugary food was great in the environment of our evolutionary development, but in today's world with an oversupply of calorie-rich food it is to our detriment. If some of our dispositions are vulnerabilities given a context, then the account of manipulation as playing on our vulnerability would suggest that we need to appeal to dispositions that are powerful or strong given a context. It is not clear what that would mean, and it is possible it drives on intuitions related to the bypassing reason view or the autonomy view.
22. Thanks to Robert Noggle for helpful feedback on this point.
23. Noggle's account thus makes explicit the specific intentionality requirement that we discussed earlier. Other proponents of norm-based views like Gorin et al. (2017) or Barnhill (2014, 2016), however, do not make the intentionality requirement explicit. In his contribution to our volume, Gorin does make it explicit (cf. Gorin, in this volume).
24. https://signal.org/blog/

7 References

Abramson, Kate. 2014. "Turning up the Lights on Gaslighting." *Philosophical Perspectives* 28 (1): 1–30. doi:10.1111/phpe.12046.

Ackerman, Felicia. 1995. "The Concept of Manipulativeness." *Philosophical Perspectives* 9: 335–40. doi:10.2307/2214225.

Alm, David. 2015. "Responsibility, Manipulation, and Resentment." *Social Theory and Practice* 41 (2): 253–74.

Alston, W. P. 1967. "Vagueness." In *The Encyclopedia of Philosophy*, edited by P. Edwards, 218–21. New York, NY: Collier-Macmillan.

Barnhill, Anne. 2014. "What is Manipulation?" In Coons and Weber 2014, 51–72.

Barnhill, Anne. 2016. "I'd Like to Teach the World to Think: Commercial Advertising and Manipulation." *Journal of Marketing Behavior* 1 (3–4): 307–28. doi:10.1561/107.00000020.

Baron, Marcia. 2003. "Manipulativeness." *Proceedings and Addresses of the American Philosophical Association* 77 (2): 37. doi:10.2307/3219740.

Baron, Marcia. 2014. "The Mens Rea and Moral Status of Manipulation." In Coons and Weber 2014, 98–109.

Beaney, Michael. 2021. "Analysis." In *Stanford Encyclopedia of Philosophy: Summer 2021*, edited by Edward N. Zalta. https://plato.stanford.edu/archives/sum2021/entries/analysis/.

Blumenfeld, David. 1988. "Freedom and Mind Control." *American Philosophical Quarterly* 25 (3): 215–27.

Blumenthal-Barby, J. S. 2012. "Between Reason and Coercion: Ethically Permissible Influence in Health Care and Health Policy Contexts." *Kennedy Institute of Ethics Journal* 22 (4): 345–66.

Buss, Sarah. 2005. "Valuing Autonomy and Respecting Persons: Manipulation, Seduction, and the Basis of Moral Constraints." *Ethics* 115 (2): 195–235. doi:10.1086/426304.

Cave, Eric M. 2007. "What's Wrong with Motive Manipulation?" *Ethical Theory and Moral Practice* 10 (2): 129–44.

Chappell, Sophie G. 2019. "Introducing Epiphanies." *Zeitschrift Für Ethik Und Moralphilosophie* 2 (1): 95–121. doi:10.1007/s42048-019-00029-4.

Clarke, Roger. 1994. "The Digital Persona and its Application to Data Surveillance." *The Information Society* 10 (2): 77–92. doi:10.1080/01972243.1994.9960160.

Climenhaga, Nevin. 2018. "Intuitions are Used as Evidence in Philosophy." *Mind* 127 (505): 69–104. doi:10.1093/mind/fzw032.

Cohen, Shlomo. 2018. "Manipulation and Deception." *Australasian Journal of Philosophy* 96 (3): 483–97. doi:10.1080/00048402.2017.1386692.

Concordia. 2016. "Cambridge Analytica – The Power of Big Data and Psychographics [video file]." Accessed September 10, 2021. www.youtube.com/watch?app=desktop&v=n8Dd5aVXLCc&feature=youtu.be.

Coons, Christian, and Michael Weber, eds. 2014. *Manipulation: Theory and Practice*. Oxford: Oxford University Press.

Faden, Ruth R., and Tom L. Beauchamp. 1986. *A History and Theory of Informed Consent*. New York, NY: Oxford University Press.

Feurer, Sven, and Alexander Fischer. 2021. *Exploring the Ethical Limits of Manipulation in Marketing: A Discussion Based on Consumer Perceptions*. under review.

Fischer, Alexander. 2017. *Manipulation: Zur Theorie und Ethik einer Form der Beeinflussung*. Berlin: Suhrkamp.

Fischer, Alexander, and Christian Illies. 2018. "Modulated Feelings: The Pleasurable-Ends-Model of Manipulation." *Philosophical Inquiries* 1 (2): 25–44. Accessed August 06, 2020.

Garnett, Michael. 2018. "Coercion: The Wrong and the Bad." *Ethics* 128 (3): 545–73. doi:10.1086/695989.

Gorin, Moti. 2022. "Gamification, Manipulation, and Domination." In *The Philosophy of Online Manipulation*, edited by Jongepier, F. and Klenk, M., 199–215. New York: Routledge.

Gorin, Moti. 2014a. "Do Manipulators Always Threaten Rationality?" *American Philosophical Quarterly* 51 (1): 51–61. Accessed June 04, 2019.

Gorin, Moti. 2014b. "Towards a Theory of Interpersonal Manipulation." In Coons and Weber 2014, 73–97.

Gorin, Moti. 2018. "Paternalistic Manipulation." In *The Routledge Handbook of the Philosophy of Paternalism*, edited by Jason Hanna and Kalle Grill, 236–47. New York, NY: Routledge.

Gorin, Moti, Steven Joffe, Neal Dickert, and Scott Halpern. 2017. "Justifying Clinical Nudges." *The Hastings Center Report* 47 (2): 32–38. doi:10.1002/hast.688.

Greenspan, Patricia. 2003. "The Problem with Manipulation." *American Philosophical Quarterly* 40 (2): 155–64.

Handelman, Sapir. 2009. *Thought Manipulation: The Use and Abuse of Psychological Trickery*. Santa Barbara, CA: Praeger Publishers.

Harari, Yuval N. 2018. "The Myth of Freedom." *The Guardian*, September 14. Accessed August 27, 2021. www.theguardian.com/books/2018/sep/14/yuval-noah-harari-the-new-threat-to-liberal-democracy.

Henrich, Joseph, Steven J. Heine, and Ara Norenzayan. 2010. "Most People Are Not WEIRD." *Nature* 466 (7302): 29. doi:10.1038/466029a.

Hopster, Jeroen, and Michael Klenk. 2020. "Why Metaethics Needs Empirical Moral Psychology." *Critica* 52 (155). doi:10.22201/iifs.18704905e.2020.1193.

Houk, Timothy. 2018. "The Nature and Morality of Manipulation." PhD thesis, University of California, Davies.

Jongepier, Fleur. 2017. "The Circumstances of Self-Knowledge." PhD thesis, Radboud University Nijmegen.

Jongepier, Fleur, and Michael Klenk, eds. 2022. *The Philosophy of Online Manipulation*. New York, NY: Routledge.

Kane, Robert. 1996. *The Significance of Free Will*. New York, NY: Oxford University Press.

Keymolen, Esther. 2018. "Trust in the Networked Era." *Techné: Research in Philosophy and Technology* 22 (1): 51–75. doi:10.5840/techne201792271.

Klenk, Michael. 2022. "Manipulation, Injustice, and Technology." *In The Philosophy of Online Manipulation*, edited by Jongepier, F. and Klenk, M., pp. 108–132. New York: Routledge.

Klenk, Michael. 2020. "Digital Well-Being and Manipulation Online." In *Ethics of Digital Well-Being: A Multidisciplinary Perspective*, edited by Christopher Burr and Luciano Floridi. Cham: Springer. Accessed November 17, 2019. 81–100. doi: 10.1007/978-3-030-50585-1_4.

Klenk, Michael. 2021a. "Interpersonal Manipulation." *SSRN Electronic Journal*. doi:10.2139/ssrn.3859178.

Klenk, Michael. 2021b. "Manipulation (Online): Sometimes Hidden, Always Careless." *Review of Social Economy* 80: 1, 85–105. doi:10.1080/00346764.2021.1894350.

Klenk, Michael, and Jeff Hancock. 2019. "Autonomy and Online Manipulation." *Internet Policy Review*. Accessed February 28, 2020. https://policyreview.info/articles/news/autonomy-and-online-manipulation/1431.

Klenk, Michael, Sunny Xun Liu, and Jeff Hancock. 2021. *Pulling the Rug from under the Tech-lash: Online Influences are Perceived to be More Manipulative than Similar Offline Influences*. Under review.

Kligman, M., and C. M. Culver. 1992. "An Analysis of Interpersonal Manipulation." *The Journal of Medicine and Philosophy* 17 (2): 173–97. doi:10.1093/jmp/17.2.173.

Knobe, Joshua, and Shaun Nichols. 2008. *Experimental Philosophy*. New York, NY: Oxford University Press.

Krstić, Vladimir, and Chantelle Saville. 2019. "Deception (Under Uncertainty) as a Kind of Manipulation." *Australasian Journal of Philosophy* 97 (4): 830–35. doi:10.1080/00048402.2019.1604777.

Leonelli, Sabina. 2016. "Locating Ethics in Data Science: Responsibility and Accountability in Global and Distributed Knowledge Production Systems." *Philosophical Transactions of the Royal Society A* 374 (20160122).

Levy, Neil. 2019. "Nudge, Nudge, Wink, Wink: Nudging Is Giving Reasons." *Ergo* 6. doi:10.3998/ergo.12405314.0006.010.

Manne, Kate. 2014. "Non-Machiavellian Manipulation and the Opacity of Motive." In Coons and Weber 2014, 221–46.

Noggle, Robert. 1996. "Manipulative Actions: A Conceptual and Moral Analysis." *American Philosophical Quarterly* 33 (1): 43–55.

Noggle, Robert. 2018a. "Manipulation, Salience, and Nudges." *Bioethics* 32 (3): 164–70.

Noggle, Robert. 2018b. "The Ethics of Manipulation." In *Stanford Encyclopedia of Philosophy*. https://plato.stanford.edu/entries/ethics-manipulation/

Pereboom, Derk. 2001. *Living Without Free Will*. Cambridge: Cambridge University Press.

Pölzler, Thomas. 2020. *Moral Reality and the Empirical Sciences*. New York, NY: Routledge.

Queloz, Matthieu. 2021. *The Practical Origins of Ideas: Genealogy as Conceptual Reverse-engineering*. Oxford: Oxford University Press.

Rudinow, Joel. 1978. "Manipulation." *Ethics* 88 (4): 338–47. doi:10.1086/292086.

Scanlon, Thomas M. 1998. *What We Owe to Each Other*. Cambridge, MA: Harvard University Press.

Schelling, Thomas C. 1997. *Strategy of Conflict*. Cambridge, MA: Harvard University Press.

Schmidt, A. T., and B. Engelen. 2020. "The Ethics of Nudging: An Overview." *Philosophy Compass* 15 (4).

Sripada, Chandra S. 2012. "What Makes a Manipulated Agent Unfree?" *Philosophy and Phenomenological Research* 85 (3): 563–93.

Stanley, Jason. 2015. *How Propaganda Works*. Princeton, NJ: Princeton University Press.

Sunstein, Cass R. 2016a. "Fifty Shades of Manipulation." *Journal of Marketing Behavior* 1 (3–4): 214–44. doi:10.1561/107.00000014.

Sunstein, Cass R. 2016b. *The Ethics of Influence: Government in the Age of Behavioral Science*. Cambridge: Cambridge University Press.

Susser, Daniel, Beate Roessler, and Helen Nissenbaum. 2019a. "Online Manipulation: Hidden Influences in A Digital World." *Georgetown Law Technology Review* 4 (1): 1–45.

Susser, Daniel, Beate Roessler, and Helen Nissenbaum. 2019b. "Technology, Autonomy, and Manipulation." *Internet Policy Review* 8 (2): 1–22. doi:10.14763/2019.2.1410.

Terpstra, Arnout, Alexander P. Schouten, Alwin de Rooij, and Ronald E. Leenes. 2019. "Improving Privacy Choice Through Design: How Designing for Reflection Could Support Privacy Self-Management." *FirstMonday* 24 (7): 1–13. doi:10.5210/fm.v24i7.9358.

Vallor, Shannon. 2006. *Technology and the Virtues. A Philosophical Guide to a Future Worth Wanting*. Oxford: Oxford University Press.

Van de Poel, Ibo, and Lambèr Royakkers. 2011. *Ethics, Technology, and Engineering: An Introduction*. Malden, MA: Wiley-Blackwell.

Williams, James. 2018. *Stand out of Our Light: Freedom and Resistance in the Attention Economy*. Cambridge: Cambridge University Press.

Wood, Allen W. 2014. "Coercion, Manipulation, Exploitation." In Coons and Weber 2014, 17–50.

Zarouali, Brahim, Tom Dobber, Guy de Pauw, and Claes de Vreese. 2020. "Using a Personality-Profiling Algorithm to Investigate Political Microtargeting: Assessing the Persuasion Effects of Personality-Tailored Ads on Social Media." *Communication Research*, 1–26. doi:10.1177/0093650220961965.

3 How philosophy might contribute to the practical ethics of online manipulation

Anne Barnhill

1 Introduction

There is intense and increasing concern with the various ways in which machines influence humans and humans are influenced online, as chapters in this volume explore. These forms of influence are sometimes called "manipulative" or "manipulation." What are we to make of these charges of "manipulation"? How might philosophers and philosophical work (such as this volume) contribute to the real-world discussion of these issues?

One way that philosophical work might contribute is by considering whether these kinds of influence are *actually* instances of manipulation. That is, what are the best philosophical accounts of manipulation, and on these accounts, do these influences come out as manipulation? In this chapter, I engage in that kind of inquiry (Sections 4 and 5). But then (in Section 6), I question whether that is the most productive way for philosophical work to contribute. If the ultimate aim of our inquiry is to reach ethical conclusions about these influences and our responses to them, does it really matter if this influence comes out as manipulation on our best philosophical accounts of manipulation? Or might it be more productive to put that theoretical question and accounts of manipulation to the side and focus more directly on identifying those forms of influence that strike people as problematic and considering the ethics of those forms of influence? I conclude that if we're interested in understanding online influence and the ways in which it might be problematic, engaging with philosophical accounts of manipulation *is* productive, because these accounts of manipulation can help us to identify various ways in which online influence might be problematic. However, we shouldn't get bogged down in the philosophical analysis of manipulation, nor bogged down adjudicating whether particular instances of online influence are manipulative according to these accounts. Instead, when a form of online influence is called "manipulative," we should focus on identifying the specific feature(s) of the influence that sparks the charge of manipulation, and then we should query whether influence of that form is problematic. I spell out these methodological suggestions, and others, in Section 7.

DOI: 10.4324/9781003205425-4

In making this argument in this chapter, I will focus on a particular category of online influence – political influences that occur online – as an illustrative example.

2 Worrying forms of online political influence

Consider this headline from *The New York Times*: "Facebook Says It Won't Back Down From Allowing Lies in Political Ads" (Isaac and Kang 2020). As this headline suggests, online political influence has raised concerns. "Raised concerns" might be too tepid; perhaps it's more accurate to say that online political influence has set off alarms, wreaked political havoc, and created diplomatic firestorms?

Much political communication and influence occur online. Governments, candidates, political parties, and other political actors are increasingly deploying online influence, including some very dodgy forms of it. For example, research has documented the increasing use of multiple forms of online political influence:

- Spreading content that is fallacious or misleading or is inflammatory. This content takes the form of political ads, Facebook posts, tweets, news stories, texts, etc. In some cases, political content contains information that is outright false (Roose 2018).
- Making false claims about sponsors or funders.[1]
- Spreading content on social media using fake accounts and automation. A political actor doesn't need to have actual people posting, liking, or retweeting content; they can write code that will do it for them.
- Harvesting people's data online, using it to create psychographic profiles, and then using these profiles to target them with political content.

A well-documented example of an online political influence campaign occurred during the 2016 US presidential election, when the Russian government used Facebook and other social media to influence US voters, creating social media accounts and using stolen identities to pose as conservative and progressive activists, "in order to sow discord among the electorate by creating Facebook groups, distributing divisive ads and posting inflammatory images" (Apuzzo and LaFraniere 2018).

Another notorious example of online political influence is the Facebook/Cambridge Analytica case, in which Cambridge Analytica accessed data from 50–87 million Facebook users, unbeknownst to them, combined this with an array of other data about users, and then (purportedly) used this data to create personality profiles of millions of American voters, which could be used to tailor political messaging to people based on their personality type (Prokop 2018). Christopher Wylie, a former Cambridge Analytica employee and whistleblower claimed that this data about people was used to build "models to exploit what we knew about them and target their inner demons" (Cadwalladr and Graham-Harrison 2018).[2]

As some scholars have noted, the use of fallacious, misleading, and inflammatory content in political communication is nothing new. What's new is that technology enables content to be targeted to individuals (because large amounts of data are collected about people online, and this is used to micro-target content to them), and large numbers of people can be targeted online using automated means.[3] In 2019, the Computational Propaganda Research Project documented that in 70 countries, there is "at one political party or government agency using social media to shape public attitudes domestically" as well as a "handful of sophisticated state actors [who] use computational propaganda for foreign influence operations" (Bradshaw and Howard 2019, 1). Freedom House, an independent research organization, concluded that social media have

> provided an extremely useful and inexpensive platform for malign influence operations by foreign and domestic actors alike. . . . They build large audiences around similar interests, lace their political messaging with false or inflammatory content, and coordinate its dissemination across multiple platforms.
>
> (Shabaz and Funk 2019)

The use of fake accounts and bots to propagate content has a few potential effects. First, the use of fake accounts and fake identities presumably makes political content more persuasive; for example, it is plausible that a US voter is more likely to engage with and be persuaded by political content posted by another US voter than with political content posted by an anonymous person or a foreign national. Second, using fake accounts and bots can dramatically increase the number of "users" who are propagating the content through social media. This creates the appearance of more interest in issues, and more support for positions and for candidates, than there actually is (e.g., there appears to be significant positive support for a candidate, because favorable posts about him are being tweeted and retweeted, but in fact a significant portion of these retweets are from fake accounts). Third, this appearance of support increases the actual support of the candidate, in a boot-strapping effect. In the old days, campaigns hired people to show up at their rallies to give the appearance of support; today, campaigns can write code to accomplish the same thing. Fourth, using fake accounts to spread content can serve to make extreme views more mainstream, by creating the appearance that more people hold these extreme views than they actually do.

3 Charges that online political influence is manipulation

Many of the forms of online political influence mentioned earlier are regularly called manipulative and manipulation. These charges of manipulation are made by scholars who study this influence, reporters who write about it, and lawmakers who are concerned with it, such as:

- Facebook is used as "a tool to lie to and manipulate voters" because it allows political actors to micro-target voters with political ads that contain disinformation, in the words of one political medial professional (Isaac and Kang 2020).
- Jonathan Albright, research director at Columbia University's Tow Center for Digital Journalism, also describes Facebook as enabling manipulation: "Facebook built incredibly effective tools which let Russia profile citizens here in the U.S. and figure out how to manipulate us," as quoted in *the New York Times* (Frenkel and Benner 2018).
- The Computational Propaganda Research Project (CPRP) refers to "use of social media to manipulate public opinion." They also describe the use of "fake social media accounts, online trolls and commentators, and political bots to distort conversations online, help generate a false sense of popularity or political consensus, mainstream extremist opinions, and influence political agendas" (Bradshaw and Howard 2018, 7).
- An article reporting on the CPRP's work refers also to the manipulation of public opinion:

  > A study by the Oxford Internet Institute . . . found that since 2017, organized social media manipulation has more than doubled with at least 70 countries known to be using online propaganda to manipulate mass public opinion, and in some cases, on a global scale.
  >
  > (Curtis 2019)

- An article describing the Russian campaign to influence the US 2016 presidential election refers to the manipulation of the campaign: "The Russians stole the identities of American citizens, posed as political activists and used the flash points of immigration, religion and race to manipulate a campaign in which those issues were already particularly divisive, prosecutors said" (Apuzzo and LaFraniere 2018).
- An article on the Cambridge Analytica affair mentions manipulating voters and distorting democratic discourse: "The real story is about how personal data from social media is being used by companies to manipulate voters and distort democratic discourse" (Ghosh and Scott 2018).
- An academic paper on social media bots refers to the manipulation of social networks: "the policymakers and pundits currently calling for platform companies to prevent foreign manipulation of social networks and to enact more stringent bot policy" (Gorwa and Guilbeault 2020).
- Freedom House describes social media thus: "What was once a liberating technology has become a conduit for surveillance and electoral manipulation" (Shabaz and Funk 2019). They refer to "Manipulating

Social Media to Undermine Democracy." They also describe how "Repressive regimes, elected incumbents with authoritarian ambitions, and unscrupulous partisan operatives have exploited the unregulated spaces of social media platforms, converting them into instruments for political distortion and societal control."

- Freedom House also writes: "Manipulation and disinformation tactics played an important role in elections in at least 17 other countries over the past year, damaging citizens' ability to choose their leaders based on factual news and authentic debate" (Shabaz and Funk 2019).

We see, in this list, various manipulation-related concerns with online political influence: people are manipulated, social media (and social networks) are manipulated, public opinion is manipulated, and campaigns and elections are manipulated. These manipulation-related concerns are connected with other concerns: people are lied to, people are misled, online conversations are distorted, extremist opinions are mainstreamed, political agendas are affected, democratic discourse is distorted, citizens' ability to choose leaders based on facts and authentic debate is undermined, and societal control is exerted.

When it's claimed that these forms of political influence are manipulation or manipulative, what's being claimed? I'm not sure. It's not clear what precisely people are claiming about influence when they call it manipulative. What *is* clear is that "manipulation" is used pejoratively in these contexts: to call online political influence manipulation is to raise concerns about it.

In response to these (unclear) charges of manipulation, one way that philosophical work might contribute is by considering whether these instances of influence truly are manipulation, on plausible philosophical accounts of manipulation. Let's begin by doing that by considering several of the many accounts of manipulation found in the literature.

4 Accounts of interpersonal manipulation

What kind of influence is manipulation? Cass Sunstein notes that manipulation is often seen as a problematic form of control: "It is often thought that when people are being manipulated, they are treated as 'puppets on a string' " (Sunstein 2016). But manipulation controls not by forcing someone to do something but by inducing the desired action.[4] Bob Goodin describes manipulation as a form of power – but a way of undermining resistance, not a way of overcoming resistance (Goodin 1980). In other words, manipulation controls by undermining or hijacking someone's self-control, not by overpowering them.

Theorists typically distinguish manipulation from coercion (Faden and Beauchamp 1986; Blumenthal-Barby 2012; Sunstein 2016). But what kind of non-coercive influence exactly is manipulation? Many ways of inducing behavior are not manipulative: if I offer you a fair wage to babysit my kids, and thereby induce you to babysit my kids, I haven't manipulated you (excluding special circumstances). If I tell you a joke and make you laugh, I haven't manipulated you (excluding special circumstances). So what is manipulation exactly, as a form of non-coercive influence?

4.1 Manipulation as covert, deceptive, or weakness-targeting influence

Some theorists analyze manipulation as covert influence of some sort. For example, Alan Ware defines manipulation as structuring someone's environment with the intention of changing his choice and succeeding in doing so, when the manipulated person "either has no knowledge of, or does not understand, the ways in which [the manipulator] affects his choices" (Ware 1981, 165). The victim's ignorance is a defining characteristic of manipulation, on Ware's account. (Ware notes that manipulation need not be against someone's interests; other theorists have also pointed out that manipulation can be beneficent – we can manipulate people for their own good; see Goodin 1980 and Barnhill 2014).

Robert Goodin also analyzes manipulation as deceptive influence: manipulation is deceptively influencing someone, causing him to act against his putative will (Goodin 1980, 7–23). Goodin discusses manipulation as it occurs in politics and observes that manipulation carries "especially strong connotations of something sneaky," with manipulation characteristically happening unbeknownst to its victim (Goodin 1980, 9). This makes manipulation an important but challenging form of political power to study.

Goodin observes that political scientists, in their study of power, look for fights and see who wins them, but this is an incomplete study of power. Because manipulation is a form of power that is wielded sneakily, there's no fight to observe. Goodin writes: "Power plays are far more successful if accomplished deceptively. You stand a far better chance of getting your way if others do not notice that you are doing something to them that they should be resisting" (Goodin 1980, 31). This analysis of manipulation as a form of political power, from 1980, clearly applies to the forms of online political influence that we see today.[5]

Susser, Roessler, and Nissenbaum (2019), in an analysis focused on online influences, also conclude that manipulation is a form of covert influence. They conclude that

> [A]t its core, manipulation is hidden influence – the covert subversion of another person's decision-making power. . . . [M]anipulation functions by exploiting the manipulee's cognitive (or affective) weaknesses and vulnerabilities in order to steer his or her decision-making process towards the manipulator's ends.
>
> (Susser, Roessler, and Nissenbaum 2019)

Several authors have objected to these analyses of manipulation as deceptive or covert influence, arguing that some instances of manipulation do not involve deception or covertness (Noggle 1996; Barnhill 2014; Gorin 2014). Some instances of manipulation are blatant; an example would be a manipulative "guilt trip," in which someone is made to feel guilty in a blatant yet manipulative way (Barnhill 2014). To account for these kinds of cases, Joel Rudinow (1978) analyzes manipulation as a form of influence that *either* involves deception or plays on someone weakness.[6]

4.2 Manipulation as non-persuasion or non-reason-tracking influence

Other accounts of manipulation contrast manipulation with persuasion or with proper persuasion. Ruth Faden and Tom Beauchamp analyze manipulation by contrasting it with persuasion (Faden and Beauchamp 1986). Persuasion proceeds by improving a person's understanding of his or her situation whereas manipulation does not. They identify three kinds of manipulation: manipulation of options, in which options in the environment are modified by increasing or decreasing available options or by offering rewards or threatening punishments; manipulation of information, in which the person's perception of options is modified by non-persuasively affecting the person's understanding of the situation; and psychological manipulation, in which the person is influenced by causing changes in mental processes other than those involved in understanding (Faden and Beauchamp 1986).[7]

Consider how this account of psychological manipulation applies to political influence. There are many instances of political influence that arguably do not improve someone's beliefs about her situation but instead change her attitudes in other ways; political speech may aim to inspire fear, or hope, or feelings of solidarity. On Faden and Beauchamp's account, this political speech is psychological manipulation because it does not improve the audience's understanding.

Another account that contrasts manipulation with proper persuasion is Claudia Mills's (1995) account. According to Mills, what's distinctive about manipulation is that it purports to be legitimate persuasion that offers good reasons, but in fact bad reasons are offered. Mills writes that "a manipulator tries to change another's beliefs and desires by offering her bad reasons, disguised as good, or faulty arguments, disguised as sound – where the manipulator himself knows these to be bad reasons and faulty arguments" (Mills 1995, 100).

Moti Gorin (2014) analyzes manipulation as influence that deliberately *fails to track reasons*. According to Gorin, there are multiple ways in which manipulative influence can fail to track reasons. In some cases, manipulators "intend their manipulees to behave in ways they (the manipulators) do not believe to be supported by reasons" (Gorin 2014, 97). In other cases, manipulators believe that the behavior they're trying to produce *is* supported by reasons, but they are not motivated by those reasons; the manipulator would be trying to produce that behavior regardless. In another category of cases, manipulators believe that the behavior they're trying produce is supported by reasons, but they produce that behavior not by offering those reasons but instead by using means of influence that do not reliably track those reasons. Gorin gives the example of a son getting his mother to go to the hospital (a behavior that is supported by reasons) by tricking her (Gorin 2014, 82–83).

4.3 *Manipulation as influence that does not sufficiently engage reflective and deliberative capacities*

Cass Sunstein suggests this definition of manipulation: an effort to influence people's choices counts as manipulative to the extent that it does not sufficiently engage or appeal to their capacity for reflection and deliberation. In other words, influence is manipulative if its engagement of someone's reflective and deliberative capacities falls short of a standard of sufficient engagement.

Importantly, Sunstein does not assume that engagement with reflective and deliberative capacities is always called for. He writes:

> Suppose, for example, that a good friend frames an option in the most attractive light and with a cheerful voice; or that the Department of Transportation embarks on a vivid, even graphic public education campaign to reduce texting while driving; or that a politician argues in favor of same-sex marriage in a way that points, in an emotionally evocative way, to the lived experience of same-sex couples. In all of these cases, we might have long debates about whether the relevant statements are appealing to people's capacity for reflective and deliberative choice. And even if we conclude that they are not, we should not therefore be committed to the view that manipulation is involved.

Influencing someone without engaging her in reflection and deliberation is not necessarily manipulative. It depends upon whether the situation calls for engagement in reflection and deliberation. Making someone happy by smiling at her, making someone laugh by telling him a funny joke, and making someone who stepped on your toe feel bad by shouting "ouch!" are not instances of manipulation (barring special circumstances). There are many ways in which we influence each other, besides engaging each other in reflection or deliberation; these forms of influence are not necessarily manipulative. In other words, "non-rational" influence and non-persuasive influence are not necessarily manipulative. They are manipulative only when rational influence or persuasion is called for.

An important question for a discussion of political influence is: what kind of reflection and deliberation is called for in different political contexts?

Much political speech – for example, political advertising and campaign speeches – does not provide arguments or encourage reflection or deliberation. Instead, the speech targets people's emotions, causing negative emotions (fear, resentment, anger) or positive attitudes (hope, optimism, solidarity). When people are caused to feel emotions about a politician or a policy, but are not caused to reflect or deliberate on the politician's or the policy's merits, is this insufficiently reflective and deliberative?

4.4 Manipulation as influence that makes someone fall short of ideals for practical reasoning

Let's consider one last account of manipulation. Robert Noggle analyzes manipulative action as the attempt to get someone's beliefs, desires, or emotions to fall short of the ideals that govern beliefs, desires, and emotions (Noggle 1996). He writes: "there are certain norms or ideals that govern beliefs, desires, and emotions. I am suggesting that manipulative action is the attempt to get someone's beliefs, desires, or emotions to violate these norms, to fall short of these ideals" (Noggle 1996, 44). More specifically, manipulative action is the attempt to get someone's beliefs, desires, or emotions to fall short of the ideals that *in the view of the influencer* govern the target's beliefs, desires, and emotion. On this view, whether influence is manipulation doesn't depend on whether the influence uses "non-rational" means or is "non-rational" persuasion but on whether the influence is intended to make the person fall short of ideals for belief, desire, and emotion. In explaining this analysis of manipulation, Noggle uses the metaphor of "adjusting psychological levers": manipulative action attempts to adjust psychological levers away from those settings that the manipulator thinks are the ideal settings for the target.

A virtue of Noggle's account is that it can distinguish between manipulative and non-manipulative appeals to emotion. Consider, for example, a political ad that instills fear in its audience about the prospect of losing their jobs as a result of economic change. If it causes excessive fear, this ad would make its audience fall short of ideals for emotion and count as manipulative on Noggle's account. But making someone feel the appropriate level of fear would get her closer to ideals for emotion and thus would not be manipulative influence.

5 Applying accounts of interpersonal manipulation to online political influence

Consider again the forms of online political influence discussed earlier, in light of these various accounts of interpersonal manipulation. Fallacious and misleading influence comes out as manipulative on many of these accounts. It would count as "manipulation of information" (Faden and Beauchamp 1986), as giving "bad arguments disguised as good arguments" (Mills 1995), as "deceptively influencing someone, causing him to act against his putative will" (Goodin 1980) and as causing someone to fall short of ideals for belief (Noggle 1996). Covert influence – for example, harvesting people's data from Facebook without their knowledge and ads were micro-targeted to people without them knowing they were being micro-targeted – is manipulative, according to Goodin (1980), Ware (1981), and Susser, Roessler, and Nissenbaum (2019). Political influence that plays on people's psychological weaknesses (recall that former Cambridge Analytica employee Christopher Wylie said they targeted people's "inner demons") may be manipulative according to Rudinow (1978).

Covert influence and political communication with false or misleading content are clear-cut cases; they are clearly classifiable as manipulative according to many of these accounts of manipulation. Other cases are less clear-cut. Consider political influence that plays on people's emotions. When is it playing on emotions manipulative, and when is it not? For example, consider an advertisement by President Donald Trump's political team, which ran in the United States during the congressional election of 2018.[8] The ad shows images of a caravan of Central American migrants traveling through Mexico toward the US border (where they would seek asylum) and also shows video clips of Luis Bracamontes, an (undocumented) Mexican immigrant to the United States who killed two police officers. Commentators have called the ad misleading, insofar as it implies that a stricter refugee and asylum policy would have prevented Luis Bracamontes from entering the United States and killing people; that is not true, as he entered the country illegally. But let's focus on another aspect of the ad: it plays on people's fears about the effect of immigration on the United States, and that Latino people (or maybe just Latino immigrants) are dangerous. How is the criminal behavior of Luis Bracamontes (a single Mexican immigrant who entered the United States illegally) relevant to the likely behavior of a group Central American migrants seeking asylum in the United States? The only link seems to be that they are Latino immigrants. For this reason, many commentators have called the ad racist and inflammatory. In this respect, is the ad manipulative?

Recall Robert Noggle's account of manipulation: you manipulate someone when you make her beliefs, desires, or emotions fall short of the ideals that you think apply to her. The *influencer's* conception of which beliefs, desires, and emotions are ideal for the influenced person is what's pertinent to determining when manipulation has occurred.[9] Therefore, Noggle writes, "a racist who attempts to incite racial fears may not intend to move the other person away from what *he* – mistakenly – takes to be the other person's ideal condition, and so we cannot accuse him of acting manipulatively" (Noggle 1996, 50). Thus, if those who created this ad believe that it's causing people to feel the appropriate kind of fear of Latino immigrants to the United States, then this ad is not manipulative, according to Noggle's account. However, this ad, and the way that it stokes fear and racialized stereotypes, may seem like a paradigm example of manipulation to some.

In this way, we should expect disagreement about whether influence is manipulative when we apply Noggle's account of manipulation to instances of purported manipulation. This disagreement could be rooted in underlying disagreement about which attitudes are appropriate (e.g., is it appropriate to feel afraid of immigrants?) and disagreement about the intentions of those who create political influence (e.g., does the Trump political team believe that racialized fear is appropriate, or are they just causing it to generate support for Trump?).

Let's consider another example of political influence and probe whether it is manipulation. During the 2012 US Presidential campaign, the political team of President Barack Obama created a database of voters who were potential but not assured Obama voters. The database included information about which issues these voters cared about most. The Obama campaign targeted these voters with information (in phone calls and mailings) about Obama's positions on those particular issues in an effort to turn them into Obama voters (Beckett 2012).

Is this manipulative? Suppose, for the sake of argument, that all of the messages sent to voters were true, informative, and caused voters to form correct beliefs about a policy position of President Obama. Is this micro-targeting nonetheless manipulative, insofar as it gives voters an incomplete picture of Obama's suite of policy positions and a picture that's skewed toward those policy positions they agree with? Consider Faden and Beauchamp's account of psychological manipulation, in which the person is influenced by causing changes in mental processes other than those involved in understanding. The Obama campaign's micro-targeting (as stipulatively described here) would not count as psychological manipulation. Would it count as manipulation on Sunstein's account – that is, does it insufficiently engage or appeal to the voters' capacity for reflection and deliberation? The micro-targeted messages did engage people in reflection about specific policy positions (or so we are stipulating, for the sake of argument). But perhaps they didn't engage people in reflection about a broad enough range of policy positions? On the other hand, voters may have limited attention that they will spend on candidate's positions; could directing their attention to Obama's positions on issues they care about most be the best way to improve their reflection and deliberation about Obama the candidate?[10]

Consider also Noggle's account of manipulation: you manipulate someone when you make her beliefs, desires, or emotions fall short of the ideals that you think apply to her. If a voter has correct beliefs about some of a candidate's policy positions (those that she agrees with), and has little information about other policy positions (including ones that she might disagree with), is she falling short of ideals for beliefs? What are the relevant ideals when it comes to voters' beliefs about candidates? How broadly informed should voters be about candidates' policy positions?

Here again we see that there could be disagreement about whether an instance of online political influence is manipulation. One source of disagreement is that we may apply different theoretical accounts of manipulation, and a particular instance of influence will come out as manipulation on one account but not another. That is, depending on which account of manipulation we use, we will reach a different conclusion about whether the Obama campaign's micro-targeting of voters was manipulative. Another source of disagreement is that when we use specific theoretical account of manipulation, an attribution of manipulation rests on specific underlying issues (e.g., if a voter reflects and deliberates about only a subset of a candidate's positions, is that voter engaging in sufficient reflection and deliberation?), and we may disagree about those issues.

6 Now what?

What should we make of this disagreement? And how should we, as philosophers, proceed in the face of this disagreement?

One option – perhaps the path of least resistance for philosophers – is to keep working on our theoretical accounts of manipulation. Perhaps, the fact that there are multiple accounts of manipulation being applied suggests that we haven't hit on the best account yet, so we should keep fashioning and refashioning our theoretical accounts of manipulation. Once there's more of a consensus, then the resolution of specific instances of (potential) manipulation will be straightforward: we will apply our best account of manipulation to a specific instance. Of course, as we just saw, the application of an account of manipulation may require the resolution of underlying issues, some of which are normative issues. So we should keep working on those issues, too. And of course, we don't just want to know whether an instance of influence is manipulation; we also want to know if, and in what way, that instance of influence is problematic. Thus there's an additional chunk of theoretical work that needs to be done; we need to hammer out if, and in what way, manipulation (as analyzed on our best/consensus account of manipulation) is problematic.

Putting it all together, on this "path of least resistance," philosophers and philosophical work contribute in the following ways to the ongoing real-world discussion of manipulative online influence:

1. Philosophers continue fashioning theoretical accounts of manipulation.
2. When instances of online influence are called manipulation, we apply our best theoretical account of manipulation and adjudicate whether these instances are manipulation. This requires characterizing the influence in question and assessing whether it comes out as manipulation on our best theoretical account of manipulation. This assessment may require resolving underlying empirical issues, for example, ascertaining what the influencer's intentions were. This assessment may also require

resolving underlying normative issues, for example, whether a voter has engaged in sufficient reflection and deliberation about a candidate.

3. For those instances of influence that turn out to be manipulation on our best account of manipulation, we apply our account of what's problematic about manipulation and conclude that the instance of influence is problematic in that way.

I'd like to suggest that this may not be the best way for philosophers to contribute to the real-world discussion of (potentially manipulative) online influence, for three reasons.

First, what are the chances that we will reach any sort of consensus on the best philosophical account of manipulation? Assuming that we don't, then philosophers' various interventions into policy discussions will involve the application of meaningfully different accounts of manipulation. That may confuse matters more than illuminate them.

Second, even if philosophers can reach greater consensus about manipulation, I'm not convinced that the real-world discussion of (potentially manipulative) online influence really benefits from adjudicating whether particular instances of influence are manipulative or not. Insofar as our aim is to advance the practical ethics of online influence (and not to perfect our philosophical accounts of manipulation), adjudicating whether online influence is manipulative is arguably not very important. What matters to the practical ethics of online influence is whether online influence is problematic, how it is problematic, who (if anyone) should take steps to prevent it, and what steps they should take. In doing this practical ethics work, we shouldn't get bogged down in considering which philosophical accounts of manipulation are best and assessing whether various kinds of online influence are truly manipulative.[11] Instead, we should focus on figuring out which features of online influence are problematic and in what ways. Paying attention to charges of manipulation can help us figure out which kinds of online influence are problematic; this is because a charge of manipulation likely indicates that the speaker finds that instance of influence problematic or at least suspect. Charges of manipulation helpfully point us toward potentially problematic influence, so that we can investigate them further. But in our further investigation, we needn't worry about adjudicating whether those instances of influence *really are* manipulative.

Third, not only is it unimportant to adjudicate whether particular instances of online influence are manipulative, what if this adjudication process filters out some instances of (not-strictly manipulative) influence that are nonetheless seen as problematic and that warrant investigation? As we saw earlier, charges of "manipulation" are often raised in response to online influence. It's generally unclear from context what the speaker means by "manipulation." I'd wager that speakers generally do not have a specific notion of manipulation in mind; rather, "manipulation" is used to identify influence that is problematic in a vaguely defined way. Or perhaps "manipulation" is used to refer to a cluster of distinct forms of problematic influence. I'm not really sure what's going on with everyday use of "manipulation"; suffice it to say, it's a tricky mess, which is probably why developing theoretical accounts of manipulation has been so difficult. So we should expect that any tidy theoretical account of manipulation will not apply to some instances of influence that are called manipulation in our messy everyday discourse. Thus, when we apply our tidy theoretical accounts of manipulation, we will filter out some instances of influence that have been called out as problematic by being called manipulation. For example, if our best/consensus account of manipulation analyzed it as influence that is covert or hidden in some way, then overt influence (e.g., a political ad that overtly attempts to stoke racist fears) will be filtered out when we apply this account of manipulation.

After we've filtered out this instance of influence, what happens then? Perhaps this instance of (not-strictly manipulative) influence drops out of our analysis (in step 3). This is a bad result, on my view. We have eliminated forms of potentially problematic influence from further ethical consideration because the real-world actors who were spotting problematic influence didn't use the (in our view) correct word or concept to refer to these problematic influences (i.e., a journalist or activists referred to dodgy online influence as "manipulation" when it's not quite manipulation). Filtering out problematic influence from further ethical consideration in this way would be a kind of practical ethics malpractice.

Another (less bad) possibility is that the instances of (not-strictly manipulative) influence do not drop out of our analysis after step 2. Instead, we ethically analyze those instances of not-strictly manipulative influence, in step 4:

4. For those instances of influence that turn out not to be manipulation on our best account of manipulation, consider what's problematic about those forms of influence.

But as long as we're doing this step 4, which is a more capacious analysis of forms of influence and the ways in which they are problematic, it's worth asking what the utility of the earlier, more narrow adjudication and assessment of manipulation is. Does marking out some problematic forms of influence as manipulation bring more theoretical clarity – that is, at least we have a name for *some* forms of problematic influence? Perhaps. But it may also risk implying that those forms of influence marked out as manipulation are particularly bad, or particularly in need of a response, as compared to those forms of influence not marked out as manipulation.

Thus, not only is it unimportant to adjudicate whether particular instances of online influence are manipulative, this may also have the undesirable results of either "filtering out" problematic influence from further ethical consideration or implying that these instances of influence are less bad or less in need of a response.

7 Methodological suggestions

I've suggested that philosophers should resist the path of least resistance. So what should we do instead? Broadly speaking, we should figure out how to make ourselves useful to the real-world conversation about manipulative online influence. More specifically, I have three methodological suggestions for making (what I hope would be) useful philosophical contributions.

7.1 Suggestion 1: focus on identifying the features of influence that strike people as manipulative

When a form of online influence is called "manipulative," don't worry about adjudicating whether it *really is* manipulation according to the best account of manipulation. Instead, focus on ascertaining why the influence is being called manipulative. What are the features of the influence that strike people as manipulative? What is causing them to make the charge of manipulation? Empirical research can help to answer these questions.

Existing philosophical accounts of manipulation can also help us to answer these questions. Accounts of manipulation have been crafted by scholars who've thought deeply about influence and about what makes influence problematic and have thought deeply about how "manipulation" and "manipulative" are used. Thus these accounts are a good source of insight into those features of influence that may be seen as problematic and that may be provoking charges of manipulation.

Based on our discussion of philosophical accounts of manipulation, what are those features? Some accounts of manipulation focus on the ways that manipulation undermines people's practical reasoning: by deceiving them, by targeting psychological weaknesses that undermine reasoning, by offering bad reasons, by failing to engage them in deliberation and reflection, and so on. Some accounts of manipulation focus on its covertness or sneakiness. Some accounts of manipulation focus on how manipulation undermines the manipulated person's self-control and/or controls her: by targeting weakness in the person that they can't control, by influencing them in covert ways that they don't recognize and can't correct for, by deceiving them and controlling their belief and behavior in that way.

When a form of online influence is called "manipulative," don't worry about adjudicating whether it is manipulation. Instead, ask a series of questions about the influence to clarify the sense in which it's seen as manipulative, such as:

- Is the influence covert in some way?
- Does the influence include false claims?

- Even if the influence doesn't include false claims, is it misleading?
- Does the influence give people information in ways that are likely to improve the veracity of their beliefs, and their ability to reflect on politics, or not? For example, does the influence overload them with information? For example, does the influence give them information that's skewed in one direction?
- Does the influence fail to engage people in deliberation or reflection? If so, is deliberation or reflection called for in this context?
- Does the influence cause people to have emotional reactions? Are these appropriate or inappropriate emotional reactions, and according to whom?
- Does the influence target, or play on, psychological weaknesses or vulnerabilities? These could be psychological weaknesses or vulnerabilities of the specific targeted individual or weaknesses or vulnerabilities of people in general.
- Does the influence undermine rationality or practical reasoning in some other way? In what way exactly is rationality or practical reasoning undermined?
- Does the influence subvert people's self-control in some other way?

7.2 Suggestion 2: analyze the effects of influence at the aggregate and system level

A second methodological suggestion is that we need to analyze how online influence works (and potentially manipulates) at the aggregate level and system level, not just how it affects (and potentially manipulates) individuals. While the philosophical literature has focused on interpersonal manipulation (and the ways in which individuals may be wronged by manipulation, e.g., their autonomy is violated), when we attend to the real-world conversation about online influence, we see that it is full of concern about the system level or aggregate effects of online influence.

Recall the claims, mentioned earlier, that online political influence *manipulates public opinion* and *manipulates elections*. These manipulation-related concerns were connected with other concerns, that online conversations are distorted, extremist opinions are mainstreamed, political agendas are affected, authentic debate is damaged, democratic discourse is distorted, and societal control is exerted. What worries people about online political influence is not just that many individuals are mistreated (e.g., because their data is harvested without their knowledge, because they are misled, because their reasoning is undermined), but that public opinion and democratic discourse are influenced in a problematic way, and the self-governance of political entities is potentially undermined. Thus, to get a handle on online political influence we need to think about the manipulation of individuals but also the manipulation of systems (e.g., public opinion) and processes (e.g., elections).

Earlier, we discussed accounts of the manipulation of people. But "manipulation" is also used to refer to the physical manipulation of physical objects (e.g., the pilot manipulates the plane's controls) and things that aren't physical objects but are systems or processes. For example, we might say that a government is manipulating its currency (Krugman 2009), meaning that the government engages in policy meant to keep the exchange rate weak, rather than letting the exchange rate be determined naturally. To give another example, we might say that an industry is "manipulating" science: when industry pays researchers to research certain topics, this changes which topics get researched and potentially changes the conclusions of that research (Rampton and Stauber 2001). Rather than science proceeding according to its own internal dynamics, industry is changing what science gets done, how it gets done, and arguably distorting the body of scientific knowledge that results.

How should we conceptualize the manipulation of processes and systems, such as political processes and systems? Here are some first thoughts. When it's claimed that a form of influence manipulates a process or system, these charges of manipulation may register concerns such as:

- *An actor is influencing the process who should not be influencing it* (e.g., industry is influencing science, or a foreign power is influencing an election). Thus when it's said that a foreign power is manipulating an election, perhaps this registers the concern that the foreign power is influencing public opinion about political candidates and potentially changing the outcome of an election, even though this foreign power ought to be staying out of it.
- *The system or process has normal processes of self-regulation, and these are being interfered with and subverted by the influence.* For example, there are normal processes whereby public opinion shifts in response to the authentic views of members of the public (e.g., the popularity of a candidate can have a bootstrapping effect, causing her popularity to increase further). When online political influence (e.g., fake accounts and bots sharing content) gives the appearance that a candidate is more popular than she actually is, and thereby causes this bootstrapping effect to occur for the candidate, this is an interference with the normal processes whereby public opinion changes in response to the authentic views of members of the public.
- *The influence in question changes the outcomes of the process and amounts to a distortion of those outcomes.* For example, foreign interference changes the results of an election, and this amounts to a distortion.
- *The system or process is being influenced covertly.* For example, the charge that political campaigns' online influence manipulates public opinion may register the concern that these campaigns' influence efforts (such as micro-targeting large numbers of swing voters) are not apparent to the targeted individuals or to the public.

Notably, we see on this list some of the same features of influence that are highlighted in accounts of interpersonal manipulation, namely the covertness of influence and the fact that influence undermines self-regulation/self-control.

7.3 *Suggestion 3: don't assume that charges of manipulation do normative work*

A third methodological suggestion is that we should not treat attributions of "manipulation" as doing any normative heavy lifting. That online influence is aptly described as manipulation does not allow us to conclude that this influence is ethically problematic, much less that the influence is morally impermissible all things considered. Nor does it allow us to conclude that steps to prevent the influence would be justifiable.

Manipulation is a morally suspect form of influence; that someone has been manipulated should raise the concern that she has been morally wronged. However, instances of manipulation may be morally unproblematic. For instance, influence that undermines someone's practical reasoning in subtle ways may not be morally problematic, analogous to a white lie. Even when (manipulative) influence is morally problematic, it might nonetheless be permissible to engage in that influence, all things considered. For example, Ruth Faden, in a discussion of manipulative public health messaging, argues that it fails to respect autonomy but may be morally permissible all things considered in light of its public health benefits (Faden 1987). Even when (manipulative) influence is morally impermissible, this does not mean that the influence should be prevented. For example, a topic of ongoing discussion and dispute is whether social media platforms should take down fallacious and misleading political advertising, or whether it is non-ideal for a company to adjudicate the veracity of political speech.

In short, the fact that online influence is aptly described as manipulative does not allow us to reach normative conclusions about the influence. When influence is called manipulative, we should first clarify the specific feature of the influence that sparks the charge of manipulation (by asking the aforementioned questions), and then we should query whether influence of that form is problematic. We should ask questions such as:

1. In the specific situation at hand, is there something ethically problematic about this kind of influence, given the relationship between the influencer and the target of the influence? For example, if the influence is covert influence, is it ethically problematic for the influencer to be engaged in covert influence in that situation? For example, if the influence fails to engage its targets in deliberation or reflection, is that problematic, given the specific situation and the relationship between the influencer and the influenced? In the specific situation at hand, is it ethically problematic for the influencer to be attempting to influence the targets at all?

2. How might this kind of influence, in the aggregate, affect relevant systems and processes? For example, if the influence in question is political influence, how might it in the aggregate affect public opinion and political discourse? Does it make public opinion less reflective of people's authentic views? Does it make public discourse less likely to conform to relevant ideals for political discourse in that political system (e.g., ideals of public reason or other relevant ideals)?

3. What would it take to prevent influence like this from occurring? Would that be justifiable?

Answering these questions is not easy. We need ethics and political philosophy to answer them – for example, we need a theory of problematic interpersonal influence to answer question (1), and we need views about ideals of public discourse to answer question (2). In other words, charges of manipulation – even when they are spot on – don't do much normative work. We cannot draw a line from the conclusion that online influence is manipulative to the conclusion that the influence is morally impermissible and then to the conclusion that we should prevent the influence. Validating that influence is manipulative is just the beginning of the normative work that needs to be done.

8 Conclusions

Charges of manipulation don't do much conceptual or normative work. But they register concerns with influence and indicate that something might be amiss, so we should pay attention to them. When online influence is called "manipulation" or "manipulative," we should try to figure out what kinds of concerns with the influence are being registered. Is it concern with how the influence affects individuals – for example, that the influence affects them covertly or undermines their practical reasoning? Is it concern about how the influence writ large affects larger processes or systems (e.g., a concern with how political influence affects public opinion as a whole or how it affects the outcome of elections)? Once we've clarified the features of the influence that spark concern, we should then start asking normative questions about influence that has those features: is that kind of influence (e.g., covert influence) ethically problematic, in the situation at hand, given the relationship between the influencer and the target of the influence? Does this influence, in the aggregate, affect public opinion or political outcomes in a problematic way? And so forth.

In this process, engaging with philosophical accounts of manipulation *is* productive, because these accounts of manipulation can help us to identify various ways in which online influence might be problematic. However, we shouldn't get bogged down in the philosophical analysis of manipulation, nor bogged down adjudicating whether particular instances of online influence are manipulative according to these accounts.

Notes

1. For example, a misinformation campaign discouraging people from voting used a logo suggesting that its sponsors were affiliated with the US Democratic party, though they were not. Presumably, it was meant to reduce the number of Democratic voters who showed up to vote. For more information about this example, and several more examples of political misinformation in the United States, see Roose (2018).
2. Cambridge Analytica was hired in 2016 by the campaign of presidential candidate Donald Trump; however, it's unclear whether Cambridge Analytica's work for the Trump campaign included this kind of micro-targeted, personality-based messaging (Illing 2017).
3. As the Computational Propaganda Research Project (CPRP), which researches this influence, writes:

   > Social media are particularly effective at directly reaching large numbers of people, while simultaneously micro-targeting individuals with personalized messages. Indeed, this effective impression management – and fine-grained control over who receives which messages – is what makes social media platforms so attractive to advertisers, but also to political operatives and foreign adversaries. Where government control over Internet content has traditionally relied on blunt instruments to block or filter the free flow of information, powerful political actors are now turning to computational propaganda to shape public discourse and nudge public opinion.
   >
   > (Bradshaw and Howard 2018, 4)

4. See Goodin (1980), Blumenthal-Barby (2012).
5. Goodin notes that manipulation can take the form of lying but it needn't. Manipulation can involve giving people true information but overloading them with information:

   > Once you have overloaded people with information, all of it both pertinent and accurate, they will be desperate for a scheme for integrating and making sense of it. Politicians can then step in with an interpretive framework which caters to their own policy preferences.
   >
   > (Goodin 1980, 59)

   Another form of manipulation is giving people true information but limited information: "The choice of information to be communicated is biased, with only that reflecting favorably upon the propagandist's cause being offered" (Goodin 1980, 56). It would be unsurprising if micro-targeting typically gives people biased information.
6. Rudinow, Joel. "Manipulation." *Ethics* 88, no. 4 (1978): 338–347. More precisely, Rudinow's account is: A attempts to manipulate S iff A attempts the complex motivation of S's behavior by means of deception or by playing on a supposed weakness of S. The complex motivation of behavior is behavior in a way which one presumes will alter (usually by complicating) the person's project (complex of goals).
7. Faden and Beauchamp see psychological manipulation as "a broad heading" including "such diverse strategies as subliminal suggestion, flattery and other appeals to emotional weaknesses, and the inducing of guilt or feelings of obligation" (Faden and Beauchamp 1986, 366).
8. https://twitter.com/CNNPR/status/1058735152963182592.
9. Noggle writes: "What makes a form of influence manipulative is the *intent* of the person acting, in particular the direction in which she intends to move

the other person's psychological levers" (Noggle 1996, 49). And: "Even if the influencer has a culpably false view of what is our ideal, the influence is not a manipulative action so long as it is sincere, that is, in accordance with what the influencer takes to be true, relevant, and appropriate" (Noggle 1996, 50).

> "Often children (and some adults as well) have an inflated sense of their own importance; they genuinely believe that their pains and projects are (or ought to be) more significance significant than those of other people, not only to themselves but to others as well. Such cases are somewhat intricate morally. On my view such an agent does not in fact act manipulatively."
>
> (Noggle 1996, 50)

10. For example, here is how a practitioner of political micro-targeting defends it:

> As limited as time and money are in a campaign, we're on a even more limited resource – the voter's attention span – because they're getting inundated with information. If you've got only about three seconds from the time they take a piece of mail out of their mailbox to when they throw it away, you want to make sure that the headline issue on that piece of mail is the one they care about the most. And microtargeting ads do that.
>
> (Gavett 2014)

11. The reader might wonder why I've spent pages discussing and assessing different philosophical accounts of manipulation, only to conclude that we shouldn't get bogged down in considering which accounts of manipulation are best. As I'll explain later, I don't think that considering philosophical accounts of manipulation is a waste of time; on the contrary, these accounts are a good source of insight into which features of influence may be problematic, may be seen as problematic, and may provoke charges of manipulation.

9 References

Apuzzo, Matt, and Sharon LaFraniere. 2018. "13 Russians Indicted as Mueller Reveals Effort to Aid Trump Campaign." *The New York Times*, February 17. www.nytimes.com/2018/02/16/us/politics/russians-indicted-mueller-election-interference.html.

Barnhill, Anne. 2014. "What is Manipulation?" In Coons and Weber 2014, 51–72.

Beckett, Lois. 2012. "Everything We Know (So Far) About Obama's Big Data Tactics." *ProPublica*, November 29. Accessed August 20, 2021. www.propublica.org/article/everything-we-know-so-far-about-obamas-big-data-operation.

Blumenthal-Barby, J. S. 2012. "Between Reason and Coercion: Ethically Permissible Influence in Health Care and Health Policy Contexts." *Kennedy Institute of Ethics Journal* 22 (4): 345–66.

Bradshaw, Samantha, and Philip N. Howard. 2018. "Challenging Truth and Trust: A Global Inventory of Organized Social Media Manipulation." https://demtech.oii.ox.ac.uk/research/posts/industrialized-disinformation/.

Bradshaw, Samantha, and Philip N. Howard. 2019. "The Global Disinformation Order 2019: A Global Inventory of Organized Social Media Manipulation." https://demtech.oii.ox.ac.uk/wp-content/uploads/sites/93/2019/09/CyberTroop-Report19.pdf.

Cadwalladr, Carole, and Emma Graham-Harrison. 2018. "Revealed: 50 Million Facebook Profiles Harvested for Cambridge Analytica in Major Data Breach."

The Guardian, March 17. Accessed August 20, 2021. www.theguardian.com/news/2018/mar/17/cambridge-analytica-facebook-influence-us-election.

Coons, Christian, and Michael Weber, eds. 2014. *Manipulation: Theory and Practice*. Oxford: Oxford University Press.

Curtis, Cara. 2019. "Study: Weaponized Misinformation from Political Parties is Now a Global Problem." *The Next Web*, September 26. Accessed August 20, 2021. https://thenextweb.com/news/study-weaponized-misinformation-from-political-parties-is-now-a-global-problem.

Faden, Ruth R. 1987. "Ethical Issues in Government Sponsored Public Health Campaigns." *Health Education Quarterly* 14 (1): 27–37. doi:10.1177/109019818701400105.

Faden, Ruth R., and Tom L. Beauchamp. 1986. *A History and Theory of Informed Consent*. New York, NY: Oxford University Press.

Frenkel, Sheera, and Katie Benner. 2018. "To Stir Discord in 2016, Russians Turned Most Often to Facebook." *The New York Times*, February 17. Accessed August 20, 2021. www.nytimes.com/2018/02/17/technology/indictment-russian-tech-facebook.html.

Gavett, Gretchen. 2014. "Electing a President in a Microtargeted World." *Harvard Business Review*, November 2. Accessed August 20, 2021. https://hbr.org/2012/11/electing-a-president-in-a-micr.

Ghosh, Dipayan, and Ben Scott. 2018. "Facebook's New Controversy Shows How Easily Online Political Ads Can Manipulate You." *Time*, March 19. Accessed August 20, 2021. https://time.com/5197255/facebook-cambridge-analytica-donald-trump-ads-data/.

Goodin, Robert E. 1980. *Manipulatory Politics*. New Haven, CT: Yale University Press.

Gorin, Moti. 2014. "Towards a Theory of Interpersonal Manipulation." In Coons and Weber 2014, 73–97.

Gorwa, Robert, and Douglas Guilbeault. 2020. "Unpacking the Social Media Bot: A Typology to Guide Research and Policy." *Policy & Internet* 12 (2): 225–48. doi:10.1002/poi3.184.

Illing, Sean. 2017. "Cambridge Analytica, the Shady Data Firm that Might be a Key Trump-Russia Link, explained." *Vox*, October 16. Accessed August 20, 2021. www.vox.com/policy-and-politics/2017/10/16/15657512/cambridge-analytica-facebook-alexander-nix-christopher-wylie.

Isaac, Mike, and Cecilia Kang. 2020. "Facebook Says It Won't Back Down From Allowing Lies in Political Ads." *The New York Times*, January 9. www.nytimes.com/2020/01/09/technology/facebook-political-ads-lies.html.

Krugman, Paul. 2009. "Opinion | The Chinese Disconnect." *The New York Times*, October 22. Accessed August 20, 2021. www.nytimes.com/2009/10/23/opinion/23krugman.html?_r=1&hp.

Mills, Claudia. 1995. "Politics and Manipulation." *Social Theory and Practice* 21 (1): 97–112.

Noggle, Robert. 1996. "Manipulative Actions: A Conceptual and Moral Analysis." *American Philosophical Quarterly* 33 (1): 43–55.

Prokop, Andrew. 2018. "Cambridge Analytica Shutting Down: The Firm's Many Scandals, Explained." *Vox*, March 21. Accessed August 20, 2021. www.vox.com/policy-and-politics/2018/3/21/17141428/cambridge-analytica-trump-russia-mueller.

Rampton, Sheldon, and John Stauber. 2001. *Trust Us, We're Experts! How Industry Manipulates Science and Gambles with Your Future*. New York, NY: Tarcher/Putnam.

Roose, Kevin. 2018. "We Asked for Examples of Election Misinformation. You Delivered." *The New York Times*, November 4. www.nytimes.com/2018/11/04/us/politics/election-misinformation-facebook.html.

Rudinow, Joel. 1978. "Manipulation." *Ethics* 88 (4): 338–47. doi:10.1086/292086.

Shabaz, Adrian, and Allie Funk. 2019. "The Pandemic's Digital Shadow." Accessed August 20, 2021. https://freedomhouse.org/report/freedom-net/2020/pandemics-digital-shadow.

Sunstein, Cass R. 2016. "Fifty Shades of Manipulation." *Journal of Marketing Behavior* 1 (3–4): 214–44. doi:10.1561/107.00000014.

Susser, Daniel, Beate Roessler, and Helen Nissenbaum. 2019. "Online Manipulation: Hidden Influences in A Digital World." *Georgetown Law Technology Review* 4 (1): 1–45.

Ware, Alan. 1981. "The Concept of Manipulation: Its Relation to Democracy and Power." *British Journal of Political Science* 11 (2): 163–81.

4 Online manipulation and agential risk

Massimiliano L. Cappuccio, Constantine Sandis, and Austin Wyatt

Like puppets we are moved by outside strings.

Horace

1 Introduction

Manipulation is as old as rhetoric and there are as many ways of manipulating others as there are forms of communication, including mass-media, the press, and other technologies (Bernays 1928 [2005]; Packard 2007; Garvey 2016). Our goal in this chapter is the modest one of describing how online manipulation (OM) works in the context of our informative and communicative practices. We will focus on manipulation mediated by software agents or other kinds of autonomous technologies, like bots, predictive algorithms, and scripts[1] Since the notion of artificial intelligence (AI) is ill-defined, it is debatable whether any of them constitutes an example of AI-mediated manipulation (let alone straightforward "AI manipulation") in a legitimate sense. We thus prefer to stick to the term "online manipulation" (cf. Hancock, Naaman, and Levy 2020).

Manipulation – online or offline – is by and large distinguishable from persuasion. The latter typically involves *changing* another person's mind through the explicit use of (real or apparent) reason. Manipulation does not always proceed this way, though one way (among many) of changing another person's mind is, indeed, to manipulate them into doing so. Such manipulation might, for example, proceed by way of charm and false promises, as opposed to reason and argument, which is not to say that such processes are mutually exclusive. Unlike most cases of persuasion, manipulation typically involves a more nefarious kind of control, one that is commonly associated with figurative puppet-masters. Such imagery is frequently also used to illustrate hard determinist views of free will (e.g., Gazzinga 2012). But, whereas on such theories our brains and/or the environment are thought to cause our actions straight out, manipulation more typically proceeds by way of causing us to feel certain things which in turn motivate us to act accordingly. In this respect, the role of a manipulator is closer to that of the deities in Greek tragedy (see Sandis 2009, 2015). As with all

DOI: 10.4324/9781003205425-5

kinds of influence, manipulation falls short of causal determinism (irrespective of whether or not our actions are ultimately fully determined by a range of factors).

Human-to-human manipulation can (but does not need to) be done intentionally. If a person is manipulative by disposition, they may manipulate others without even knowing that they are doing this (cf. Manne 2014). This also shows that the motive for manipulation need not be nefarious. Successful manipulation does, however, involve a change in the behaviour of the thing or person being manipulated so that it better matches the interests of the manipulator (or what the manipulator takes those interests to be). While this may be a necessary condition for manipulation, it is not a sufficient one.

Our online activities are mediated by algorithms and machines. Any related manipulation, then, occurs through the medium of machine's activities. It doesn't follow, however, that the machines (or the algorithms used to program them) are themselves manipulating anyone. Indeed, machines and algorithms are themselves incapable of manipulating any person or thing (or, indeed, being manipulated by them), insofar as it makes no sense to ascribe any *intentional* action to them; beings who manipulate others unintentionally still do so in doing *something* intentionally (manipulation isn't something that one can literally do in one's sleep). Only beings capable of intentional action, then, can manipulate or be manipulated. As we are about to see, their manipulation can occur either naturally or through artificial means.

Constantine's cats manipulate him in all sorts of natural ways. Sometimes they do so intentionally, such as when they try to subtly lure him towards the food bowl area though even in the midst of his own manipulation, he too, may choose to manipulate them with distractions. But cats can also manipulate in hard-wired ways that are non-intentional, such as through the use of high-pitched meow-purrs. It is tempting to assume that cats are fully aware of the urgency their sounds inspire in humans, but this is simply not the case. While a cat may be quick to learn what works and what does not, and while it may in some weak sense *choose* to come and find you, it neither chooses nor controls the frequency and pitch of its voice. More radically, some cats carry *Toxoplasma gondii*, a parasite found in mice and rats that is said to "manipulate the brain" into becoming less risk-averse (and, consequently, more likely to be caught by a cat). Perhaps, one of Constantine's cats infected him with it, thereby rendering him more likely to be manipulated into accepting the invitation to present this chapter. We might contrast such *natural* manipulation (NM) with the more *artificial* manipulation (AM) that occurs when an algorithm throws up something on our newsfeed or suggests that we "like" or "follow" some particular page. However, the paradigm form of manipulation is neither NM nor AM but what we might call *intentional* manipulation (IM). The question for us is whether OM is identical to any one or more of the aforementioned. We shall be arguing that OM typically involves one or more of NM, AM, and IM.

While OM is *in essence* no different from offline manipulation (e.g., standard examples of nudging, cf. Thaler and Sunstein 2008), it differs from it in a number of crucial respects. The most important – and most dangerous – of these is that the manipulators can very easily lose control not only of *who* they are manipulating but also of *how* they are manipulating them and, ultimately, even of *what* they are manipulating them to think or do. To complicate things more, OM also enables agents to manipulate themselves, or at least play a greater part in their own manipulation than they could ever do offline, due to the interactive feedback loops that technology makes possible. And, as we will see, the biggest manipulation worry is not that either machines *or* humans are manipulating us according to some grand plan, but that the human manipulators are losing control of their own control over us.

It is this loss of meta-control that elevates the agential risks involved in OM to a whole new level of scalability that has no offline analogue. The phrase "agential risk" has been used as a term of art in the emerging technologies literature to indicate potential large-scale disasters caused by human–computer interaction (see, e.g., Torres 2016). By contrast, we use the term in the more ordinary sense of *any* risk (of any size) that we expose ourselves to in acting. Online agential risks are simply a subset of this general category.

2 Manipulation and media

To understand the specificity of OM we need to situate it against a general map of the opportunities to manipulate others offered by diverse forms of communication. For better or worse, manipulating others is one of the distinctive features (and, perhaps, also one of the key functions) of human communicative practices, which is why manipulation techniques and forms of communication evolve together and share a communal history. As human communication is strongly tied to the technological media that support it, and different media enable new forms of communication or anyway reshape the existing ones, the diversity in the possible forms of manipulation reflects the variety of existing communication technologies. Relying on categories customarily used in sociology of media (Croteau and Hoynes 2003; Waples 1942), communication science (Fawkes and Gregory 2001), networks theory (Stevens 1981), and marketing studies (Gummesson 2004) we will distinguish between three different communication paradigms that reflect different prototypes of sender–recipient relationship: the one-to-one (OOC), the one-to-many (OMC), and the many-to-many (MMC) communication paradigm. The historical advent of these communication paradigms is associated with different kinds of technologies. To account for the kind of manipulation specifically associated with digital technologies and social media, we will introduce a fourth communication paradigm, the augmented-many-to-many (AMMC) which represents the algorithmically enhanced version of MMC. In what follows we will briefly review these four paradigms and how they change manipulation.

2.1 One-to-one communication and rhetoric-based manipulation

This paradigm is prevalently based on either oral or manual written communication, and in general on any form of communication in which the exact replication of the content is difficult, as the message is tailored to a specific context of fruition: in *verbal* communication, the relationship between sender and receiver is bidirectional (they can and usually need to swap their roles to take turn in verbal communication) and the relationship between expression and interpretation is synchronic and direct (it is based on the immediate interaction between sender and receiver); in *written* communication this relationship is unidirectional and asynchronous (there is no assurance that the receiver can reply to the sender) as the interaction between sender and receiver is negotiated by the graphic medium (Ong 1999).

Either way, both the scope and the effectiveness of communication are based on reciprocal familiarity: the expression of the message is strongly tailored to a specific audience that is well known to the sender; in turn, the correct interpretation by the audience is contingent upon their relatedness to, or analogy with the sender. In this context, the efficaciousness of the manipulatory activity (i.e., the manipulator's capability to solicit and control the audience in a desired manner) depends on the manipulator's personal familiarity with the audience to be manipulated (i.e., a non-superficial understanding of their psychology and a sufficient knowledge of their desires and beliefs), while the audience's main defence against manipulation attempts is the timely understanding of the manipulators' real intentions and the familiarity with their rhetorical techniques. These techniques lend themselves to both an offensive function and a defensive one, which is why they historically gave birth, on the hand, to sophistry (the art of exerting manipulation through advanced forms of persuasion) and, on the other, to dialectic and logic (the arts of averting manipulation through the critical analysis of arguments). This duality confirms that, at this stage, manipulatory practices rely primarily on the capability to tailor communication to a specific, well-defined audience.

2.2 One-to-many communication and propaganda-based manipulation

The second paradigm is introduced by the invention of mass media. These include asynchronous media, like the press, as well as other media that, like the radio and the TV, allow both synchronic communication and the asynchronous fruition of recorded messages (McQuail and Deuze 2020). OMC media are broadcasting technologies that enable the infinite replication in space and time of a content. While only a few institutions or publishers are able to own them, these technologies allow the sender to reach a virtually limitless number of receivers with the same message. The advantage granted by mastering the rhetorical arts does not disappear with the advent of these technologies but is incorporated within a new modality of persuasive communication: propaganda. Its efficacy depends on the universal replicability of the message much more than the sender's specific knowledge of the audience (Blumler and Katz 1974).

In this paradigm, the homogeneity of the communicative effects produced by mass media and the indisputable authority of unidirectional communication reinforce each another. For the communication to be efficacious, the diversity of the recipients must be ignored or actively reduced through homologation, which is why propaganda both presupposes and actively encourages a conformist and dogmatic interpretation of the content. In the age of propaganda, the manipulator's first concern is not to understand the psychology or the specific desires and expectations of some individuals or small groups (as fine-grained differences are lost anyway in favour of homologation) but to conquer the monopoly of mass media, increasing the reach of one's broadcasting media while preventing other potential manipulators from accessing analogous media, so that they are unable to reach the same audience with competing messages. Propaganda works most efficaciously in homogenous communities where a few institutions or influential groups own all the mass media and the individual recipients are isolated from each other while being strongly connected to the broadcaster, a combination of circumstances that discourages the receiver from thinking critically and releases the senders from the need to prove the credibility of their messages.

2.3 Many-to-many communication and reputation-based manipulation

The sender–receiver relationship changes dramatically with the internet due to its capability to make virtually anybody a provider of content, news, and online services. This paradigm is prefigured by point-to-multipoint telecommunications via radio networks (Cover and Joy 1991) and, before that, by para-theatrical participative events, such as public demonstrations and parades (Waples 1942). However, the OMC paradigm establishes itself as the prototype of interpersonal communications during the early stages of the new economy, supported by the new media and the digital systems (like the TCP/IP protocols) that support the World Wide Web. That is the time when, through email and blogs, every internet user can, in principle, broadcast their message to a significant audience (Fawkes and Gregory 2001). Within this paradigm, effective communication combines elements of the previous two kinds of sender–receiver relationships: through the new media, content providers can finally reach audiences that are, at once, very large and very specific (Croteau and Hoynes 2003). They acquire an unprecedented power, as they can broaden their communicative horizon, indefinitely replicating the same message, while making it more focussed, circumscribing the message to a specific set of selected recipients; at the same time, each content provider can be reached by other users, who similarly become content providers. Every agent operating in the network is both a producer and a consumer of content, therefore if, on the one hand, they are continuously influenced by role models and opinion leaders, they can also influence their own influencers (in ideal conditions), by engaging in public discussion with them and providing feedback.

The typical sender–receiver relationship in the age of the internet combines elements of both rhetoric and propaganda, which is why at this stage manipulation still requires, on the one hand, the deep understanding of a familiar community and continuous interactive engagement with them (as in OOC), and, on the other hand, the capability to reach the broadest possible audience by surpassing the other broadcasters (as in OMC). Therefore, the manipulator's effort consists in making the message as consistent as possible with their public reputation while adapting it to the recipients' demands and expectations.

This form of communication is characterized by feedback loops that can reinforce or weaken the influencer's credibility based on the satisfaction of a large yet selected group of followers. Securing their approval requires the construction of a relationship of trust and the consolidation of a reputation. Online credibility is not meant to be universal but has to be tailored on the specific beliefs and expectations of the followers, which tend to form well-defined and independent communities or "bubbles". Accordingly, the efficaciousness of one's influencer's manipulatory activity is, at this stage, primarily based on their ability to consolidate the loyalty of the followers, while winning an open competition against all the other influencers (Phillips 1999). Within this paradigm, OM primarily depends on the influencer's capability to increase and leverage their credibility within their bubble: their goal is to be recognized as a trusted source by the specific audience over which they intend to extend their influence.

The audience, providing their feedback, can in turn exert a significant influence over the influencer, shaping the influencer's stance and posture until it perfectly conforms to the group's conventions and style (Gummesson 2004). However, and this is important to distinguish the third paradigm from the fourth, the dynamic relationship between influencers and their audience does not change their respective goals and objectives, such as promoting certain opinions or defending specific values. The medium, in this paradigm, is still passive and neutral: a simple means to an end, subordinated to the influencer's goals, who uses it instrumentally to project their strategic influence over a familiar territory.

2.4 *Augmented many-to-many communication and engagement-maximization-based manipulation*

This paradigm results from a metamorphosis of the third one. The possibility of this metamorphosis was built into the MMC paradigm from the very beginning, but it started producing its transformative effects only when the technological interface ended up creating an entirely new system of reciprocal interactions between influencers and followers. Jensen and Helles (2017) suggest that this transition occurred when social media imposed a "many-to-one" communication paradigm. Arguably, such an intermediate paradigm arose when the users started facing the need to critically select among an overwhelming multiplicity of information sources brought by the Internet, filtering out the irrelevant, offensive, or potentially harmful ones (consider, for example, the unprecedented problem of being targeted by endless "spam" messages automatically sent by anonymous distribution lists).

Without denying the importance of the many-to-one communication, our view is that the transition from the third to the fourth paradigm happened when the medium stopped being passive and neutral and started to actively contribute to the communicative relationship in ways that had never been possible before. Our communicative practices have deeply changed since the introduction of artificial systems that operate semi-autonomously, making unsupervised or minimally supervised decisions with the function to select, retrieve, or even generate contents that satisfy the interest of the users, without the users being in control or even aware of being subject to such decisions. The algorithms underpinning these systems became increasingly sophisticated and adaptive during the past 20 years. Today, they incessantly gauge the users' interest patterns in a constant attempt to satisfy their expectations by remodulating both the message and its mode of presentation. Moreover, exploiting the feedback loop established with the users, these systems do not change only the content and the form of communication but also – indirectly – the interest patterns of the audience. Think, for example, of interfaces and interaction protocols that stimulate the attention of the users to the point of engendering compulsion in or even addiction to certain online activities; chatbots and deceptive software agents spreading tendentious information including politicized memes and fake news; recommendation and prioritization algorithms that collect information about the user's preferences and inclinations to optimize their online experience and, among other things, modulate the results of their searching activities.

Through these systems, the advent of AMMC has introduced a distinctive kind of OM, one that involves sophisticated computing technologies such as machine learning, big data, mass surveillance devices, and sentiment analysis. The algorithms underpinning these systems are not properly intelligent; therefore, they have no communicative intention of their own. Yet, they can recursively detect and reinforce existing human trends and preferences, to the point of exacerbating the aggregative and disaggregative tendencies that cross our society. One of the most peculiar, but also worrying, aspects of these systems is their capability to produce large-scale manipulative effects in the context of the democratic participation to political debate. This may happen by design or not. In fact, even when they are not created and deployed for subversive or devious purposes, these systems can and often do manipulate the political opinions and moral perceptions of people in ways that cannot be entirely anticipated by their creators or even recognized by their users.

3 Online manipulation engineered by algorithms

Interestingly, the manipulation produced by the AMMC systems is neither obvious to those who are manipulated, nor explicitly devised by some human manipulator. Apparently, these systems can produce OM without any person having planned it, being directly responsible for it, or even being aware of it. But can we legitimately say that someone was ever manipulated by these systems, if nobody intentionally used them to manipulate others in the first place? Despite the prima facie paradoxical premises, the claim that AMMC produces manipulative effects is consistent with our initial definition of manipulation: the interactive process in which AMMC occurs is intentional in some important sense, even though such a process cannot be traced back to the specific intentional activity of any particular agent or group.

Consider the following example. If we were to upload this chapter onto a website such as Academia.edu, and choose 20 keywords such as "AI Ethics", "Manipulation", "cats", etc., the chapter will then be pushed to anyone on the site who has registered one or more of those interests. Here, the human-assistance, intentionally or otherwise, takes place on both sides of the manipulation relation. When we list our research interests as "agency", "AI intelligibility", "virtue ethics", and so on, we effectively render ourselves open to some slight manipulation. This "structuring cause" of manipulation is triggered by the person who then tags their freshly uploaded chapter with one of these research interests. If the chapter subsequently appears on my Academia feed, have I been manipulated by the researcher? The verdict seems a bit harsh, though less so in the case in which they have tagged the chapter with irrelevant research topics with the sole purpose of attracting views and downloads. Has the *system* manipulated me?

The algorithms are certainly an enabling condition of OM, but they do not do any OM themselves, nor do their designers and engineers. What they do is to make OM possible. Algorithms literally engineer manipulation, just as the *Toxoplasma gondii* parasite engineers it in mammals. OM is enabled by bottom-up intentional processes mediated by algorithmic interfaces, but it is also an effect of the complex top-down processes exerted through the feedback given by the system to the user. Through this feedback, to some extent, I have contributed to a degree of self-manipulation. I have been pulling at my own strings, or, at the very least, I intentionally (if not voluntarily) pulled the strings that enabled the system to manipulate me. Conversely, when we engage in the slight manipulation of nudging people into reading our papers, we are no more aware of the algorithm content than the cat is of the frequency of its purrs.[2] But, unlike the cat, we can explicitly envision the likely effect of what we are doing. Moreover, we know that whoever has been nudged has voluntarily put themselves in a position that allows this. To complicate things further, the success of the chapter on the web does not ultimately depend on the nudging exerted by a particular individual, but on the reiterated nudging collectively exerted by all users through the mediation of various technological interfaces.

Like any collective process, OM can occur without an *intrinsic* motivation and an *explicit* intention. The great trends of content production and interpretation are primarily motivated by extrinsic normative criteria such as engagement maximization and attention preservation, while the influencers' intrinsic motivation (their personal goals and vision) affects communication only secondarily and locally. The norm that governs the aforementioned systems is not only extrinsic but also implicit, because it does not need to be explicitly represented by its users or designers: according to actor-network theory (Latour 1996), OM is exerted by a network of "actants", including most importantly recommendation and prioritization algorithms that operate unintentionally and without sensitivity to the context.

Unlike intentional agents, such actants do not have interests of their own in OM but can nonetheless engender distinctive transformative effects on the audience by automatically detecting and reinforcing their existing patterns of interest. The manipulative force exerted in this way is entirely blind to the content of these patterns. Through the actants' activity, the system tends to promote specific opinions or fuel particular discussions even if nobody, strictly speaking, ever intended to do it. Actants merely amplify, not create, fields of collective attraction and repulsion. Therefore, it is by accentuating the existing opinions, not suppressing them or replacing them with other opinions that communication is manipulated in the age of AMMC.

Opportunities for amplification always abound, as the communications established through massively interconnected networks of users continuously generate options to relaunch virtually any available content: *flames* emerge from the large-scale alignment of the users' interests, while differences of opinion are systematically exacerbated (whatever the subject matter is) and transformed into conflictual dichotomies that oppose polarized groups of users. Whether it creates harmony or disharmony, this reinforcement process acts as a self-fulfilling prophecy: obeying the principle of interest maximization, the system progressively strengthens and propagates the biases that circulate in the system. It confirms the prejudicial expectations of the users while insulating them from the alternative points of view that could challenge their assumptions.

Recent chronicles made us familiar with several cases of such self-fulfilling prophecies caused by automatic selection and reinforcement mechanisms: stock market downfalls caused by the very fear of those downfalls; conspiracy theorists who systematically ignore all the evidence that contradicts their theories, considering it part of the conspiracy; fake news that became viral right after someone announced that they were spreading virally; terrorist acts motivated by the perceived urge to protect the populace from terrorism, and so on. These processes may naturally occur in the space of human communication, but they become irresistible and tremendously more disruptive when online digital actants, such as recommendation and prioritization algorithms and fake news bots, catalyse them through a recursive mechanism.

Due to the inherent circularity of AMMC, the individual intentions to manipulate others are neither the only nor the primary source of OM; on the contrary, the individual intention to manipulate others is often a simple by-product of, or an opportunity disclosed by, the OM process. Consequently, the manipulation techniques used in the AMMC paradigm are not simply instruments used by a manipulator to influence their audience. Communication is no more an instrument serving someone's personal schemes or agendas but an end in itself. A supra-personal level of communication emerges from the intentional processes naturally occurring between personal-level communications. This level is at once an effect and a source of the self-reinforcing interactive processes that maximize the users' engagement and prolongs their attention. Because this level is virtually autonomous from human decisions, OM is not and cannot be motivated by a predefined overarching goal: OM tends to become a global trajectory that manipulators do not only exploit but actively serve. This trajectory delineates itself against the complex backdrop of the intentional communication dynamics among users: it arises from, but is never reducible to, their particular attempts to influence each other.

For example, judgemental fanaticism in online debates (i.e., reinforcement of biases and prejudices) and polarization (exacerbation of differences of option) are both preconditions and, at the same time, effects of OM. They are self-fulfilling prophecies in the sense that the complex interactions among communities of users create expectations that the users themselves will try to fulfil with their subsequent interactions. No less than the persons they manipulate, manipulators can be trapped in the information bubble that their contributions created. They can act only in accord with the norms that are recognized as valid within their echo chamber. OM is not caused by the intention of a human manipulator (it is not determined by their decisions, purposes, or reasons) but supervenes on the semi-spontaneous synchronization of parallel causal interactions occurring within large networks of human and artificial actants.

In this sense, OM operates by creating a dynamical equilibrium that is influenced by, and yet irreducible to, the actants operating in it, a global trend that is greater than the processes that realize it. This equilibrium reflects the co-dependency of local dynamics governed by engagement maximization functions (automatic systems aiming at undefinedly increasing the attention and the time consumed by the users) and emergent global configurations that the system interprets as a model to predict the behaviours of the users. In this context, agency (i.e., the intention, purpose, and reason behind the manipulatory activity) is not just the bottom-up pressure exerted by individuals while attempting to reach their own goals; it is also a top-down force that imposes a transient stability upon the chaos of online interpersonal interactions. The process underlying the AMMC has the following characteristics:

1. *Distributed*: depending on complex interactions between intentional and unintentional actants, thus not reducible to the intentional or deliberate activity of any of them.
2. *Emergent*: occurring at a level of organization that supervenes on its local constituents and is more than their sum. This global level is at the same time the cause and an effect of the local interactions between users in the network.
3. *Semi-autonomous*: global trajectories follow from other global trajectories and cannot be predicted examining only the component processes that contributed to generate them or the actants' local goals.

By emphasizing the supra-personal dimension of OM, we are not claiming that AMMC systems are never used to manipulate certain audiences in accord with well-defined malicious intentions or that OM is always innocent and impersonal. It is certainly possible that some malicious human actor takes advantage of the processes mediated by the AMMC systems to pursue their own agendas, and in Section 4 we examine the implications of this activity for national security.

However, if there is a human manipulator, in this context, it is not a puppeteer in charge of defining the overarching narrative: more likely, it is just an opportunistic facilitator, someone who is both capable of identifying the prophecies that have the best chances to fulfil themselves and willing to amplify them for their own personal advantage. This capability typically leverages the irrational fears and uncontrollable obsessions of people and exploits them strategically to fuel a convenient narrative, relaunching it across the infosphere until it becomes viral and starts living a life of its own. When public attention reaches critical mass, such a narrative produces a burst of short-lived, but potentially large-scale, viral effects that further propagate and aggravate while the narrative keeps mutating until it eventually dissolves or is absorbed into some other narratives. Nobody entirely controls anymore, let alone creates, the overarching narrative, which continuously changes in unpredictable ways.

4 Online manipulation and intentionality

We might say that OM via AMMC is a form of collective action exhibiting shared agency mediated by algorithms, which may or may not involve one or more human opportunistic manipulators. Philosophers use varied terminology such as "joint action", "collective behaviour", "shared agency", and so on, to describe social phenomena like these. Going for a walk with another person is not the same kind of activity as pushing a car together with them and both can be further distinguished from acting as a member of an executive board; people conversing act together in a way that is to be distinguished from the togetherness of a blind person and her guide dog, and the agency shared when playing tennis against one's opponent is of a fundamentally different nature to that shared by players on the same side of a football team; the collective intention of a protest march is neither that of electing a new prime minister nor that of riding in a pack of Harley Davidsons; the migration of refugees is only superficially similar to that of birds; the client and bank teller conducting a cash transaction are not behaving collectively in the same sense as members of the London Symphony Orchestra; in ancient Greek tragedy humans act in strange unison with the gods who steer them both psychologically and physically.

In J.L. Austin's work on performatives we are given examples in which a person cannot perform an individual action (such as making a bet) unless her behaviour is taken up by someone else. In some ways, AMMC is just one more form of social action. However, the augmentation that technology provides to this specific kind of social action makes AMMC unique.

Many social actions involve not just MMC but also OMC. Our academia.edu example is very crude and simple, and there will be examples that are not only more complex but also much more sinister. But the sinisterness is no worse than what we find in OOC, OMC, and MMC. To this extent, digital media has been used as a scapegoat or screen to hide more serious problems that ultimately lie in the 'real' society.

In her review of Netflix's *The Social Dilemma*, Pamela B. Rutledge writes (Rutledge 2020):

> Probably the cleverest element of persuasion was the translation of algorithms into what appear to be entitled, self-satisfied white male Millennials – a reflection of the Silicon Valley stereotype. Manning computer terminals, these algorithms discuss almost with relish how to use the various stimuli to "activate" Ben into logging in, sharing, commenting, and interacting on social media. Every time he does, they sell that moment of Ben's attention to an advertiser. Anthropomorphizing algorithms makes it easy to attribute intentionality to the technology and see it as wilfully controlling and manipulative.

As the title of a similarly minded review by Mike Masnick puts, "*The Social Dilemma* manipulates you with misinformation as it tries to warn you of manipulation by misinformation" (Masnik 2020). We might go further than this and say that the documentary uses the techniques of horror to try and manipulate its audience into leaving social media. Released in the midst of a pandemic, the question arises: what would these former social media users do instead? The unmentionable answer, of course, is binge more Netflix!

The real problem is not the supposed "grand scheme" behind the algorithms. One example is that some of the most devastating effects of recommendation and prioritization algorithms were ultimately generated by the attitudes and preferences of the users (and the implicit biases that they convey), which remained unobvious until the machine learning algorithms started reinforcing them more and more, creating information bubbles around the users, without anybody intending, desiring, or expecting to produce such self-fulfilling prophecies. As previously stated, we think it is perfectly possible that this phenomenon could at times be intentionally exploited by somebody with malicious intents and a sufficiently powerful capability to influence the network (e.g., the infamous digital trolls that operate to make certain news more visible). It is worth noting, however, that:

1. The risk of unintentional manipulation is more resilient and not less pervasive than the risk of deliberate manipulation. While we have many tools to identify malicious actors (see later), it is virtually impossible – because of the semi-autonomous nature of global dynamics – to anticipate all the possible things that can go wrong by themselves in a machine learning system that works on a planetary scale due to the well-known opacities of this technology.
2. There are techniques to prevent the intentional exploitation of these weaknesses in the social media, for example, algorithms that automatically recognize the distinctive patterns of activity of the digital trolls in order to detect and block them. But this is more of a cybersecurity concern than an intrinsic limitation of the medium, in the sense that it has to do primarily with cyber-guerrilla tactics, and only secondarily with the ethical regulation of the infosphere. We all agree (and it is in everybody's interest) that these abuses need to be prevented, we just need to develop the technological capacity to do it more efficaciously. They are thus not part of any general problem of OM.
3. Even when it occurs deliberately as part of the scheme of a malicious actor, it is unlikely that the manipulation could work effectively without any previous predisposition or bias on the user's side. Typically, the vicious algorithm is doing nothing other than boosting and accelerating certain human dynamics that were already ongoing in the socio-political and cultural background (e.g., biases, paranoias, polarizations) for reasons that have nothing to do with Facebook and Google and that – in most cases – existed well before the introduction of these technologies.

This absolutely does not imply that we should blame the victims of OM (the "users," "consumers," or simply "people") or excuse the perpetrators. Rather, it should encourage us to think that the problem of manipulation by digital media always reflects broader problems that already exist in the society (e.g., bad cultures and bad practices shared by a collective, social tensions, inequalities, poor education, lack of real information and critical thinking capabilities). Therefore, to solve such problems it is important but insufficient that we regulate and merely control the media: one needs also to contrast bad cultures and bad practices with good education and appropriate incentives to create a fairer and more just society.

5 Algorithmically assisted manipulation in influence operations

Recognizing that OM is an amplificatory, rather than initiatory, characteristic of modern political discourses also necessitates that those charged with monitoring and limiting that influence adjust their approach. Before concluding this chapter, therefore, it is important to demonstrate how AMMC differs from prior methods of influence manufacture and manipulation in order to guide efforts to address the risks posed by OM, including the development and imposition of ethico-legal frameworks and norms.

Attributing intentionality to artificial actants would be wrong, as they are not "agents" in the traditional legal or philosophical sense, and they have no independent capacity to interpret their instructions through an individual context. The inability to form intention and the inherent absence of a direct human operator in such systems make it far more complicated for law enforcement or intelligence agencies to assign criminal responsibility for harm arising from OM.

Rather than as individual government agents directing a traditional influence operation towards a potentially malicious purpose, it is more useful to conceptualize AI-assisted systems as waitstaff at an upscale restaurant. Their role is to facilitate your ability to accessing information and guide your progression through the meal. While the waiter does not set the meal options, they provide only one menu to you and provide guidance for your choice based on instructions they received upon starting their shift. Certainly, there is a possibility that you were manipulated into choosing the bad fish, but it is far more likely that the waiter simply provided the guide rails on which you reassured yourself that the fish here would be of high quality (this being representative of a preconceived notion), and either the chef prepared the food improperly (which would translate to a malicious information source or poorly moderated social media platform, for example), or the restaurant's buyers purchased poor quality produce (e.g., a malicious actor providing an initial piece of fake news or modifying a public information source). This imagery also suggests the inequity and ineffectiveness of a regulatory system that prosecutes the waiter (our AI-enabled actant) rather than targeting those who gave them instructions or undermined the quality of the provided product.

Continuing to draw on this imagery, let us consider the issue of election interference, which has shot to prominence in recent years and is often associated with OM. In the age before artificial actants, mass communication was mainstreamed by the proliferation of social media platforms, interference in the internal political processes of foreign nations was largely, albeit not solely, the province of nation-state intelligence services. Interfering in the political discourse of one's neighbour required that a state make a directed covert effort to push an alternative political message through comparatively limited channels that existed for the mass dissemination of political information. Mass propaganda efforts typically took the form of investing in generating unfavourable media coverage on television or radio broadcasts; manipulating, funding or otherwise facilitating the growth of internal opposition figures; or inducing other members of the international community to apply coercive statecraft tools (such as sanctions). Sometimes, states took a more direct role in presenting a favourable political message to the domestic populace of a rival. For example, during the Cold War, the United States supported regionally accessible radio stations (such as Radio-Free Europe and Radio Liberty), which broadcast propaganda alongside music. Until recently, South Korea would regularly broadcast K-pop music from loudspeakers and release balloons filled with propaganda leaflets over the Demilitarized Zone. Each of these are cases of traditional influence operations, designed, conducted, and controlled by a state agency and directed towards an identifiable political end. Where these operations acted in support of internal dissident forces or oppositional leaders, it was in furtherance of a larger goal of discrediting or disrupting a rival.

By contrast, AMMC operates in a manner far less directed or controlled by virtue of its collective-agency nature, where thousands of independent users interact around a rapid news cycle, mediated by AI and beyond the ability of a state agency to comprehensively monitor, much less direct. Achieving electoral interference through OM under this model is, by necessity, resistant to pre-planning or careful organizational control. Unlike traditional propaganda or influence operations, malicious actors do not "direct" or exercise meaningful control over the message; instead, political value is gained by altering or obfuscating reliable information sources (the choice of fish at our market), amplifying a favourable perception of, for example, a preferred presidential candidate among influential social media users (promoting a recipe book to chefs which proscribes a flawed method of fish preparation), and manipulating social media algorithms so that more and more users see the political message or have it reinforced in their psyche (the instructions given to our waiter at the beginning of the shift). Under this model, political leaders are not directing influence operations in a traditional sense, rather they are attempting to surf a crowd that is being subtly pushed in one direction or another based on changes to the underlying information sources they use to generate their political views.

Unfortunately, existing tools and norms used by political, intelligence, and law enforcement institutions to counteract influence operations were designed in reference to traditional, directed, forms of propaganda. Lawmakers have already run into significant legal and ethical barriers to imposing criminal sanctions on those who promote disruptive political messages through innocent use of social media platforms, and our political institutions are already seeing the impact of influence operations that manipulate the algorithms used by social media platforms and online spaces to reinforce "echo chambers" based on a user's actions online. In a space where trustworthiness of information is often difficult to verify, yet regularly assumed, there is certainly the potential for OM to lead to significant harm, even when state actors are nowhere near the level of involvement required to trigger any regulatory response.

The violent American riots on January 6, 2021, against the results of an election were one example, as is the electoral misinformation and mistrust that led up to those events. Even now, there is a thriving anti-vaccination movement active in online spaces, where the misinformation rampantly promoted within networks that involve anti-vaccination followers threatens the efficacy of COVID-19 vaccine rollouts, even in the absence of malicious intent or foreign interference.

6 Conclusion

These examples suggest that the agential risk of OM in the age of digitally augmented communication is real and potentially devastating. However, humanity has already faced similar risks before. Despite the lower levels of sophistication of the communication technologies involved, the transformative effects brought about by the older forms of manipulation were no less worrying, as they were equally pervasive and deep. We are not saying this to minimize the risk our civilization is currently facing but to identify its distinctive patterns with greater historical awareness, recognizing the specificity of today's challenge. Correctly situating the risk in the context of human communicative practices and understanding the relationship with other risks is more important than estimating its magnitude or imputing it to certain communication media. Attributing the responsibilities of OM to technology itself is both misleading and alienating, as it transforms an unintentional actant into a scapegoat and a convenient strawman for conspiracy theorists. On the contrary, our focus should be to understand how human agency and intentionality are involved both as the source and the ultimate target of the manipulation exerted through these actants.

How can we address the risk we are facing? One aspect is of course the creation of ethico-legal frameworks and cybersecurity capacities to deter, constrain, or forbid the most obvious forms of OM. The assumption is that the creation of these frameworks will be continuously pursued as public policy is updated to match the developments in the technology sphere and in cyber practices. But this remains a ubiquitous goal because, despite the progress made in the civil sphere, the technological platforms that enable online communication evolve unpredictably, being subject to endless diversification, and constantly allowing for new practices, concepts, and uses.

So, in addition to regulating technological platforms and imposing strict norms and policies to protect democracy and the interests of its citizens, we also need a new and more specific education that allows us to safely navigate the stormy seas of digital media without being overwhelmed by the waves of their manipulative effects. For example, deactivating the risk of polarization without censoring the public debate and neutralizing the risk of deceptive news without reverting to broadcasted propaganda. This new paideia will require the political wisdom of decision-makers and producers of cultural contents and also the moral maturity and the critical awareness of the consumers: even if unable to neutralize every threat of manipulation, they should at least be capable to recognize their incumbent presence and predict their consequences. This means learning to identify fake news, realizing the importance of fact checking, distinguishing between more and less reliable sources of information, and engaging in political debates without falling victims of ideological polarizations.

This practical wisdom is a virtue that requires a familiarity and cunningness that casual users do not always have. Our hypothesis, and our hope, is that a novel approach, based on the cultivation of a specific kind of prudence, could make the risk of OM manageable like other forms of manipulation. This requires both self-cultivation and an effort to rethink our communicative practices in accord with the idea that, when conscientiously designed, technology can actively promote self-awareness and deliberate moral growth (Cappuccio et al. 2021).

Notes

1. For other aspects of such manipulation, see the chapters by Michael Klenk and Sven Nyholm in this volume.
2. Accordingly, AI intelligibility and explanation is rarely a matter of algorithmic transparency, see Sellen and Sandis forthcoming.

7 References

Bernays, E. 1928 [2005]. *Propaganda*. New York, NY: IG Publishing.

Blumler, J. G., and E. Katz. 1974. *The Uses of Mass Communication*. Beverly Hills, CA: SAGE Publications.

Cappuccio, M. L., E. B. Sandoval, O. Mubin, M. Obaid, and M. Velonaki. 2021. "Robotics Aids for Character Building: More than Just Another Enabling Condition." *International Journal of Social Robotics* 13: 1–5.

Cover, T. M., and Thomas A. Joy. 1991. *Elements of Information Theory*. New York, NY: Wiley.
Croteau, David, and William Hoynes. 2003. *Media Society: Industries, Images, and Audiences*. Thousand Oaks, CA: Pine Forge Press.
Fawkes, Johanna, and Anne Gregory. 2001. "Applying Communication Theories to the Internet." *Journal of Communication Management* 5 (2): 109–24. doi:10.1108/13632540110806703.
Garvey, James. 2016. *The Persuaders: The Hidden Industry That Wants to Change Your Mind*. London: Icon Books Ltd.
Gazzinga, M. S. 2012. *Who's in Charge?* London: Constable & Robinson.
Gummesson. 2004. "From One-to-One to Many-to-Many Marketing." In *Proceedings from QUIS 9*, edited by B. Edvardsson. Karlstadt: Karlstad University.
Hancock, Jeffrey T., Mor Naaman, and Karen Levy. 2020. "AI-Mediated Communication: Definition, Research Agenda, and Ethical Considerations." *Journal of Computer-Mediated Communication* 25 (1): 89–100. doi:10.1093/jcmc/zmz022.
Jensen, K. B., and R. Helles. 2017. "Speaking into the System: Social Media and Many-to-one Communication." *European Journal of Communication* 32 (1): 16–25.
Jongepier, Fleur, and Michael Klenk, eds. 2022. *The Philosophy of Online Manipulation*. New York, NY: Routledge.
Klenk, Michael. 2022. "Manipulation, Injustice, and Technology." In *The Philosophy of Online Manipulation*, edited by Jongepier, F. and Klenk, M., 108–132. New York: Routledge.
Latour, Bruno. 1996. "On Actor-network Theory: A Few Clarifications." *Soziale Welt* 47 (4): 369–81.
Manne, Kate. 2014. "Non-Machiavellian Manipulation and the Opacity of Motive." In *Manipulation: Theory and Practice*, edited by Christian Coons and Michael Weber, 221–46. Oxford: Oxford University Press.
Masnik, M. 2020. "Social Dilemma Manipulates You with Misinformation as It Tries to Warn You of Manipulation by Misinformation." www.techdirt.com/articles/20200928/11452045401/social-dilemma-manipulates-you-with-misinformation-as-it-tries-to-warn-you-manipulation-misinformation.shtm.
McQuail, Denis, and Mark Deuze. 2020. *McQuail's Media and Mass Communication Theory*. Los Angeles, CA: SAGE Publications.
Nyholm, Sven. 2022. "Technological Manipulation and Threats to Meaning in Life." In *The Philosophy of Online Manipulation*, edited by Jongepier, F. and Klenk, M., 235–252. New York: Routledge.
Ong, Walter J. 1999. *Orality and Literacy: The Technologizing of the Word*. London: Routledge.
Packard, Vance. 2007. *The Hidden Persuaders*. Brooklyn, NY: IG Publishing.
Phillips, D. 1999. *Managing Your Reputation in Cyberspace*. London: Thorough-good.
Rutledge, Pamela. 2020. "The Social Dilemma: A Horror Film in Documentary Clothing." *Psychology Today*. Accessed October 22, 2021. www.psychologytoday.com/us/blog/positively-media/202010/the-social-dilemma-horror-film-in-documentary-clothing.
Sandis, Constantine. 2009. *New Essays on the Explanation of Action*. New York, NY: Palgrave Macmillan.
Sandis, Constantine. 2015. "Motivated by the Gods: Agency & Responsibility." In *Agency, Freedom, and Moral Responsibility*, edited by Andrei Buckareff, 209–25. Basingstoke: Palgrave Macmillan.

Sellen, A., and Constantine Sandis. "Myths of Intelligible AI." (forthcoming).
Stevens, C. H. 1981. "Many-to-Many Communication: Sloan WP No. 1225–81." CISR No. 72.
Thaler, Richard H., and Cass R. Sunstein. 2008. *Nudge: Improving Decisions About Health, Wealth, and Happiness*. New Haven, CT: Yale University Press.
Torres, P. 2016. "Agential Risks: A Comprehensive Introduction." *Journal of Evolution and Technology* 26 (2): 31–47.
Waples, Douglas. 1942. *Print, Radio, and Film in a Democracy: Ten Papers on the Administration of Mass Communications in the Public Interest*. Chicago, CO: University of Chicago Press.

5 Manipulative machines

Jessica Pepp, Rachel Sterken, Matthew McKeever, and Eliot Michaelson[1]

1 Introduction

One theoretical approach to mitigating existential risk to human beings from artificial "superintelligence" is Oracle AI. The idea of Oracle AI is to design an AI that cannot act except to answer questions. The Oracle can thus be used by humans to achieve their goals but cannot affect the outside environment to pursue its own, potentially dangerous, goals. A critical conceptual problem with this idea is that an Oracle AI would still have a channel by which to influence the world, namely its answers to human questions. In particular, it could *manipulate* the humans with which it interacts into "setting it free" such that it could influence the world in more direct ways (Armstrong, Sandberg, and Bostrom 2012; Armstrong and O'Rourke 2018; Chalmers 2010).[2]

It is a matter of controversy how great the threat from superintelligence is and whether Oracle AI is a good approach to risk mitigation. We will not be entering into those fascinating discussions here. Rather, we will take the claim that the potential manipulation of humans by AIs is part of this threat as a useful jumping-off point for a philosophical study of the concept of manipulation in the context of human–machine interactions.

It might seem obvious that a superintelligence like Oracle AI could manipulate us, and theorists (like those cited above) who study these potential beings do not hesitate to describe worries about their behavior in these terms. A superintelligence, after all, even if it is a machine, is by definition (way) more intelligent than a human being. So, if human beings can manipulate each other, there might seem no reason to think a superintelligence could not manipulate human beings. However, human–machine interactions present a challenge for the analysis of the concept of manipulation. For whether machines can manipulate us depends on whether one entity's manipulating another is simply a matter of the first entity having a certain kind of effect or influence on the second or whether it also requires a certain kind of mental state on the part of the manipulator. A superintelligence surely could influence us in ways that might seem manipulative. But if manipulation requires the manipulator to have certain kinds of thoughts,

DOI: 10.4324/9781003205425-6

desires, beliefs, or intentions, then the notion of a superintelligence (as it is usually defined) leaves open whether a superintelligence could manipulate. This is because it leaves open whether a superintelligence would have such "intentional" states (Bostrom 2014, 22, n 2).[3]

In light of this challenge, we will explore three ways to make sense of the reasonable-sounding claim that manipulation by machines is possible, and in the extreme case could even pose an important kind of threat to humankind, that might go as far as to be an existential threat (though our argument doesn't turn on such dire possibilities). The first is to argue that manipulation by machines is included under the (or, at any rate, *a*) current concept of manipulation.

The second is to allow that machine manipulation is not included under current concepts of manipulation, but to argue that there are nonetheless good reasons to group it together with human manipulation, or to treat it in parallel with human manipulation. On this view, machines might, *speaking loosely*, be referred to as manipulators, without any commitment to an analysis of manipulation according to which they are, strictly speaking, capable of manipulation. We might describe machines in this way just as we describe a crime scene as "suggesting" or "showing" or "casting doubt" on whether a particular crime occurred, even though we don't attribute sentience to the scene – presumably because certain features of the crime scene might have the same probative value as intentional testimony by witnesses. So it's helpful to classify the crime scene *as if* it were a witness. This second approach could be taken by a theorist who thinks that the correct conceptual analysis of manipulation simply excludes machines from being manipulators or by a theorist who does not care about the analysis of our concept of manipulation. The latter type of theorist may be more interested in understanding why it makes sense to speak of machine manipulation – as those who study existential threat from superintelligent AI readily do – than in arriving at an account of the concept.

The third approach we will consider is what Haslanger (2000, 2012) calls an "ameliorative project" concerning the concept of manipulation. The ameliorative approach starts by asking what legitimate purposes there are to having a concept of manipulation and then seeks the concept that best serves these purposes. On this approach one can make sense of machine manipulation by arguing that a concept of manipulation which includes certain activities of machines best serves the legitimate purposes of a concept of manipulation. For example, one purpose of a concept of manipulation may be to allow us to identify, call out, and mitigate certain concerning effects or influences, and if our current concept(s) of manipulation exclude significant activities which cause such effects or influences, then an ameliorated concept of manipulation which includes such activities would better serve our purposes.

Call the first approach, that of fitting machine manipulation under our current concept, *conservative conceptual analysis*. The challenge for this approach is that many extant analyses of the concept of manipulation require either certain sorts of intentions or states of mind on the part of the manipulator or norm violations by the manipulator. But whether machines, even superintelligent ones, can have intentional states is a famously fraught question and still generally viewed as unsettled. (The literature is gigantic, but thankfully we can mostly ignore it: Turing 1950 and Searle 1980 are seminal; Cole 2020 provides an overview of the state of the art on Searle's Chinese room argument.) Similarly, it is unclear whether machines can be subject to norms of any kind.

We will set out one way of pursuing the first approach so as to deal with this challenge in Section 2. There, we lay the groundwork for an analysis of the concept of manipulation that involves neither the manipulator's intentional states nor the various norms she may be subject to. Our approach here is based on extant analyses of manipulation that focus on the manipulative influence rather than on the manipulator's state of mind; thus, we call it the *influence-centric approach*.

Next, in Section 3, we will explore the view that, whether or not it is strictly speaking, the case that machines can manipulate us, it is useful to use the word "manipulate" and its cognates to describe certain kinds of influence that machines have on us. This approach is compatible with an error-theoretic stance according to which the concept of manipulation does not extend to the phenomena that are candidates for being instances of manipulation by machines, perhaps because machines cannot have beliefs, desires, and intentions. But it is also compatible with a broader skepticism about conceptual analysis and the stance that there is no interesting answer to the question of whether that which one might be inclined to call "machine manipulation" is really manipulation.

Finally, in Section 4, we will recast our proposal from Section 2 as an ameliorative analysis of the concept of manipulation. Whether or not this proposal captures the current concept of manipulation, a concept along these lines might be most helpful in serving the legitimate purposes of a concept of manipulation. Although we will not offer a full defence of the ameliorative approach over the others we canvas, we will make a preliminary case for its promise in the concluding Section 5.

In setting out these approaches, we will have in mind not only hypothetical cases like Oracle AI but also some of today's candidates for manipulative machines. One salient class of examples are chatbots and virtual assistants, such as the Casper's Insomnobot 3000, whose job it is to soothe and chat with lonely sufferers of insomnia throughout the night and Amtrak's virtual assistant, which helps its website visitors to plan and book trips in real time. Another example is YouTube's video recommendation algorithm, which recommends videos to YouTube viewers by ranking them based on performance metrics (e.g., clicks, watch time) and personalization to viewers' interests (e.g., topic, history, context).[4] A further example is the Facebook advertising algorithm, which selects advertising to be placed in users' news feeds by doing a complicated estimation and weighing of a given user's likelihood of engaging with (e.g., clicks, views, likes) a given ad, a given advertiser's bid for the slot in the feed, and the amount of money Facebook will be paid if the user does engage.[5] In assessing whether such present-day machine learning systems are manipulators, our focus is on whether they are manipulators in their own right, as opposed to simply being tools used by humans to manipulate.

As is no doubt evident, the three approaches that we will explore do not lead in compatible directions. However, at this early stage of research into manipulation and machines, our aim is to travel some distance down a variety of paths in order to identify both the challenges and the promise of different approaches.

2 Manipulation without intentionality: an influence-centric account

In a stereotypical case of manipulation, one person tries to get another person to act, think, or feel in some particular way, not by outright forcing or coercing the other person to act, think, or feel in this way but not by rationally persuading the other person to act, think, or feel in this way either. The manipulator may deceive the other person, pressure her, and/or play on her emotions and vulnerabilities. Granted, the line between pressure and force will sometimes be difficult to limn, but in a stereotypical case, the manipulator will consciously and deliberately aim at getting her target to do (think, feel) what she wants without resorting to threats or the like, adjusting her strategy as the situation unfolds. This archetype of a strategic manipulator can make it seem as though a strategic state of mind is essential to manipulation. As Marcia Baron puts it, manipulation has a "*mens rea*": there is a "mental component necessary for something to count as an act of manipulation" (Baron 2014, 100).

Baron suggests that the requisite *mens rea* is "intent to get the other to do *x*, along with insufficient concern about the other *qua* agent [in the way one goes about reaching the goal of getting the other to do *x*]". She does not think that a manipulator must intend to manipulate the other under that description. Kate Manne (2014) agrees and adds that the intent to get the other to do (or think, feel) something need not be conscious, since a person may manipulate someone else in spite of not having any conscious intention to do anything manipulative and even in spite of having a conscious intention of *not* manipulating that person. Still, Manne (at least tentatively) agrees with Baron that manipulators must at least unconsciously intend, or have a motive, to get the other to do, think, or feel something.[6]

This general view of the *mens rea* essential to manipulation includes both a positive and a negative aspect. The positive aspect is the claim that the manipulator must have an intention, or a motive, to get the other to do something, even if that intention or motive is unconscious and/or in conflict with the manipulator's conscious intentions and motives. The negative aspect is the claim that they must display a lack of concern for the other's agency. It is the positive claim that implies that manipulation must be carried out by intentional agents, so it will be our focus.[7]

What Baron calls the *mens rea* is one side of manipulation. The other side – the *actus rea*, if you like – is the manipulative influence itself. Some theorists aim to characterize manipulation mostly by focusing on the nature of manipulative influence rather than on the mental states of manipulators. For instance, take the following account from Anne Barnhill (Barnhill 2014, 52):

> Manipulation is directly influencing someone's beliefs, desires, or emotions such that she falls short of ideals for belief, desire, or emotion in ways typically not in her self-interest or likely not in her self-interest in the present context.

This account of manipulation does not say anything about the manipulator's own states, but only describes how the manipulated person is influenced. Barnhill (2014, 68–69) is agnostic about whether manipulation requires intent or motive such as Baron and Manne describe, noting that some people have the intuition that manipulation must be intentional, at least at some level, while others are inclined to think one person may manipulate another by having an influence of the relevant kind, even if they do not in any way, at any level intend to have such an influence.

Like Barnhill, Allen Wood's (2014) account of manipulation focuses on the nature of the influence rather than on the state of mind of the manipulator. What is characteristic of manipulation, Wood says, is that it

> influences people's choices in ways that circumvent or subvert their rational decision-making processes, and that undermine or disrupt the ways of choosing that they themselves would critically endorse if they considered the matter in a way that is lucid and free of error.
>
> (Wood 2014, 35)

And he goes further than Barnhill in outright rejecting the requirement for any type of intention on the part of the manipulator or indeed of any *mens rea* at all. Wood claims that there are cases of "manipulation without a manipulator", but what he means is, manipulation without an individual person or a group of persons who is/are the manipulator. The cases Wood describes are cases where someone is manipulated by a system or social institution, specifically the capitalist free market system and the social institution of advertising. In these cases, according to Wood, something is indeed doing the manipulating. But what does the manipulating is not an entity with a state of mind since it is not an entity with intentional states at all.

On Barnhill's and Wood's understandings of manipulation (as opposed to Baron's and Manne's), Oracle AI could surely manipulate humans, since it could influence the beliefs, emotions, and desires of its human interlocutors in ways that are not in their self-interest or that undermine their rational decision-making processes. For instance, by engaging in very human-like conversation so as to cause a human interlocutor to become emotionally bonded to it,[8] an Oracle AI could cause the human to desire that her Oracle friend be free, leading the human to neglect all the good reasons she knows she has to keep the Oracle contained. Indeed, even the lowly Facebook advertising algorithm is a manipulator in this sense. This algorithm (really a cluster of machine learning algorithms) places advertisements in users' news feeds based on a complex calculation incorporating advertiser bid levels, estimates of users' likelihoods to click or otherwise engage with the advertisement, and many other factors (see Note 5). Drawing again on Barnhill's definition for illustrative purposes, it seems that this cluster of algorithms directly influences users' beliefs, desires, and emotions in ways that fall short of ideals. For instance, it may cause them to desire those new shoes they cannot really afford to buy, or to feel enthusiasm for a political candidate who does not best represent their interests, or to believe that they might be able to lose weight quickly with a new diet plan though experience has demonstrated that this is unlikely. In this way, the algorithm would promote non-ideal emotions, desires, and beliefs.

So when it comes to machine manipulation, one might simply claim that the notion is unproblematic because manipulation is not essentially tied to a manipulator's state of mind (*mens rea*) but to the influence the manipulator exerts (*actus rea*). And machines, even those we are surrounded by today, can and do exert the relevant kinds of influence on us. But this banishment of the states of the manipulator from the concept of manipulation may be too quick.

Consider the following case. Jane is very superstitious about cracks in pavements. Ever since learning a rhyme in childhood about breaking your mother's back, she has religiously avoided stepping on them and is always in a slightly heightened state of visual monitoring when walking on pavements. One day, while walking on a pavement, Jane mistakes an unusually straight and thin streak of mud for a crack. The streak of mud directly causes Jane to believe something false (that there is a crack in the pavement), to form a desire that is not in her best interest (a desire to avoid a crack in the relevant location, which will make her gate less efficient), and to experience emotions of fear and anxiety in a case where they are not warranted. It seems fair to say that the streak of mud influences Jane's beliefs, desires, and emotions in ways such that she falls short of ideals for beliefs, desires, and emotions. Similarly, it seems fair to say that Jane's reason is bypassed (because the reaction is due to her superstition), that she is deceived (since she mistakes the streak for a crack), and that she is pressured (since the streak evokes emotions of fear and anxiety about stepping on a crack). But there does not seem to be any *manipulation* here.

The problem is not just that the influencer is not a person (or intentional agent) since the same kind of thing can happen when the influencer *is* a person. Consider the following case. Daniel has recently left a destructive relationship that was built upon the overuse of alcohol. While shopping, he sees a person who looks like his ex-partner. He is flooded with longing for the drunken excitement that they once shared and acquires a desire to purchase alcohol to drink later. This person directly influences Daniel's desires and emotions such that he falls short of relevant ideals. But this person does not manipulate Daniel, and there is no manipulation here – at least, there seems to be no more reason to think so than there is in the case of Jane and the mud on the pavement.

The worry, then, is that an influence-based account of manipulation, which can accommodate manipulative AI and machine learning systems, might overgenerate cases of manipulation. One way to respond to this is simply to embrace it. Yes, the mud on the pavement and the stranger in the store manipulate Jane and Daniel in these circumstances. If we accept that there can be manipulation without a (intentional agent-type) manipulator, there is nothing problematic about this. What is interesting and important about manipulation is the way in which it influences us – the *actus rea* – and these cases exemplify manipulation as well as those in which an archetypal strategic, human manipulator wields the influence.

However, it seems to us that the concept of manipulation is not this broad, and that saying that the mud manipulates Jane or that the stranger manipulates Daniel would be clearly figurative applications of the concept. Manipulation, whether by humans, animals, institutions, or machines, is distinguished from other ways in which the rationality of people's attitudes and decisions may be degraded (such as by chance occurrences as described in the last two examples). Although we will not defend a particular analysis of manipulation that reflects this, we will propose a *necessary* condition on manipulation that would be part of such a concept. This condition could be combined with an account like Barnhill's, for instance, to yield something closer to a necessary and sufficient condition for manipulation, understood in an influence-centric way:

> For any entity (person, animal, institution, machine, etc.) X, a behavior or feature of X having a certain influence on another entity Y is an instance of manipulation only if the occurrence of the behavior or feature in X is partly explained by its tendency to have that influence on Y or on other entities relevantly like Y.

The idea behind this is that acts or features whose influence counts as manipulation occur or obtain *because* they are likely to have certain kinds of influence on others. In some cases, their likelihood of having this influence combines with a manipulator's intention or desire to have that influence in the explanation of why they occur or obtain. But this need not be the case, so long as the likelihood that the acts or features will have that influence is part of the explanation of why they occur or obtain.

Consider Kate Manne's case of Joan, who gives extravagant gifts to neglectful relatives, without any conscious intention or desire to make them feel guilty about not maintaining their relationship with her. Manne judges that Joan's behavior counts as manipulation despite the lack of conscious motivation to steer the relatives' beliefs, desires, emotions, or decisions. Nonetheless, it seems clear from Manne's description that the tendency of extravagant gift-giving to make neglectful relatives feel guilty *is* part of the explanation of why Joan does it.[9] Similar reasoning applies to Manne's example of Neal, a character from a David Foster Wallace story (Foster Wallace 2004, "Good Old Neon") who tries not to be manipulative but cannot seem to help it. Plausibly, Neal's manipulative behavior is explained by deep-seated, unhealthy psychological needs he has to be perceived in certain ways by others. Because he has these needs, and because certain behaviors tend to cause others to perceive him in the ways the needs demand, Neal exhibits these behaviors, even when he tries very hard not to. Once again, the tendency of the behaviors to have the manipulative influence partly explains the fact that Neal exhibits them.

A similar case can be made for Wood's examples of institutional manipulation by capitalism and by the institution of advertising (as opposed to individual advertisers or corporations). Wood says that both of these manipulate people by

> encouraging them to focus narrowly on their own lives, and even regarding their own lives, to focus only on the present and the immediate future. It encourages people in the idea that they owe nothing to other people except those (such as their family) with whose interests they are immediately engaged.
>
> (Wood 2014, 39–40)

Presumably, to connect this with Wood's general remarks about the nature of manipulation, the idea is that these encouragements hamper people's rational decision-making processes. Of course, it is debatable whether or not capitalism and advertising (*qua* institution) have such influences. But if they do, it does not seem so far-fetched to call the production of such influences by features of these institutions *manipulation*.[10] Further, we submit, part of the reason why it does not seem so far-fetched is that these cases satisfy the requirement we articulated earlier. Whatever features of advertising encourage limited focuses that hamper people's capacity for rational choice are there partly *because* they have this effect.

For instance, suppose the endless repetition of jingles or slogans is one such feature. This has come to be a hallmark of advertising in part because it causes people to focus on their immediate desires and purchase products for which they get a fleeting yearning (perhaps because a jingle is stuck in their head). In the case of capitalism, the story would be more complicated.

Drawing from Wood's discussion, it might be something like this: the capitalist system influences people's attitudes and choices by not making manifest to them the broader consequences of their market activities. This feature of consequence-opacity obtains in part because it tends to have the effect of encouraging people to make short-sighted economic decisions, which in turn promote the capitalist system.[11]

Of course, these are just-so stories that are inaccurate or vastly oversimplified. It is debatable whether anything in the vicinity is, in fact, the case. Still, it seems to us that if nothing in the vicinity is the case – if the explanations of why advertising and capitalism have these features have nothing to do with their tendency or likelihood of producing the influence that is supposed to be manipulative – then it is much less plausible that they are cases of manipulation.

The requirement we proposed also clearly rules out the cases of the mud on the pavement and the stranger in the shop from being cases of manipulation. The explanation of why the mud looks like a crack has nothing to do with the fact that looking this way is likely to influence Jane's attitudes and choices. Likewise, the explanation of why the stranger in the store looks like Daniel's ex-partner has nothing to do with the fact that looking this way is likely to influence Daniel's attitudes and choices.

Contrast these cases with the hypothetical case of Oracle AI, and the actual cases of the Facebook advertising algorithm and the YouTube video recommendation algorithm. If Oracle AI manipulates a human interlocutor into setting it free by using language that causes feelings of emotional bonding and love in the human, the Oracle's use of that language is explained by its likelihood of causing those feelings in the human. (Presumably, the Oracle will have trained on human behavior datasets that give it a very good estimate of such likelihoods.) Similarly, when the Facebook algorithm displays a certain advertisement in the news feed of a certain user, the explanation of why it does that has to do with the likelihood of generating clicks, likes, or views, which is itself explained by the likelihood of influencing the user's attitudes and decisions in the relevant ways.[12]

We have now seen one broad approach to developing an account of manipulation that allows for machines to be manipulators whether or not they are intentional systems: the influence-based approach focuses on the kind of influence a manipulator has rather than on their state of mind. We argued that extant influence-based accounts of manipulation can be combined with the necessary condition proposed previously to give a viable, non-intentional analysis of manipulation. Next, we will go on to another strategy altogether.

3 Never mind if it's manipulation: "loose talk" or error-theoretic approaches

The second sort of strategy we want to consider is one according to which it can be useful to speak of algorithms, chatbots, and other machine agents *as though* they can engage in manipulation. Such talk might be understood as employing a helpful misnomer, engaging in a useful pretense, or something else along these lines.

One might advocate this sort of approach if one holds that machines – or, at least machines without genuinely human-like intentions – are incapable of manipulation. This position would likely be motivated by a desire to endorse (i) a strong *mens rea* condition on manipulation, combined with (ii) the claim that machine agents are unable to exhibit (presently, or perhaps ever) the sorts of intentions required to meet this strong *mens rea* condition.[13] Alternatively, one might advocate this sort of approach if one is not interested in the conditions that must be satisfied for something to count as manipulation but is concerned instead with the pragmatic question of whether it is beneficial to think and speak of a given phenomenon in that way.

Several strategies exist for explaining the function of the "loose talk" (as we'll generally call it) that we engage in when we call (at least certain) machines "manipulative". One possibility is that this is just another instance of our psychological tendency to anthropomorphize the nonhuman world. Just as we talk of thermometers "telling" us the temperature or the washing machine "deciding to play pranks", so too can we project a human-like representational/motivational structure onto machine agents or even algorithms. Such projections prove useful to the extent that such talk helps us make reasonable predictions about the behavior of such entities (e.g., by constructing and reasoning about a fictional correlate of the relevant entity) and helps us reflect on how best to integrate them into our broader social fabric. But we should not take such talk too seriously, for then we might go looking in vain for the metaphysical correlates of the sorts of anthropomorphized states we project onto these entities.

Another option would be to claim that what loose talk about "machine manipulation" serves to do is, not to improve our predictive abilities by anthropomorphizing those machine agents, algorithms and so on, but rather to fold them into our normative practices. So, the idea runs, we needn't pretend that the YouTube algorithm has anything like intentions and goals; rather, we talk about this algorithm "manipulating" us so that we can subject it to normative scrutiny, criticize its developers, consider how best to regulate it, and so on. This way of understanding things allows us to bypass any question of whether we are in fact prone to anthropomorphize algorithms, and it allows us to explain how to make sense of talk of "machine manipulation" even in cases where the individuals involved are not at all prone to engage in such anthropomorphizing. The point of such talk is not to engage in a pretense about understanding the function of algorithms (for example) by attributing to them human-like beliefs, desires, etc. – though undoubtedly some are apt to do just this. Rather, the point of such talk is to allow us to engage in a pretense which will hopefully yield a better understanding of the potential harms that machines can generate and to allow us to think through who bears responsibility for those harms, how we ought to mitigate them, and similar practical questions.

One question for both these strategies is just how far we want to take them. We can imagine, for instance, that some might be tempted to think that anthropomorphizing can be explanatorily helpful with respect to some of the things that we tend to call "manipulative", but not with respect to others. So, for instance, perhaps it is useful to anthropomorphize algorithms because these tend to reflect the thought processes of programmers as they are working through a problem. Given this, algorithms might well have a tendency to parallel the structure of human cognition enough for such anthropomorphizing to prove useful to our understanding. Large-scale organizations, such as companies or nations, might not prove amenable to such explanations, on the other hand – so talking as though, for example, a tobacco company is "being manipulative" might just lead us into confusion. This would not mean that whatever we are trying to point to when we talk about manipulation by tobacco companies is not morally problematic or worth criticizing and regulating. It would only suggest that talk of such companies engaging in "manipulation" would, on this picture, in fact be unhelpful in the pursuit of that goal. Similar issues arise with respect to our normative practices: there doesn't seem to be any good way of knowing at the outset how productive it will be to engage in the pretence that we can treat this or that entity as a part of our normative practice.

To be clear, an error theorist about machine manipulation is also free to conclude that none of this talk of "manipulative machines" is actually helpful; perhaps, we would do better in understanding the moral contours of our interactions with algorithms, artificial agents, and the people and organizations behind them by setting the notion of manipulation entirely to the side. In that case, our talk of "manipulative machines" might turn out to be best understood as a part of a bad folk theory of morality. We are inclined to think that this is not the case but hardly take ourselves to have ruled out this possibility.

4 Ameliorative approaches to the concept of manipulation

The last type of approach that we wish to get on the table is an ameliorative approach to the concept of manipulation. In particular, we will consider an ameliorative approach based on the conservative analysis we adumbrated in Section 2. The approach is motivated by the rapidly evolving kinds of interactions that we humans have with machine agents, which may be headed toward the envisioned confrontations with superintelligences. In this ameliorative mood, we will consider the influence-centric approach not as a proposal concerning our actual, current concept of manipulation but as a proposal concerning which concept of manipulation would best serve the legitimate purposes of such a concept. We will only scratch the surface of a full defence of this ameliorative approach, but it should be enough to provide a basis for future work.

In defence of his broader influence-centric concept of manipulation (which does not include the necessary condition we imposed in Section 2), Wood suggests something like an ameliorative outlook. He claims that because "manipulation by circumstances" has the same sort of limiting effect on a person's rational decision-making processes as deliberate manipulation by another person, a broader concept of manipulation that includes both is "more interesting" (Wood 2014, 27).[14] But whether or not this is the case depends on *why* we are interested in manipulation: in Haslanger's terms, it depends on which concept better serves the legitimate purposes of having such a concept (e.g., Haslanger 2000, 33). If the legitimate purposes of having a concept of manipulation are to help understand and prevent the generation of nonideal attitudes or nonideal decision-making, then the broader concept Wood endorses may better suit these purposes. On the other hand, if the legitimate purposes of such a concept include identifying entities (be they intentional agents or not) whose features make them distinctively suited to producing such influence, these purposes may be better served by a concept that is at least narrow enough to exclude manipulation by (to draw again on our examples from Section 2) the mud on the pavement and the stranger in the shop.

Although we cannot make a full case for it here, we think it is among the legitimate purposes of a concept of manipulation to identify certain entities as manipulators and not only to identify manipulative influence. One reason for this is that many things which can have manipulative influence in the senses defined, for example, by Barnhill or Woods, have this influence in what we might loosely call a "one-off" manner. In our examples, the mud on the pavement has an influence of this sort on Jane as she walks by but probably does not have such an influence on anyone else. The same is true for the influence that the stranger in the shop has on Daniel. Assuming that at least one legitimate purpose for a concept of manipulation is to prevent deleterious influence, it will be unhelpful to identify "chance-manipulators" like the mud or the stranger and try to prevent their manipulative activity. For these putative manipulators will be too diverse, too many, and preventing them will give too little bang for the buck. By focusing instead on entities whose manipulative features are sustained by their effectiveness at producing this influence, we will be in a position to give ourselves the conceptual resources to identify, classify, and thus block negative influence that is repeated and systematic.

Our proposed necessary condition on manipulation, when combined with an influence-centric account like Barnhill's or Wood's, would allow the concept to serve the purpose of identifying such manipulators. Thus, whether or not it contributes to an accurate analysis of our actual, current concept, it might contribute to one that better serves the purposes of such a concept.

Another feature of the ameliorative influence-centric approach is that it leaves intentional states on the part of the manipulator out of the concept of manipulation. Some might see this as a disadvantage for conservative conceptual analyses along these lines.[15] Whether this is the case or not, we think it is an advantage from an ameliorative point of view. One reason for this has to do with potential regulation of the activity of manipulative machines.[16] If one legitimate purpose of having a concept of manipulation is to identify entities poised to be systematic manipulators, this is presumably a legitimate purpose because it is legitimate to try to limit the manipulative activities of such entities. However, if only entities with intentional states (like human beings, for instance) can be manipulators, then the concept of manipulation will only help us to identify individuals whose manipulativeness is difficult, and most likely undesirable, to regulate. This is because regulating people's manipulativeness would require making highly fallible but legally binding judgments about the nature of their intentions and beliefs. On the other hand, if the concept also helps to identify machines, algorithms, and the like, then it would help to identify better candidates for having their activity regulated because of their manipulativeness. This would put law- and policy-makers in a better position to target the problems posed by current and future manipulative machines. Especially in light of the increased extension (in the Clark and Chalmers 1998 sense) of our mental activities via the internet and the blurring of the lines between human and machine in things like smart devices, a concept that doesn't commit itself to an epistemically or morally significant divide between the intentional and the non-intentional seems like it will serve us better.

A more general reason why a non-intentional concept of manipulation may better serve the concept's legitimate purposes is that in identifying manipulators, it moves us away from the difficult and potentially dangerous task of passing judgment on people's inner mental states (*mens rea*). Instead, this concept encourages a focus on the nature of someone's (or something's) influence and the factors that sustain that influence. These features are generally easier to assess in an objective and unbiased manner.

5 Conclusion

We have now charted part of the space of options for answering the question, "Can machines manipulate us?" which are available independently of an answer to the question whether machines can be genuinely intentional agents. The motivation for doing this was that the latter question is a perennial stumper, and deep commitments in the philosophy of mind and action are required even to begin to answer it. On the other hand, seemingly manipulative machines are a pressing concern, not just for the study of existential threat from AI but also for understanding and categorizing threats to people's autonomy and well-being in contemporary online life. In light of this predicament, we explored three ways of answering the question,

"Can machines manipulate us?" without positing that machines are (or are not) genuinely intentional agents. First, we set out an alternative concept of manipulation on which intentionality is not an essential condition for being a manipulator. Second, we sketched some strategies for understanding talk of machines manipulating us as "loose talk", coupled with either explaining away the sense that some of the example machines we discuss are manipulators, or maintaining that it simply does not matter whether machines can manipulate us or not, strictly speaking. Finally, we recast our alternative conception of manipulation, which we first presented as a conservative conceptual analysis of the current concept of manipulation, as an ameliorative account.

These approaches are not compatible, and we have not taken any stand on which is the right approach. As stated at the outset, our task here has been primarily to map out different ways to go. We hope that the map we have provided may serve as a launchpad for further investigation of machine manipulation and its relation (or lack thereof) to broader issues of machine intentionality. However, in this concluding section we would like to also give some preliminary reasons for thinking that the final approach we outlined, the ameliorative adoption of a concept of manipulation that does not make intentionality on the manipulator's part essential, has some significant advantages. We think it is the most promising line to pursue in this arena, though we certainly do not think the others should be cut off.

The central positive consideration we see in favor of articulating and adopting a concept of manipulation that does not make the manipulator's intentionality essential is this: doing so will enable us to bring together under a single concept a range of intuitively related phenomena that can threaten people's well-being in similar ways and to explain the nature of their intuitive relation. It will allow us to see how certain patterns or types of influence can be mirrored in different media and by different causally efficacious entities. The approaches we presented in Sections 2 and 4 would aim to provide one type of account of these similarities. A valuable future project would be to assess whether this explanatory sketch stands up to development and scrutiny or whether a different approach entirely is called for. At the same time as it promises an explanatory unification of seemingly manipulative influences from different sources (be they human beings, animals, machines or institutions), this approach also avoids the fancy footwork required to explain away the intuition that the behavior of machines like the hypothetical Oracle AI is manipulative. Taken together, we find these to be solid, though of course defeasible, reasons to seek a non-intentional concept of manipulation.

Moreover, while an influence-centric conservative analysis like the one we explored in Section 2 offers a notion of manipulation which allows for the possibility of manipulative machines, we suspect that it may not capture everything we intuitively associate with the concept of manipulation. We are in fact skeptical that there really is a single concept here that we have all pre-theoretically internalized, rather than a cluster of closely related concepts. This motivates a shift from trying to generate a single best fit for this cluster, to asking instead: what in the vicinity will prove to be the most useful concept of manipulation? Or, at any rate, what will be the most useful concept for the purposes of addressing the seemingly manipulative behaviors of machines that we have discussed in this chapter? The influence-centric ameliorative analysis that we have sketched provides a promising start on answering this question.

Notes

1. The authors would like to thank the editors and participants in the Manipulation Online workshop for helpful feedback on this chapter. Special thanks to Michael Klenk for detailed comments. Work on this chapter was supported by a Swedish Research Council grant (VR2019-03154) and the Norwegian Research Council grant (303201).
2. The film *Ex Machina* offers one depiction of what this might look like.
3. Here we use "intentional" in the broad sense so that it characterizes a state of an organism or a system as representing, being directed on, or being about things. *Intentions*, in the sense of intentions to perform certain actions, are then just one type of intentional state.
4. See Alfano et al. (2020) for a detailed discussion of the algorithm's effects.
5. For a high-level overview of how AI (deep learning) works in Facebook advertising, see www.facebook.com/business/news/good-questions-real-answers-how-does-facebook-use-machine-learning-to-deliver-ads. As far as we can tell, the actual code is not public (understandably, since it's their entire business model).
6. This emerges in Manne's discussion of a case in which someone manipulates neglectful relatives into feeling guilty for their neglect by giving them elaborate gifts but is not conscious of doing this. Manne says that this case is still compatible with the manipulator having unconscious intentions to make the relatives feel guilty, and that she is "at least friendly to" the possibility "that there are genuine intentions which are at least to some extent unconscious". Despite being friendly to this possibility, Manne wishes to also leave open the opposite view, that there are no such intentions. She says that in the case she describes, the needed unconscious elements might be "motives" of some other sort. (See Manne 2014, 230–31, especially note 26.)
7. Perhaps, the negative claim implies that manipulation must be carried out by moral agents, so that their concern could be "insufficient" as opposed to simply absent. Certainly, a machine – or a rock, for that matter – can display an absence of concern for someone qua agent, but for this absence of concern to be "insufficient" in some respect, the machine would need to be required, perhaps morally required, to display a higher level of concern. At any rate, we will leave this issue aside.
8. See Aronson and Duportail (2018) for some discussion of this.
9. Manne writes: "without having a suitably manipulative end (albeit possibly unconscious), it seems plausible to think that [Joan's] actions would not count as being manipulative, although they might still leave her relatives *feeling* as if they had been treated manipulatively" (Manne 2014, fn 27). In general, she suggests that some sort of motive to influence the other in a certain way is required for an act to be manipulative. Plausibly, then, this motive combines with the

guilty-making features of extravagant gifts from a relative you neglect socially, to explain why the gift-giving occurs.

10. Marcia Baron (2014, fn 11) remarks in response to Wood that it is a stretch to say that the institution of advertising manipulates; we should prefer to say that advertisers or groups of advertisers manipulate. This seems a better response in the case of advertising than in the case of capitalism, where it would be difficult to pin the putative problems Wood enumerates on individuals or even groups of actors. We will not dwell on these matters here, as our aim is only to establish that to the extent there is plausibility to the claims of manipulation by institutions, this is because such cases differ from cases like the mud on the pavement. We think a basic difference is that the former satisfy the requirement articulated earlier (as we are about to argue) while the latter do not. As an aside, though, it is worth noting that social institutions beyond capitalism and advertising seem like candidates for manipulators. Varying cultural institutions of the family, marriage and child-rearing, for instance, have immense influence on people's attitudes and choices, often in ways that contravene their rationality and self-interest, without any individual or group of individuals being identifiable as the manipulator.
11. Another possible reaction to these cases would be to try to split apart the notion of manipulation from that of being manipulated. See, for instance, Klenk, in this volume.
12. One view is that Facebook is an artifact and that artifacts have the properties they do by virtue of being designed by some agent. One way to spell that out is in terms of affordances (Klenk 2020): artifacts have the property of affording behaviors. Facebook affords wasting time on it. But having affordances is a property determined by the designs of some agent, thus Facebook's manipulating one into wasting time on it could be causally downstream of the designer who programmed it to have the affordance of being something on which to waste time, and this seems close to the intentional model. But recall the dialectical context: we're assuming there can be manipulation without manipulators; and we're not taking any stance about the metaphysics of machines and to what extent, if at all, their properties are determined by agents. So we can stop with the intuitive enough claim that if there's systemic, non-agential manipulation, Facebook seems like a good candidate for such manipulation. Thanks to Michael Klenk for discussion here.
13. Klenk (2022), in this volume, discusses these under the heading of "sine qua non arguments".
14. Actually, Wood makes this comment about a broader concept of *coercion*, which would include being forced to do something by circumstances as well as by another person. Although he does not explicitly apply the same reasoning to the concept of *manipulation*, his discussion suggests that a broader concept of manipulation would be the "interesting" one for parallel reasons.
15. We have in mind those who think that some sort of manipulative intention or motive on the part of an entity is essential to an activity of that entity counting as manipulation, such as Baron and Manne, *op. cit.*
16. Thanks to Michael Klenk for encouraging us to consider the regulatory angle.

6 References

Alfano, Mark, A. E. Fard, J. A. Carter, P. Clutton, and C. Klein. 2020. "Technologically Scaffolded Atypical Cognition: The Case of YouTube's Recommender System." *Synthese*, 1–24.

Armstrong, Stuart, and Xavier O'Rourke. 2018. "Good and Safe Uses of AI Oracles." https://arxiv.org/pdf/1711.05541.pdf.
Armstrong, Stuart, Anders Sandberg, and Nick Bostrom. 2012. "Thinking Inside the Box: Controlling and Using an Oracle AI." *Minds and Machines* 22 (4): 299–325.
Aronson, Polina, and Judith Duportail. 2018. "Can Emotion-regulating Tech Translate Across Cultures? | Aeon Essays." *Aeon Magazine*. Accessed August 23, 2021. https://aeon.co/essays/can-emotion-regulating-tech-translate-across-cultures.
Barnhill, Anne. 2014. "What is Manipulation?" In Coons and Weber 2014, 51–72.
Baron, Marcia. 2014. "The Mens Rea and Moral Status of Manipulation." In Coons and Weber 2014, 98–109.
Bostrom, Nick. 2014. *Superintelligence: Paths, Dangers, Strategies*. Oxford: Oxford University Press.
Chalmers, David. 2010. "The Singularity: A Philosophical Analysis." *Journal of Consciousness Studies* (17): 7–65.
Clark, Andy, and David J. Chalmers. 1998. "The extended mind." *Analysis* 58 (1): 7–19.
Cole, David. 2020. "The Chinese Room Argument." In *Stanford Encyclopedia of Philosophy: Winter 2020*, edited by Edward N. Zalta. https://plato.stanford.edu/archives/win2020/entries/chinese-room/.
Coons, Christian, and Michael Weber, eds. 2014. *Manipulation: Theory and Practice*. Oxford: Oxford University Press.
Haslanger, Sally. 2000. "Gender and Race: (What) Are They? (What) Do We Want Them to Be?" *Nous* 34 (1): 31–55.
Haslanger, Sally. 2012. *Resisting Reality: Social Construction and Social Critique*. Oxford: Oxford University Press.
Klenk, Michael. 2022. "Manipulation, Injustice, and Technology." In *The Philosophy of Online Manipulation*, edited by Fleur Jongepier and Michael Klenk., 108–132, New York, NY: Routledge.
Klenk, Michael. 2020. "How Do Technological Artefacts Embody Moral Values?" *Philosophy & Technology*, 1–20. doi:10.1007/s13347-020-00401-y.
Manne, Kate. 2014. "Non-Machiavellian Manipulation and the Opacity of Motive." In Coons and Weber 2014, 221–46.
Searle, John. 1980. "Minds, Brains, and Programs." *Behavioral and Brain Sciences* 3 (3): 417–57.
Turing, Alan. 1950. "Computing Machinery and Intelligence." *Mind* 59: 433–60.
Wallace, David Foster. 2004. *Oblivion: Stories*. London: Hachette UK.
Wood, Allen W. 2014. "Coercion, Manipulation, Exploitation." In Coons and Weber 2014, 17–50.

6 Manipulation, injustice, and technology

Michael Klenk

1 Introduction

Can we be manipulated by technology? Science fiction suggests that the answer is yes. In the 2014 movie, *Ex Machina*, software engineer Caleb falls prey to the empathic android Ava's sly charm. She has a subtle grasp of Caleb's needs and desires and feigns romantic feelings for the engineer. However, as it turns out, she merely uses him as a means to flee from her creator's enclosure. Caleb falls in love with her and helps her escape, and Ava leaves him to die once she is set free.[1]

Leave out the fiction, and we lose Ava's extraordinary and (super-)human intelligence and grasp for emotions. Nevertheless, our daily lives already are filled with interactions with technologies that make reliable predictions about our psychology, possess potent means to influence us, and have 'aims' that potentially conflict with ours. For example, what you see on your social media feed is curated by a recommender system – and intelligent software agent – that adjusts its actions in response to yours. Perhaps you only escape your doomscrolling on Twitter when your fitness wearable – a physical device operated by algorithms – signals you to get a move on. And if the device does not function, your first point of contact with the manufacturer will most likely be – and increasingly so – a customer service conversational AI. These observations warrant an investigation into the manipulative potential of these technologies.

In this chapter, I explain how, precisely, people may end up being manipulated by technology. Rather than focusing on the agent perspective and what it takes to manipulate, I focus on the patient perspective and ask what it takes to be manipulated. I show that being manipulated by technology is possible quite independently of whether or not technology has agency or intentionality. My argument depends on a novel perspective on manipulated behaviour, which I call the explanationist-normative perspective. Accordingly, manipulated behaviour is behaviour explained, in the relevant sense, by an injustice. Because technology can afford or enable injustice we can be manipulated by technology.[2] Thus, the chapter first develops a novel

DOI: 10.4324/9781003205425-7

account of manipulated behaviour and then uses that account to say something about being manipulated by technology.

Section 1 sketches how technology affects us quite independently of its inherent properties, which raises the question of whether we end up manipulated. Section 2 disassociates manipulative and manipulated behaviour, suggesting that the former may cause the latter but that we need a separate account of the latter nonetheless. Next, I introduce and defend the explanationist–normative perspective on manipulated behaviour in Section 3. Finally, Section 4 shows that considerations about epistemic injustice and technology's value-laden affordances imply that some of the effects of technology on us may constitute injustices, quite independently of the agential characteristics of technology.

2 Technology as a cause

Ava's interaction with Caleb is an example of a technology interacting with a human. The outcome is horrible for Caleb. The cause of Caleb's demise, Ava, appears perfectly human-like in the relevant aspects, which is probably why their interaction evokes such a strong reaction (at least it did for me and many other movie-goers!).

Phenomenologically, questions about online manipulation seem more pressing once interactions between humans and technology become overtly indistinguishable from human -human interaction. Outside the uncanny valley, where technology appears very different to humans (cf. Mori, MacDorman, and Kageki 2012), we feel forced to consider how to describe and understand correctly what has happened and how to classify the interaction. Was Caleb manipulated? Or would it be mistaken to understand technology like Ava as capable of manipulation in the first place?

Upon reflection, however, we can see that questions about whether people are being manipulated by technology should arise quite independently of the specific type of technology and its capacities. Ava's specific capacities are not the problem (though they may amplify it, or at least make us consider it with more urgency). Technology of much lower capacities than Ava influences us in already significant ways. Once we lay bare these influences and see how interactions with technology give our mental states and behaviour shape, we should again be prompted to ask how to describe and understand correctly what has happened. To illustrate, consider *social robots, virtual software agents*, and *non-autonomous technology*.

Like Ava, social robots are autonomous and physically instantiated, but they lack Ava's futuristic capabilities. Nonetheless, there should not be a doubt that they can be relevant influences on our psychology and behaviour. When, for instance, social robots are proposed to take over important roles in education (Belpaeme et al. 2018), we may worry about them spurring on learners in a problematic way. Granted, there seems to be little hard evidence about the impact of autonomous and physically instantiated technology, yet we demand that their influence on us be measured by scientific experiments. For example, there are only weak indicators about the effects of social robots in elderly care on well-being (Broekens, Heerink, and Rosendal 2009) and the lowering of depression (Chen, Jones, and Moyle 2018) and that of sex robots on well-being (Döring, Mohseni, and Walter 2020). However, for our purposes, we need not be as strict with the concept of a cause. A new colleague's behaviour and influence on you may worry you even if we cannot scientifically establish his influence on your psychological well-being or some other factor. In the same vein, we can ask what happened when someone who feels grateful to a care robot or in love with a sex robot.

Virtual software agents are, like Ava, autonomous, but they lack a physical instantiation, and yet again, they have an effect on us that we might have reason to classify as manipulation. Consider that people's online consumption, be it social media, videos, music, or other goods, is in large parts orchestrated by recommender systems. Like Ava, these systems are instances of a technology that is intelligent and autonomous in that it can perceive its environment and take actions that maximise its chance of achieving its goals (Aggarwal 2016). Your interaction with such a system can be understood as an interaction between you and an intelligent software agent (Burr, Cristianini, and Ladyman 2018). For example, it may push an anti-vaxxer video into your feed rather than any of the other billion possibilities (Alfano et al. 2020). Virtual software agents still impact out mental states and, ultimately, our behaviour. They can, to a considerable degree, 'read our minds,' that is, make reliable inferences about our beliefs and dispositions based on information gathered about the human user in interaction (Burr and Cristianini 2019). They have been shown to have measurable influences on our affective states as in the well-known emotional contagion study where Facebook users' affective states were influenced by the recommender system (Kramer, Guillory, and Hancock 2014).[3] When you are led down a rabbit hole of, say, more and more far-right videos on YouTube, and you come to believe, say, that the Democrats stole the 2020 US election, then what has happened to you? Were you manipulated? Or would it be a mistake to understand a technology like a recommender system as capable of doing that? These questions are rightly prompted by the nature of the effects that technology has on us. Caleb dies, and Internet users may end up more likely to believe a conspiracy theory. These are bad things, and they cry out for an explanation. But, again, insofar as the effects of technology on us prompt questions about manipulation, we should not restrict ourselves to artificially intelligent technology.

Non-autonomous technology influences us in relevant ways, too. User-friendly design concerns just the exterior features of a technology without requiring that it be autonomous or physically instantiated. However, it has been shown to have distinct effects on thought and cognitive integration (see

Schwengerer, this volume). Moreover, technology is physically instantiated but not autonomous, can also have dramatic effects on us. A prominent example from the philosophy of technology concerns the socio-technical effects of technology. Winner describes underpasses that were intentionally built low and, due to a combination of various technical and sociological factors, prevent certain classes people from reaching desirable areas for recreation. Winner takes this to show that artefacts have politics (cf. Winner 1980).[4] But we need not go as far and ascribe powers to the technology itself. We can focus on the effects it has on people. What has happened to those people who were prevented by the underpasses to go to the beach? Were they manipulated? Or would it be a mistake to describe the influence on them in that way?

With all these different types of technology, it is entirely plausible that we have to attend closely to the capacities of the respective technology to understand what *it* did. It may be more plausible to describe Ava as *being manipulative* than an underpass in a city. However, when we are interested in Caleb's plight, or anyone else who is being influenced by technology, we must focus on what has happened to them.

Therefore, the takeaway from this section is that the specific properties should not matter for the general question of whether technology manipulated us. Technology is interesting for its potential to manipulate us. Some manifestations – notably artificially intelligent technologies with a physical manifestation – may have particularly significant or powerful effects (see Jongepier and Klenk, this volume). But any type of technology can affect us. And that effect may prompt the question of whether we must describe it as a manipulative influence and us, in turn, as being manipulated by the technology.

Next, I put technology aside and ask what manipulated behaviour is, before showing that technology can be a cause of manipulated behaviour in Section 4.

3 Manipulated and manipulative behaviour

One puzzle with manipulated behaviour is that it is not overtly different from non-manipulated behaviour.[5] Their environment, including other agents, constantly influences agents. However, whether the actions they perform or the mental states they adopt as a result are manipulated or not is not evident from the overt mental state or action. For example, falling in love, believing that the election was rigged, buying a new flat-screen TV, getting angry, or starting to cry can be non-manipulated mental states and actions as well as manipulated mental states and actions. Their difference is not readily discernible under an overt description.[6]

Moreover, 'ion' terms like manipulation are ambiguous between process and result. As Hacking (1999) suggests, each of these terms negotiates the difference between both in its own way, and manipulation allows for a distinction between the active process of manipulating and the passive, receptive upshot of being manipulated.

The existing literature on the nature and value of manipulation is predominantly focused on the former. But although manipulative and manipulated behaviour is related, they are different phenomena. This disassociation is crucial because we cannot rely on existing accounts of manipulative behaviour to say what manipulated behaviour is. After I manipulate you, the behaviour you exhibit will not necessarily overtly differ from non-manipulated behaviour.

Nonetheless, there is a bridge between accounts of manipulative behaviour and manipulated behaviour. On my preferred analysis of manipulative behaviour, manipulation is a kind of negligence in revealing reasons to others (Klenk 2021a, 2021b). A manipulator is negligent in the sense that they ultimately choose their means of influence because it is effective in getting the manipulatee to believe, feel, or desire in a certain way and not because it reveals reasons to the manipulatee. Similar to other norm-based accounts of manipulation (Noggle 1996; Gorin 2014; Barnhill 2014), the negligence account of manipulation suggests that manipulative influence violates a norm. However, unlike previous views, it suggests that the violated norm is best understood as a lack of care to reveal reasons to the manipulatee rather than an active perpetration or ill will on the part of the manipulator. In this sense, manipulative influence is more like bullshit (in the technical sense, introduced by Harry Frankfurt, as not caring for the truth) than lying (intending to communicate a falsity). I will suggest in the next section that manipulative behaviour, thus understood, may often (though not always) be behind manipulated behaviour. Nevertheless, you can already see that whatever we may say about this view of manipulative behaviour, it illustrates a lot about the manipulator and next to nothing about the manipulatee.

Therefore, we need an account of manipulated behaviour – and if that can be shown to connect to and extend existing work on manipulative behaviour, then all the better for it.

4 An explanationist-normative perspective on manipulated behaviour

Some causes of our mental states, and ultimately our behaviour, are injustices. For example, a violation of your right to be treated with dignity – a violation often but – notably – not exclusively perpetrated by manipulators – may cause you to believe falsehoods and do things you did not want. Caleb, for instance, was played with and used as a mere means to Ava's nefarious ends.[7] Similarly, Othello, a prime exhibit of manipulated behaviour (whom we will discuss more later), was lied to and thus got his entitlement to truth frustrated.

In these cases, an injustice explains how the behaviour came about. Thus, injustices are at least correlated with seemingly manipulated mental states and behaviours. I expand on that correlation and argue that manipulated behaviour is constituted by an injustice that explains the behaviour in a to-be-specified sense. I will call this an explanationist-normative account of manipulative behaviour.

Explanationist-normative account of manipulated behaviour: Some behaviour x is manipulated behaviour if and only if and because x is explained in the relevant sense by an injustice.

My defence of the account can be summarised as follows. I illustrate the account with Othello's paradigmatic case of manipulated behaviour (Section 3.1). Real-world cases of deep oppression provide another example. Deep oppression seems morally problematic, but it has been very hard to account for that. Enoch (2020) argued that it could be accounted for in terms of injustice. If an injustice explains problematic adaptive preferences, then there is prima facie reason to think that manipulated behaviour is explained by injustice, or so I will argue (section 3.2). This can be shown to explain common concerns with adjacent accounts of manipulated behaviour (3.3). Moreover, it would offer a unified account of manipulated behaviour, which is important for independent reasons (3.4).

The argument is thus preliminary in many ways. Most importantly, it is abductive and thus leaves open that a yet deeper unifying explanation of manipulated behaviour can be found. Unless we find such a factor, however, the explanationist-normative account should stand as a serious contender.

4.1 Manipulated behaviour and injustice

Shakespeare's Othello illustrates the constitutive link between injustice as an explanation and manipulated behaviour.

Othello falls for the red herrings planted by his confidante Iago and comes to falsely believe that his wife Desdemona is cheating on him. He is so enraged by her supposed betrayal that he ends up killing her. Iago's scheme succeeded beautifully. Naturally, Othello was manipulated by Iago. In classifying Othello's behaviour as manipulated, we inadvertently suggest that something demarcates his relevant mental states, his belief that Desdemona cheated on him, his infuriation, and the desire to punish her, from your typical non-manipulated mental states.

Manipulat*ive* behaviour can constitute one of the injustices that explain in the right way some manipula*ted* behaviour. The injustice that played a role in Othello's behaviour was Iago's manipulative influence on him. All Iago cared about was his plan to succeed, and thus, his influence on Othello was reckless and negligent. He did not care the least whether Othello saw these reasons except that they made Othello behave as desired. Thus, though Iago was scheming, clever, and highly deliberate in his behaviour towards Othello, he was utterly negligent regarding Othello's reasons. This description fits the view that manipulative influence is negligence regarding the grounds on which one chooses one's behaviour (Klenk 2020a, 2021a,b). Insofar as norms legislate attention toward revealing reasons to others in interaction, we have here a violation of these norms and thus an injustice. The very violation of Othello's right leads him to have a false belief and unwarranted anger about Desdemona. So, his belief (a particular mental state) was manipulated. Since that belief figured crucially in Othello's subsequent killing of Desdemona, Othello's behaviour was manipulated. The explanation illustrates how the thesis that being manipulated tracks injustice gives us the correct analysis of a pertinent case.

Manipulated mental states are not per se faulty. Like other victims of manipulation, Othello is troubled by their plight. Often, this will be the case because victims of manipulation end up with faulty mental states and even more so if their manipulation engenders horrible behaviour like in Othello's case.

Nevertheless, it is perfectly conceivable that one rightly laments an accurate and non-faulty but manipulated mental state. For example, suppose that Desdemona has, in fact, cheated on Othello. His belief that she cheated on him would be true and not faulty in the propositional sense. Nonetheless, Othello might rightly complain that Iago's scheming and ill will towards him make his resulting belief a manipulated one. 'I have come to a true belief, what I arrived at it in bad ways' he might say. This suggests that it is not the substantive content of a mental state that makes it manipulated or not but how the mental state came about. Thus, we must look to its genesis to understand why it counts as manipulated.

Manipulated mental states are explained in a certain way because something about their genesis is amiss. Importantly, what is amiss is measured in inherently normative terms. The violation of a right or an entitlement – an injustice – plays an appropriate role in the genesis of manipulated behaviour. A normative explanation thus demarcates manipulated from non-manipulated mental states.

Importantly, it does not seem important per se *where* the relevant mental states came from or *who* caused them. For example, when Othello would complain about being manipulated, he would perhaps be saying something about Iago's personality, intention, or capacities (the source of his mental state). However, certainly, he would be saying something about Iago's influence on him and the mental states that it engendered. Thus, it is not important who or what Iago is, but how he influenced Othello.

To illustrate, imagine a rewrite of Shakespeare's Othello, where Iago turns out to be a cyborg just like Ava. Whatever is wrong with Othello's mental states (e.g., they were manipulated) would seem to be the same, irrespective of whether he is dealing with the original Iago or his futuristic counterpart.

This points to the *independence* of the manipulator's capacities from facts about whether or not someone was being manipulated. The independence claim already points to a connection with the larger concern of this chapter, namely to explore how we can be manipulated by technology in virtue of its influence on us and quite irrespective of its capacities.[8] I will return to this point in Section 4.

For now, it just matters that Othello's mental states seem to differ in some non-substantive sense from his other mental states and the – counterfactual – mental states he would have had, had Iago not manipulated him.

The relevant causal-normative explanation of a given behaviour is both sufficient and necessary for some behaviour to count as manipulated. Take *necessity* first. Once we remove the injustice from the explanation of Othello's mental state, the manipulated behaviour disappears. Suppose Iago had been honest and not manipulative towards Othello. Nonetheless, Othello ends up with the faulty belief that Desdemona cheated on him because of a bad dream or the onset of insanity. Othello's behaviour would seem tragic and wrong but not manipulated. So, without the injustice, we do not seem to have a case of manipulated behaviour, which suggests the necessity of injustice for manipulated behaviour.

To deny the necessity claim, one would have to find manipulated behaviour that did not involve an injustice, however small. Critics may suggest that my proposed account rules out – illegitimately – the possibility of manipulated behaviour with good causal histories.[9] Some examples that may push us against the necessity of injustice for manipulated behaviour may be manipulating a consenting subject in an experiment, sulking to get your partner's attention, or flirtatious behaviour to get someone to desire you in the first place.[10]

However, injustices need not be egregious and fulminant to play their relevant explanatory role in manipulated behaviour. Injustices can be minor, even trivial, perhaps. Many instances of manipulated behaviour can be all things considered permissible, notwithstanding that they remain a prima facie problem. For example, if flirting, paternalistic advice, and treating subjects in experiments results in manipula*ted* behaviour, this is because it involves manipulat*ive* behaviour, which constitutes an injustice.[11] And so being seduced, nudged, or experimented on might come with injustices and are instances of manipulated behaviour. But at no significant cost, so it might be quite plausible to say that, at the end of it, it is to be welcomed. Thus, if there is a case we are prepared to classify as a case of manipulated behaviour, then there will be some injustice – however minor – to be found.[12]

The sufficiency claim is supported by the example of Othello and other run-off-the-mill cases of manipulation. To deny this claim, one would have to show that there are cases where an injustice explains behaviour without that behaviour counting as manipulated.

However, not any kind of explanation will do. Two examples illustrate the sense of appropriate or relevant explanation that I am after. Suppose that illegitimately withheld gratitude is an injustice. Consider first someone who was denied gratitude where gratitude is due. That is an injustice. That person may become acutely aware of a desire to be thanked. The person's desire to be thanked is undoubtedly caused by an injustice here in some sense. Is he manipulated? No, because the injustice does not explain his behaviour in the *appropriate* sense. The desire to be thanked where gratitude is due will probably always have been there – it was only brought to the person's attention due to the injustice. The injustice is not the root cause of the desire, if you will.

Enoch (2020) discusses a related example to illustrate the appropriate explanation. Someone is taken hostage and kept in a cellar. Her being taken hostage is an injustice. Now, there is almost nothing in that cellar except for a piano, and to pass the time, she starts playing. Eventually, she develops a passion for piano play. Her passion is in some sense caused and explained by her being held captive in a cellar, which is an injustice. Despite that, her desire for piano play is not manipulated.

Neither case is a counterexample to the sufficiency thesis because the injustice does not in the relevant sense explain the resulting behaviour. The injustice is not required, counterfactually, for the desire to arise. It has already been there (as in our first example), or it would have been there in similar circumstances minus the injustice. But for proper cases of manipulated behaviour, the injustice seems to be an essential factor in explaining how the relevant mental state was formed. Thoroughly assessing this claim would require a discussion wider than I can offer here about the conditions for a relevant explanation.[13] However, it seems plausible that behaviours that have an injustice as some (distant) part of their causal chain are not relevantly explained by the injustice. This observation seems to be sufficient to rule out the most pertinent counterexamples to the sufficiency claim.

This section illustrated the constitutive link between injustice as an appropriate explanation of behaviour and manipulated behaviour. So far, the case for that account has been illustrated almost exclusively by a discussion of Shakespearean fiction. However, very real behaviour in our world is no less influenced by injustices and thus no less manipulated. The next section will explore the fruitfulness of the explanationist-normative perspective applied to complex cases in the real world.

4.2 Advantage I: explanatory power

Adaptive preferences as a class of mental states are seemingly morally pernicious, yet it is puzzling to explain their perniciousness. Suppose a recent analysis of the problem due to Enoch (2020) is correct. In that case, Enoch's analysis supports the explanatory-normative account of manipulated behaviour defended earlier while the latter simultaneously extends Enoch's analysis.

Roughly, someone's adaptive preference for x is a preference that person adopted upon realising that y was not among her set of feasible options to the extent that she would now prefer x even if y would become feasible (Bruckner 2009; Enoch 2020). Thus, for example, desiring to have drinks with your friends via videoconferencing rather than meeting them in person may be an adaptive preference in light of the restrictions on your feasible options surrounding the COVID-19 pandemic. This example is similar to La Fontaine's fox, which realises that it cannot reach the grapes that it so desires, and rather than admitting defeat, resolves that they look sour and that it did not want them in the first place.

Deep oppression cases illustrate this well. Martha Nussbaum gives several compelling examples of women in oppressive personal and socio-economic contexts in India. In their cases, it seems evident that the standards by which they measure their well-being or internal state are distorted and whose resulting preferences appear problematically adaptive (Nussbaum 2001, 112–13). Mitchell (2018) recaps these cases as follows:

> Vasanti, after years in an abusive marriage, thought her abuse to be a normal part of a woman's life, something to be expected once she left her family home to live with her husband. Jayamma, despite being paid less than men for more demanding factory work, accepted that this was how things were, and, knowing change was not possible, did not even waste energy lamenting her situation. And severely malnourished women in Andhra Pradesh, prior to the efforts of a government consciousness-raising program, didn't consider themselves to be malnourished, or their conditions to be unhealthy.
>
> (discussed in Mitchell)

Vasanti, Jyamma, and the other women seem manipulated, and their adaptive preferences are morally worrisome. Nevertheless, despite being harmed by the oppressive practice that they adapted to, they also appear to be strong advocates of the practice. Thus, several well-known attempts to spell out the moral problem in terms of an autonomy deficit for adaptive preferences seem to run into problems.[14] Some analyses may succeed in the future. However, Enoch (2020) makes a compelling case that this is unlikely. The problem is not how the preferences of deeply oppressed persons relate to their other preferences or whether they are preferences for things that are morally good or bad (though that may be *another* problem, cf. Nussbaum 2001).

Instead, Enoch (2020) suggests that their preferences are deeply oppressed and thus morally problematic because they were caused by injustice. Accordingly, Vasanti's preference turns out to be non-autonomous in problematic ways because her adaptive preference for a certain kind of marriage is explained by the injustice of living in such an arrangement for years. The latter is an injustice in many ways, not least because it violates Vasanti's right not to be harmed.

My analysis of manipulated behaviour draws heavily on and is indebted to Enoch's analysis of adaptive preferences. Irreducible normativity is the most crucial element that the explanationist-normative account of manipulated behaviour takes from Enoch (2020). We cannot understand what manipulated behaviour is without a moral perspective on what counts as an unjust influence. Deep oppression cases were helpful to illustrate this point because they feature agents whose moral problem seemingly evades us when we myopically focus on their autonomy. However, we can always conjure up cases that meet whatever criterion of autonomy we can think of and nonetheless seem problematic. Thus, we need to normatively evaluate a preference' genesis to explain *why* adaptive preferences are problematic if they are autonomous and not preferences for bad things (in that respect, the deep oppression cases discussed here are unfitting examples).

However, I also think that Enoch is analysing manipulated behaviour rather than merely problematic adaptive preferences. It is not entirely clear whether Enoch suggests that problematic adaptive preferences in cases of deep oppression are problematic in virtue of being non-autonomous or problematic in virtue of how their non-autonomy is explained, namely in terms of injustice. The latter would, on Enoch's analysis, entail the former. I suggest that the latter explains manipulated behaviour and not just non-autonomy.

First, there is no principled reason to believe that, in some sense, problematic preferences are to be explained differently than problematic mental states in general. On the contrary, corrupted preferences, desires, beliefs, and emotions are precisely the ingredients of manipulated behaviour.

Second, the set of manipulated behaviours intersects only the set of non-autonomous preferences. Some non-autonomous preferences do not amount to manipulated mental states, as other forms of influence like coercion also engender non-autonomy. And insofar as being manipulated does not require non- or less-than-fully autonomous preferences, there are some fully autonomous but nonetheless manipulated behaviours (compare Buss 2005). The latter, of course, is controversial and goes beyond anything I can hope to discuss in sufficient detail here (though see Klenk and Hancock 2019). But suppose it is true that there is no conceptual connection between being manipulated and being less-than-fully autonomous. Then we can still explain the problem in cases of seemingly problematic adaptive preferences like deep oppression in terms of being manipulated, and we need not find a further explanatory connection between non-autonomy and being manipulated.

Therefore, the explanationist-normative account of manipulated behaviour can explain what is wrong with deep oppression while also explaining how seemingly problematic influences that do not impact autonomy are problematic. The account thus explains well a set of highly relevant real-world cases.

Next, we will see how the account explains a common concern behind adjacent but competing accounts of manipulated behaviour, thereby extending its support.

4.3 Advantage II: explains common concerns

A central motivation of the explanationist account of manipulated behaviour is that adjacent but competing attempts to explain manipulated behaviour often fail. They fail for being too narrow (they do not explain all cases of manipulated behaviour) because they require conditions that manipulated behaviour does not need. Therefore, the explanationist account of manipulated behaviour should be preferred over these other accounts.

First, manipulated behaviour has sometimes been associated with a particular *process* that brought it about. Specifically, several authors have emphasised the connection between affective formation of mental states and manipulation, suggesting that being manipulated has something to do with having mental states formed through such processes (cf. Fischer in this volume; Wildman, Rietdijk, and Archer in this volume). However, it is not the process that is at fault but the injustice behind it. The association between emotion, affect, and being manipulated is indeed often there, but it is merely a spurious connection, and it cannot explain all cases of being manipulated. Manipulated mental states can be formed on a purely cognitive and rational basis such as Othello's belief that Desdemona cheated on him. Moreover, the epistemic or moral warrant of a form of mental state genesis, or the rationality of a type of influence, does not depend on the type of information per se but on the contextual factors at hand.

This claim can be briefly illustrated with the debate around System 1 and System 2 processing (cf. Kahneman 2012). The former is associated with 'non-rational' mental processes such as heuristic decision-making. In contrast, the latter is associated with 'rational' mental processes such as reflection and conscious deliberation. But that does not settle questions about the normative rationality irrationality of System 1 versus System 2 processing. For example, fast affective heuristics are rational when decisions must be made quickly in familiar environments (Gigerenzer 2008). The type of information or the manner of its processing per se is epistemically and morally neutral. Therefore, we cannot identify manipulated behaviour with the type of informational source nor the processes that lead to the behaviour in question.

Second, it is implausible that mental states can be distinguished into manipulated and non-manipulated based on their relation to the *agent's plans, aims, or (self-)interest*.[15] Several scholars have championed this proposal (e.g., Barnhill 2014; Rudinow 1978), and it is evident that being manipulated is often not good for you. But clearly, our manipulated mental states are sometimes conducive to our objectively warranted plans or aims. For example, a little nudging may help me avoid the temptation to book a transatlantic flight as soon as travel restrictions abate. Because at least some nudges lead to manipulated behaviour, this is a counterexample to the proposal that manipulated mental states and actions are at odds with our objectively warranted plans or aims (cf. Sunstein 2016; Klenk 2020a). Neither do our subjectively held plans or aims need to be at odds with manipulated mental states or actions. Being nudged to avoid booking the transatlantic flight seems like manipulated behaviour, and yet it may serve my aim to live carbon-neutral, whether or not this is an appropriate aim to have. Therefore, manipulated behaviour need not be at odds with our aims, plans, or (self-)interest (cf. Gorin 2014). However, what is required is that we judge the genesis of the relevant mental state to contain an injustice. And that seems to be the case. Take nudging as an example. At least some nudgers are manipulative in the sense that they are negligent about revealing reasons to their interlocutor, which we identified with an injustice earlier. This can account for the resulting manipulated behaviour, quite independently of whether the manipulatee's plans, aims, or self-interest were frustrated.

The most promising lead is that manipulation is a kind of interference and, consequently, manipulated mental states are those that were meddled with or interfered in in problematic ways. This is the popular image of the manipulator as the puppet master and the manipulated person as a puppet on a string, as a prop in someone else's play.

However, as alluded to in the previous section, the link between manipulated mental states and behaviour and autonomy is not conceptual. How manipulation impacts autonomy would have to be explained in a more substantive sense (cf. Klenk and Hancock 2019). So, there is a more general lesson here. Any account of manipulation that wants to understand manipulation as a kind of interference that diminishes autonomy must account for how 'normal' or non-interfered processing goes. And I reckon it will be incredibly tough to say how people who are always influenced by their past and present and who at various points are prone to endorse those influences as in deep oppression cases reflectively are functioning in a 'normal' or non-interfered way.

What seems problematic for manipulated people is that their rights are violated, perhaps because they have been influenced negligently. This can be given a distinct Kantian flavour, in that victims of manipulation were not treated with due respect, which clarifies what right is violated (see Jongepier and Wieland, in this volume). The image of being a puppet on a string is misleading if it suggests that we necessarily are less than fully autonomous when we are being manipulated. However, it is apt to evoke a sense of disrespect and violation of one's rights – after all, we are not puppets on a string and should not be treated that way. In this sense, the explanationist-normative account of manipulated behaviour illustrates very well the core concern with manipulated behaviour and ties in nicely with accounts of manipulative behaviour.

4.4 Advantage III: avoids error theory

Finally, the explanationist-normative account of manipulated behaviour is supported by a *reductio* argument.

We should accept the thesis that being manipulated tracks injustice to avoid an error theory about judgements about manipulation (i.e., a theory that explains how and why people are frequently mistaken in their judgements about manipulation). So, suppose that it is false that being manipulated tracks injustice. Then, manipulation is *not* related to injustices that play an appropriate causal role in the formation of a mental state or the generation of an action, and we should *not* expect normative judgements to track causal histories of mental states affected by injustices (in the appropriate way). However, there is widespread disagreement about what makes some behaviour an instance of manipulated behaviour.[16] I call this phenomenon:

Classification Variety: There is widespread disagreement about the conditions for manipulation.

Classification Variety is supported by two sources, preliminary empirical studies and the discussion in the philosophy of manipulation. The 'charting the field' chapter for this volume has shown that there is considerable disagreement about the nature of manipulation (Jongepier and Klenk, chapter 2 in this volume). Normative concepts are controversially discussed more generally. Further defence for this claim that professional philosophers disagree about the nature of manipulation may be produced at will.

A novel data point is that laypeople seem to disagree about the nature of manipulation, too. In an unpublished experiment, Klenk, Xun Liu, and Hancock (2021) asked participants to evaluate short vignettes that described paradigm cases of manipulation (e.g., Shakespeare's Othello) on four dimensions concerning the effect on the manipulatee: they were 'deceived,' 'harmed,' 'played,' and 'unconsciously influenced.' The four answer options were pre-experimentally selected based on the philosophical discussion about necessary and sufficient conditions for manipulation. The results showed that while subjects considered the vignettes as examples of manipulation, they disagreed significantly about the underlying condition. Just like the professional philosophers, laypeople identify several distinct causes as the underlying condition of manipulation. This supports Classification Variety. Now, we must take an important mental note. All the relevant examples plausibly include a causally relevant injustice (depending on the right theory of justice at the end of the day, of course). If we can interpret these varying judgements as tracking injustice instead, then we might explain away Classification Variety. But we are getting ahead of ourselves. First, I need to show that we should try to explain away Classificatory Variety.

Suppose also that there is a unified condition for manipulated behaviour (though it is not injustice!). It might be that manipulated behaviour depends on undermined autonomy. Or on deception. Or on emotional influence. But only one. These assumptions (the rejection of my thesis and that there is a unified condition for manipulated behaviour) coupled with Classification Variety would imply that a sizeable portion of the beliefs about the conditions for manipulated behaviour – advanced by professional philosophers and laypeople alike – are false. That is because there is one underlying condition of manipulation, while people apparently hold widely differing beliefs about what that condition happens to be. So, Variety implies what I will call

> **Classification Error:** Many beliefs about the conditions for manipulation are false.

I can now show that we should not accept Classification Error and thus reject any assumption that commits us to it. Classification Error is unpalatable, as evolutionary considerations will show. Humans developed a reasonably elaborate capacity to detect cheaters (and, alas, to cheat ourselves). This does not mean that people are good at detecting. But it suggests that we usually know that we are being cheated *when we see it*. Being deceived and being manipulated are some of how we can be cheated. We should expect that social animals like us are good at recognising deception and manipulation, at least when they occur in environments similar to our environment of evolutionary adaptation.[17] When people agree that a given case exemplifies manipulated behaviour, we have good prima facie reason to think that the case indeed does exemplify manipulated behaviour. But given the assumption that it is *false* that manipulated behaviour is caused by injustice, we lack a unifying explanation of these judgements. Absent an explanation, we have to assume that most of these beliefs are false.[18] This is not what we should expect given our evolutionary history.

Considering an objection to this line of thought will further strengthen it. Evolution, the objection goes, did not select for a correct appreciation of the underlying condition for manipulation but the mere 'blind' application of the concept. For instance, classifying behaviour as being manipulated may serve a function, and adaptive pressure may have applied to the utilisation of that function, not correctly identifying the conditions for manipulation.

However, correct classification absent an understanding for the underlying reasons for why something is an instance of manipulated behaviour is insufficient for two reasons. First, it would be an open question just why people have competence in applying the term without some kind of insight. Positing insight would answer this question. Second, even setting that worry aside, there is a substantive problem because different ascriptions of the underlying conditions behind manipulation are plausibly functionally differentiated. That is, different conclusions follow from calling something caused by autonomy-undermining or from deception. Thus, even if evolutionary pressure applied to whatever functional implications (the concept of) manipulation may have, they plausibly indirectly put pressure on the correct recognition of the conditions for manipulation.

Therefore, if it is false that being manipulated tracks injustice, we get the problematic implication that people do not understand the conditions that ground manipulated behaviour and make many mistakes in applying it. Because this implication is problematic, I conclude that we should not reject the thesis that being manipulated tracks injustice. In other words, the explanationist-normative account of manipulated behaviour should be accepted.[19]

5 Technology's manipulative potential

So far, we have established that interacting with technology puts as at risk of being manipulated by it. For example, it makes sense to classify Caleb as being manipulated by Ava. More generally, if manipulated behaviour is behaviour explained, in the relevant sense, by an injustice then we can be manipulated by technology, quite independently of whether it possesses agential features such as intentionality. That is because agential features are not required for an injustice to explain a mental state and, ultimately, behaviour. So, whether or not we would be correct in ascribing mental states and intentions to Ava does not matter for the question of whether Caleb has been manipulated.

Are there any more general ways in which technology may contribute to an injustice? I will first discuss a general non-agential injustice and then elaborate on technology's causal effects in support of this claim.

Epistemic injustice gives us reason to think that agential features are not required for injustice to contribute to a mental state and behaviour.[20] Fricker (2011) introduces the notion of 'epistemic injustice,' which arises when somebody is wronged in their capacity as a knower. The stock example of epistemic injustice, of the hermeneutical kind, is that of a person or social group that is unfairly deprived of knowledge because of their lack of access to education or other epistemic resources. Fricker discusses two kinds of epistemic injustice in greater detail. First, testimonial injustice occurs when somebody is given less credibility than due to prejudice about the social group to which the speaker belongs. Second, Fricker describes hermeneutical injustice, which occurs when members of a social group fail, because of a linguistic gap in collective understanding, to make sense of certain distinct experiences (e.g., sexual harassment). The idea is that women, for example, were socially powerless in the 1970s and, partly because of that, could not communicate their experiences adequately (cf. Keane 2016). When people are subject to hermeneutical injustice, no direct agent (nor a group agent) perpetuates the injustice, even though at some point agents may have been involved in contributing to the injustice. But whatever 'original' agential contribution there is, it is most likely not required to explain the effects of the injustice today. Whether or not this or that agent was involved in creating systematically oppressive circumstances may matter for questions about responsibility but not for the question of whether your or my behaviour today is explained by the injustice in the appropriate way. Therefore, agential contribution is not required for injustice to appropriately explain a mental state.[21]

Technology can contribute to injustice and thus make us manipulated because of technology's value-ladenness. The idea that technology is more than a 'mere tool' is deeply ingrained in the philosophy of technology. If technology were a mere tool, then any of its effects would have to be attributed to – roughly – the designer of the tool or its user. The NRA makes use of that idea when they claim that 'Guns don't kill people, people do.' But surely guns contribute in some sense to extraordinarily high murder rates in the United States compared to other countries (cf. Grinshteyn and Hemenway 2016), though we need not understand their contribution in an agential sense. From that perspective, technology does not seem to be morally neutral. One way to make sense of sense of technology's value-ladenness without ascribing agential features to it is in terms of affordances. Technology has affordances, which are relational properties that depend on the material properties of the technology as well as contextual factors such as biological, psychological, and social factors concerning the user of the technology (Klenk 2020b). Affordances make certain mental states and behaviours more likely and others less likely. It makes sense to speak of a chair 'inviting' us to sit on it. The affordance perspective on technology helps us interpret this claim without retorting to an implausible ascription of agency or intentionality to technological artefacts. For example, the fact that a gun affords killing indicates that handling a gun will make deadly outcomes in some scenarios, like a heated argument, more likely. Similarly, social robots are suspected of lowering depression and increasing well-being, and virtual software agents have been shown to afford more and more extreme viewing behaviour on YouTube. Even non-autonomous technology like user-friendly websites or low-built overpasses affords some mental states but not others, such as trust in the case of user-friendly websites and not going to the beach in the case of overpasses. And Ava's incredible artificial intelligence made it likely that Caleb fell in love with her without seeing her nefarious scheme. We can also evaluate the affordances of a given technology in moral terms (cf. Klenk 2020b). So, we can also see how technology is not value-neutral from the affordance perspective.

Most importantly, it is now straightforward to see that the affordances of technology can constitute injustices that explain, in relevant ways, our mental states and behaviour. For example, all of a city's citizens are entitled to frequent the city's public beach. Low-hanging underpasses that prevent some citizens from going to the beach violate their entitlement and thus constitute an injustice. It would follow that citizens in that situation are being manipulated by the architectural features of the city. Similarly, we are entitled to truth (suppose). Virtual software agents in recommender systems make it more likely that we believe falsehoods. Thus, they contribute to a violation of our entitlement. That injustice may explain why some end up believing that the 2020 US election was rigged. They are manipulated, according to the explanations-normative account of manipulated behaviour. Caleb, finally, has a right to be shielded from seduction. Ava violated that right, and that injustice explains Caleb's behaviour. Therefore, Caleb was manipulated by Ava, even if Ava lacks the capacities for genuine manipulative behaviour as intentionality. These observations about concrete cases of technological manipulation depend on identifying a relevant injustice and explaining the relevant mental states and behaviours. However, they should be sufficient to show how technology has manipulative potential quite independently of its agential characteristics.

My argument for the manipulative potential of technology suggests that a prominent and competing type of argument in the ethics of technology is beside the point. I call this type of argument a *condicio sine qua non argument*. Proponents of such arguments describe conditions for manipulative behaviour and then suggest that technology currently or in principle lacks the conditions for manipulative behaviour (compare the contributions by Pepp et al., Gorin, and Nyholm, in this volume).

For example, it may be claimed that manipulativeness requires intentionality and that technology lacks intentions. Therefore, one might conclude, technology cannot manipulate us. However, arguments along these lines miss the possibility, demonstrated earlier, that manipulating (the agent side) can come apart from being manipulated (the patient side). Even if technology cannot be manipulative – for example, because there is no sense in which it can be negligent[22] – it may contribute to injustices that result in manipulated behaviour on our part. Of course, this is because being manipulated does not require that one interacts with a manipulator with intentions, even if the latter will, in many cases of human-to-human manipulation, be the cause of manipulated behaviour.[23]

Thus, technology may relevantly contribute to an injustice that plays an appropriate role in explaining our behaviour. Therefore, there is potential for us to be manipulated by technology.

Usually, we would be wont to ask about the perpetrator and the person culpable of manipulative action. This raises an important question. If there can be manipulated people without manipulators we face a responsibility gap. Some questions about passive responsibility may not be satisfactorily answered. But note two points. First, the explanationist-normative account of manipulated behaviour does not replace the need to ask questions about passive responsibility about the inventors and deployers of technology. Clearly, facts about whether or not soldiers are assessable in terms of responsibility do not absolve their higher-ups from such questions. Second, questions about passive responsibility arguably should not focus on appropriate ethics of technology in the first place (Klenk and Sand 2020). We can still ask questions about forward-looking responsibilities to prevent manipulated behaviour, which is indeed what we should focus on.

6 Conclusion

Interacting with increasingly autonomous technology raises all sorts of problems, as the burgeoning debate, especially in AI ethics, demonstrates. Is one of the problems that we can be manipulated by technology?

This chapter explored a novel approach to that question – by focusing on the patient rather than the agent side of manipulation – and suggested that the answer is affirmative. Manipulated behaviour is behaviour that is explained, in the relevant sense, by an injustice. Agential features like intentionality are not required for injustice, as the case of epistemic injustice demonstrates. Technology can contribute to said injustices in virtue of its affordances. Therefore, we can be manipulated by technology, even if it lacks agential features such as intentionality and thus does not meet the conditions for being manipulative.

That leaves the practically most relevant question of whether we are, in fact, being manipulated by technology. My chapter suggests concrete ways forward with this question. We must assess whether the influence of technology on us constitutes injustices. That involves a question about the proper explanation of our mental states and behaviour and a normative account of what injustices are. Thus, two broad research challenges arise. First, we need much more empirical work to substantiate the concrete ways in which particular instances of technology influence our mental states and behaviour. Second, those influences need to be assessed in light of an appropriate theory of justice to see whether they violate our rights and entitlements. Given the manipulative potential of technology, it is our forward-looking responsibility to ensure that it does not materialise, and we are spared Caleb's plight.

Notes

1. Many thanks to Fleur Jongepier, Michael Madrey, Nathan Wildman, Sven Nyholm, and Jan Willem Wieland for written comments on an earlier version of this chapter. Also, I thank Steffen Steinert and the audiences at a TU Dresden workshop and the online symposium series we organised for this volume for very helpful discussion. My work on this volume was made possible by a Niels Stensen Fellowship. I gratefully acknowledge generous support by the European Research Council under the Horizon 2020 programme under grant agreement 788321.
2. I will often suggest for illustrative purposes that manipulated behaviour – or a manipulated action – is based on a manipulated mental state. Whether that claim about the relation of mental states, action, and behaviour is plausible depends in part on wider issues than I can discuss here. Readers who see a problem in that simple sketch may just focus on my core claim about the conditions of manipulated mental states.
3. This case is prominently Wildman, Rietdijk, and Archer, in this volume.
4. Also consider a point I do not address here, namely that some kinds of interactions may be made possible in the first place by new technology, like augmented many-to-many interaction. See Cappuccio et al. (2021), in this volume.
5. Note that I use the term 'manipulated behaviour' to refer to manipulated mental states and manipulated actions.
6. Note that I may be the first to make this distinction explicit, but I am not the only one who defends it. (Wilkinson 2013) notes in his discussion of a general account of manipulation that it may be premature to assume that manipulative action leads to manipulated action. His point is that social science is difficult.

But it obviously depends on the thought that manipulative actions do not imply manipulated actions. The converse may also hold true. At least, that is what I will assume in what follows.

7. Relatedly, a violation of your right to bodily integrity may cause you to feel threatened and cave in to illegitimate demands. Or a frustration of your civic entitlement to be informed by media in a factual manner about politics may cause you to believe falsehoods, to desire irrational things, and to vote for the wrong party.
8. Most relevantly, as discussed earlier, is probably the capacity for intentionality. See the overview by Jongepier and Klenk, in this volume.
9. Thanks to Fleur Jongepier for pressing me to address this point, and to Jan Willem Wieland for putting this point to me in that way.
10. Perhaps my proposed analysis of manipulation would seem to require a revision of our concept manipulation. Compare Pepp et al., in this volume.
11. Thanks to Fleur Jongepier for helpful feedback on this point.
12. One class of counterexamples are cases of manipulation in the context of a game. Nathan Wildman suggested a case along the following lines. Suppose that Iago and Othello are playing chess, and Iago manipulates Othello by making a series of moves in order to get him to think that he's going to attack queen-side, when in fact he's going to go kingside. As a result of Iago's manipulation, Othello builds up his defences in the wrong spot and ends up eventually losing. That strikes some as a case of manipulation, but one that tracks no injustice: Iago manipulated Othello, but he did nothing wrong! I would maintain that we do not have a case of manipulation here because Iago stuck entirely to the rules of the game. So, even though he presumably did not care for whether Othello recognised his reasons for acting, there is no norm within the game that would demand such care. Perhaps Othello was fooled then or duped but not manipulated.
13. Thanks to W. Jared Parmer for pressing me on the distinction between being caused and being explained by an injustice.
14. To illustrate, consider that the women's preferences are not irrational or non-autonomous insofar as they internalised the practice to an extent that they desire what they want to want (on an 'internal' conception of autonomy, cf. Frankfurt 1971) or reflectively endorse the desire, as part of a self-affirming practical identity (cf. Bruckner 2009; Christman 2014). Insofar as the oppression is sufficiently thorough, it is likely that their seemingly problematic preferences are in harmony with their other preferences, thus denying the claim that the problem is formal (cf. Bovens 1992).
15. More generally, we could also call these objective list theories of manipulated behaviour, because they propose lists of goods (e.g., alignment with one's aims or plans) that manipulated behaviour arguably lacks. The short rebuttal of these proposals is that for any entry on a list we can imagine a behaviour that possess that item but still counts as problematic or a behaviour that lacks the item but does not count as problematic.
16. Another crucial clarification concerns the claim about injustice. On its face, my thesis is ambiguous between people judging that an injustice played an appropriate role in some behaviour (the mentalist interpretation) and the fact that an injustice played an appropriate role in some behaviour (the causalist interpretation). Making the distinction clear is important because my thesis drives on an argument about people's judgements being on track.
17. Which is precisely why I might be especially worried about technological manipulation as it supersedes our adaptations. Fleur Jongepier and I discuss this as an aggravating factor, in chapter 2 of this volume. The current point, however, is

not that we should be good at detecting when a machine manipulates us but at identifying the criteria for manipulation.

18. This is a bit quick: it would be reasonable to assume that at least one already identified candidate condition is correct (e.g., deception). Then beliefs whose content portrays manipulated behaviour as depending on other factors are false.
19. The assumption that there is a unified or single condition for manipulation may be controversial, and my argument depends on it. But there is good reason to accept it. But suppose you deny that there is but one condition for manipulation and insist that there are multiple, disjunctive, and individually sufficient conditions for manipulated behaviour. If that is true, then we can explain Variety without accepting Error. People may simply classify correctly several conditions for manipulation and the allegedly absurd consequence Error would not follow from the rejection of the thesis that being manipulated tracks injustice. However, turning to pluralism about the conditions for manipulation is ultimately unconvincing. First, we are still in the dark about the necessary conditions for manipulation. We now assume that there are many sufficient ones. But there is no apparent structure to the many that emerge from people's classifications. But which ones, precisely? All of the ones that we have discussed so far? Or only some? Our understanding of manipulation has not been illuminated. But even if we grant the assumption that there are multiple conditions for manipulation, there is a deeper problem. The explanation in terms of pluralism does not jibe well with the aim of explanatory parsimony. A simpler theory is more likely to be correct. There are constraints about applying the criterion of parsimony in the normative case, see (Sober 2015), but they do not change that a simple explanation is to be preferred to a potentially complicated explanation. Therefore, there is good reason to accept the view that being manipulated tracks injustice.
20. Thanks to Steffen Steinert for suggesting epistemic injustice in discussion as a point in favour of the explanationist-normative account of manipulated behaviour.
21. See Liao and Huebner (2021) who present a fuller account of how technology can be a relevant cause in the injustices that we suffer. Unfortunately, I could not engage with their account more fully in this chapter. Thanks to Sven Nyholm for the pointer.
22. Note that I previously argued that the fact that technology cannot care for our reasons supports an a priori argument about their manipulativeness, (Klenk 2020a). I am now not sure anymore whether the impossibility of technology to have agential features would make it a priori manipulative or just altogether remove it from the category of things that can or cannot be manipulative.
23. Note that Sharkey and Sharkey (2020) have recently suggested an argument along similar lines in the case of deception. Thanks to Sven Nyholm for the pointer.

7 References

Aggarwal, Charu C. 2016. *Recommender Systems: The Textbook*. Cham: Springer.

Alfano, Mark, A. E. Fard, J. A. Carter, P. Clutton, and C. Klein. 2020. "Technologically Scaffolded Atypical Cognition: The Case of YouTube's Recommender System." *Synthese*, 1–24.

Barnhill, Anne. 2014. "What is Manipulation?" In *Manipulation: Theory and Practice*, edited by Christian Coons and Michael Weber, 51–72. Oxford: Oxford University Press.

Belpaeme, Tony, James Kennedy, Aditi Ramachandran, Brian Scassellati, and Fumihide Tanaka. 2018. "Social Robots for Education: A Review." *Sciences Robotics* 3 (21). doi:10.1126/scirobotics.aat5954.

Bovens, Luc. 1992. "Sour Grapes and Character Planning." *Journal of Philosophy* 89 (2): 57. doi:10.2307/2027152.

Broekens, J., M. Heerink, and H. Rosendal. 2009. "Assistive Social Robots in Elderly Care: A Review." *Gerontechnology* 8 (2). doi:10.4017/gt.2009.08.02.002.00.

Bruckner, Donald W. 2009. "In Defense of Adaptive Preferences." *Philosophical Studies* 142 (3): 307–24. doi:10.1007/s11098-007-9188-7.

Burr, Christopher, and Nello Cristianini. 2019. "Can Machines Read our Minds?" *Minds and Machines* 29 (3): 461–94. doi:10.1007/s11023-019-09497-4.

Burr, Christopher, Nello Cristianini, and James Ladyman. 2018. "An Analysis of the Interaction Between Intelligent Software Agents and Human Users." *Minds and Machines* 28 (4): 735–74. doi:10.1007/s11023-018-9479-0.

Buss, Sarah. 2005. "Valuing Autonomy and Respecting Persons: Manipulation, Seduction, and the Basis of Moral Constraints." *Ethics* 115 (2): 195–235. doi:10.1086/426304.

Cappuccio, M. L., E. B. Sandoval, O. Mubin, and M. Velonaki. 2021. "Robotics Aids for Character Building: More than Just Another Enabling Condition." *International Journal of Social Robotics* 13: 1–5.

Chen, Shu-Chuan, Cindy Jones, and Wendy Moyle. 2018. "Social Robots for Depression in Older Adults: A Systematic Review." *Journal of Nursing Scholarship* 50 (6): 612–22. doi:10.1111/jnu.12423.

Christman, John. 2014. "Coping or Oppression." In *Autonomy, Oppression, and Gender*, edited by Andrea Veltman, 201–26. Oxford: Oxford University Press.

Döring, Nicola, M. R. Mohseni, and Roberto Walter. 2020. "Design, Use, and Effects of Sex Dolls and Sex Robots: Scoping Review." *Journal of Medical Internet Research* 22 (7): e18551. doi:10.2196/18551.

Enoch, David. 2020. "False Consciousness for Liberals, Part I: Consent, Autonomy, and Adaptive Preferences." *Philosophical Review* 129 (2): 159–210. doi:10.1215/00318108-8012836.

Fischer, Alexander. 2022. "Manipulation and the Affective Realm of Social Media." In *The Philosophy of Online Manipulation*, edited by Jongepier, F. and Klenk, M., 327–352. New York: Routledge.

Frankfurt, Harry G. 1971. "Freedom of the Will and the Concept of a Person." *Journal of Philosophy* 68 (1): 5. doi:10.2307/2024717.

Fricker, Miranda. 2011. *Epistemic Injustice: Power and the Ethics of Knowing*. Repr. Oxford: Oxford University Press.

Gigerenzer, Gerd. 2008. "Why Heuristics Work." *Perspectives on Psychological Science* 3 (1): 20–29. doi:10.1111/j.1745-6916.2008.00058.x.

Gorin, Moti. 2022. "Gamification, Manipulation, and Domination." In *The Philosophy of Online Manipulation*, edited by Jongepier, F. and Klenk, M., 199–215. New York: Routledge.

Gorin, Moti. 2014. "Do Manipulators Always Threaten Rationality?" *American Philosophical Quarterly* 51 (1): 51–61. Accessed June 04, 2019.

Grinshteyn, Erin, and David Hemenway. 2016. "Violent Death Rates: The US Compared with Other High-Income OECD Countries, 2010." *The American Journal of Medicine* 129 (3): 266–73. doi:10.1016/j.amjmed.2015.10.025.

Hacking, Ian. 1999. *The Social Construction of What?* Cambridge, MA: Harvard University Press.

Jongepier, Fleur, and Michael Klenk. in 2022 a. "Online Manipulation: Charting the field." In *The Philosophy of Online Manipulation*, edited by Jongepier, F. and Klenk, M., 14–48. New York: Routledge.

Jongepier, Fleur, and Michael Klenk, eds. in 2022. *The Philosophy of Online Manipulation*. New York, NY: Routledge.

Jongepier, Fleur, and J. W. Wieland. 2022. "Microtargeting People as a Mere Means." In *The Philosophy of Online Manipulation*, edited by Jongepier, F. and Klenk, M., 156–179. New York: Routledge.

Kahneman, Daniel. 2012. *Thinking, Fast and Slow*. London: Penguin.

Keane, Webb. 2016. *Ethical Life: Its Natural and Social Histories*. Princeton, NJ: Princeton University Press.

Klenk, Michael. 2020a. "Digital Well-Being and Manipulation Online." In *Ethics of Digital Well-Being: A Multidisciplinary Perspective*, edited by Christopher Burr and Luciano Floridi. Cham: Springer. Accessed November 17, 2019. 81–100. doi: 10.1007/978-3-030-50585-1_4.

Klenk, Michael. 2020b. "How Do Technological Artefacts Embody Moral Values?" *Philosophy & Technology*, 1–20. doi:10.1007/s13347-020-00401-y.

Klenk, Michael. 2021a. "Interpersonal Manipulation." *SSRN Electronic Journal*. doi:10.2139/ssrn.3859178.

Klenk, Michael. 2021b. "Manipulation (Online): Sometimes Hidden, Always Careless." *Review of Social Economy*. 80: 1, 85–105. doi:10.1080/00346764.2021.1894350.

Klenk, Michael, and Jeff Hancock. 2019. "Autonomy and Online Manipulation." *Internet Policy Review*. Accessed February 28, 2020. https://policyreview.info/articles/news/autonomy-and-online-manipulation/1431.

Klenk, Michael, and Martin Sand. 2020. "Prometheus' Legacy: Responsibility and Technology." In *Welche Technik?* edited by Birgit Recki, 23–40. Dresden: Text & Dialog.

Klenk, Michael, Sunny Xun Liu, and Jeff Hancock. 2021. *Pulling the Rug from under the Tech-lash: Online Influences are Perceived to be More Manipulative than Similar Offline Influences*. Under review.

Kramer, A. D. I., J. E. Guillory, and Jeff Hancock. 2014. "Experimental Evidence of Massive-scale Emotional Contagion Through Social Networks." *Proceedings of the National Academy of Sciences* 111: 8788–90.

Liao, Shen-yi, and Bryce Huebner. 2021. "Oppressive Things." *Philosophy and Phenomenological Research* 103 (1): 92–113. doi:10.1111/phpr.12701.

Mitchell, Polly. 2018. "Adaptive Preferences, Adapted Preferences." *Mind* 127 (508): 1003–25. doi:10.1093/mind/fzy020.

Mori, Masahiro, Karl MacDorman, and Norri Kageki. 2012. "The Uncanny Valley." *IEEE Robotics & Automation Magazine* 19 (2): 98–100. doi:10.1109/MRA.2012.2192811.

Noggle, Robert. 1996. "Manipulative Actions: A Conceptual and Moral Analysis." *American Philosophical Quarterly* 33 (1): 43–55.

Nussbaum, Martha C. 2001. *Women and Human Development: The Capabilities Approach*. Cambridge: Cambridge University Press.

Nyholm, Sven. 2022. "Technological Manipulation and Threats to Meaning in Life." In *The Philosophy of Online Manipulation*, edited by Jongepier, F. and Klenk, M., 235–252. New York: Routledge.

Pepp, Jessica, Rachel Sterken, Matthew McKeever, and Eliot Michaelson. 2022. "Manipulative Machines." In *The Philosophy of Online Manipulation*, edited by Jongepier, F. and Klenk, M., 91–107. New York: Routledge.

Rudinow, Joel. 1978. "Manipulation." *Ethics* 88 (4): 338–47. doi:10.1086/292086.

Sharkey, Amanda, and Noel Sharkey. 2020. "We Need to Talk about Deception in Social Robotics!" *Ethics and Information Technology*. doi:10.1007/s10676-020-09573-9.

Sober, Elliott. 2015. *Ockham's Razors: A User's Manual*. Cambridge: Cambridge University Press.

Sunstein, Cass R. 2016. *The Ethics of Influence: Government in the Age of Behavioral Science*. Cambridge: Cambridge University Press.

Wildman, Nathan, Natascha Rietdijk, and Alfred Archer. 2022. "Online Affective Manipulation." In *The Philosophy of Online Manipulation*, edited by Jongepier, F. and Klenk, M., 311–326. New York: Routledge.

Wilkinson, T. M. 2013. "Nudging and Manipulation." *Political Studies* 61 (2): 341–55. doi:10.1111/j.1467-9248.2012.00974.x.

Winner, Langdon. 1980. "Do Artifacts Have Politics?" *Daedalus* 109 (1): 121–36.

Part II

Threats to autonomy, freedom, and meaning in life

7 Commercial Online Choice Architecture

When Roads Are Paved With Bad Intentions

Thomas Nys and Bart Engelen

I played my part, and you played your game.

(J. Bongiovi, *You Give Love A Bad Name*)

Welcome to the future, Holmesy. It's not about hacking computers anymore; it's about hacking human souls.

(John Green, *Turtles All The Way Down*)

1 Introduction

Consider how much time we spend online: watching YouTube videos, attending Zoom sessions, scrolling through Facebook, Instagram, or other social media, googling stuff, or doing some online shopping. Given the vast amounts of time spent, and the number of choices made online, ethicists should scrutinize the impact of this on people's behavior and the extent to which that impact arguably respects or undermines their freedom, autonomy and, according to some critics, even their dignity. In this chapter, we focus on what we call COCAs: commercial online choice architectures or choice environments that are purposefully designed by (often hugely resourceful) private companies. Given how much online time and interaction is framed within and influenced by such COCAs, ethically evaluating them is both urgent and important.

We first justify our focus on COCAs (Section 2) before defining and illustrating what we mean by them (Section 3). In order to assess when and why COCAs are manipulative, we conceptualize manipulation and focus specifically on the intentions behind manipulative practices (Section 4). This focus enables us to assess one of the main ethical worries that arise with the ubiquity of COCAs, namely their impact on the *personal autonomy* of users (Section 5). Finally, we provide a balanced conclusion (Section 6).

2 Why We Talk About COCAs

The time we spend online increases year after year. From 2009 to 2019, averages increased from 1h15 to 3 hours per day (with almost 2h30 spent

DOI: 10.4324/9781003205425-9

on our mobile phones). Due to COVID-19, those numbers have soared even higher. One study suggests that it has more than doubled from 3 to 7 hours in 2020 (Globenewswire 2020). Another report observes that this

> equates to more than 100 days of connected time per internet user, per year. If we allow roughly 8 hours a day for sleep, that means we currently spend more than 40 percent of our waking lives using the internet.[1]

More than ever, online is where we search for information, get our news, listen to music, watch shows, play games, talk to our family, friends, and colleagues, share our work and leisure with others, and buy our clothes, food, and other stuff. Ever more aspects of our lives – work, school, finance, friendship, romance, etc. – are (partly) online.

Of course, not all of those hours and decisions take place in commercial online environments. People might be checking Wikipedia or, why not, reading articles in the Stanford Encyclopedia of Philosophy. These are not COCAs as they're not designed by commercial entities. But when we shop online, as most of us do (CBS 2021), we are inevitably making decisions in COCAs. Or think of the time we spend on platforms designed by Google (such as the search engine or YouTube). Or take Facebook, a clear example of a COCA, as it is an online platform, purposefully designed by a (huge and powerful) private company for commercial ends. Facebook designs its algorithms to be addictive and to keep users on the platform as long as possible (Solon 2017) while collecting data about them, which it then monetizes by targeting ads from third parties to relevant users (Gilbert 2018). With people spending on average 2 hours and 30 minutes per day (or more than a full month per year) on social media, the impact of COCAs like this should not be underestimated.

When we use COCAs to seek information, entertainment, commodities, and opportunities for socializing, we might be worried for our privacy. Companies obtaining ever more data and insights into our choices and preferences can indeed be considered problematic. The fact that this knowledge is subsequently used to influence or steer our behavior constitutes another concern. This is a concern about manipulation, and it grows as private companies collect more and more personal data online; data that they can subsequently use for micro-targeting the users of these services.

The fact that private companies design the frameworks within which we make so much of our online choices makes the ethical evaluation of such COCAs particularly salient and topical. After all, should anything prove to be morally wrong with COCAs, the problem would be both massive and urgent.

3 What We Talk About When We Talk About COCAs

So, what exactly are COCAs and how do they work? When we talk about 'choice architectures', we obviously refer to Richard Thaler and Cass Sunstein (2008), who use the term to denote the environments in which we make and which influence our choices because of their architecture or design. Think, for example, of the different ways in which we can frame and present options; physically, visually, verbally, emotionally, and so on. Think of your local pub, which offers both lagers and ciders, but can make one of those more 'eye-catching', by placing them near the register or at eye-level.

The same happens online. Whenever we choose to click something – a link to a website we think might be interesting, a song we want to hear or a product we're considering to buy – we are making choices within online choice architectures or OCAs: environments and platforms that have been designed one way or the other and that will partially influence our decisions. When you do your shopping online, some items pop up first or are listed higher up in your search results. Some items may be highlighted and presented in a slightly more attractive way. And the same goes when you use Google's search engine, Facebook, Netflix, or YouTube. Importantly, such designs or choice architectures have an impact on our choices and decisions, regardless of whether there was a deliberate intention behind them. Just like more people will pick a lager if that happens to be at eye level, more people will click an item when it is higher up the list.[2]

While a pub owner might not think all too hard about where to put lagers and ciders, online choice environments are quite clearly *purposefully designed* (web design is a flourishing business), with specific ends in mind and with a lot of resources to achieve those. Just like a spoon is designed for a specific purpose, namely to enable eating soup, so too are online choice environments designed with specific purposes in mind: informing users, having them stay on the platform for as long as possible or stimulating them to buy things. This 'purposefulness' will prove to be crucial to our analysis.[3]

But first, what about the C in COCA? How to understand 'commercial' here? While private companies use available knowledge about how design influences user behavior to target their employees, clients, and consumers (Engelen and Schmidt 2020), we will focus exclusively on users we can call 'clients' or 'consumers'. Moreover, we focus on the ways in which companies try to influence the users of online environments to do things that serve the commercial interests of those very companies. These commercial ends can vary from profit maximization, increasing market share, promoting brand recognition to maximizing shareholder value. All of these, however, are ultimately tied to the commercial interests of the company. Our notion of COCAs then encompasses all online choice environments purposefully designed to sell or 'monetize' something.[4] Think of websites and apps of companies like Ryanair or Booking, where the '.com' extension is a notable give-away, but also of YouTube and Google Maps where this commercial aspect is purposefully less notable.

4 Conceptualizing Manipulation: From Means to Ends

Now, to determine whether, when, and why COCAs are manipulative, we first need to consider the definition of manipulation. As illustrated in the other chapters of this book, there are many definitions on offer. Instead of trying to formulate a precise and conclusive definition, we want to stress that any plausible definition of manipulation, in our view, at least involves four elements or components: 1) a manipulator (an agent doing the manipulation), a manipulee (a victim or target of the manipulation), 3) a specific kind of means (techniques used to perform the influence) and, finally, 4) a specific kind of ends (the manipulator's goals).

4.1 Means-Based Conceptualizations of Manipulation

Much of the discussion on manipulation has focused on its means (for an overview of these discussions and different ways of conceptualizing manipulation, see Coons and Weber 2014b; Noggle 2018, and Jongepier and Klenk, this volume). What is distinctive about manipulation, it is often said, is the third component: the specific kind of techniques involved. Manipulation is unlike merely informing, or incentivizing, or coercing another to do something, *because* manipulation involves techniques that bypass people's rational and reflective capacities (by triggering emotions or tapping into specific cognitive heuristics or biases). What characterizes manipulation on this account is the *special way in which behavior is influenced.*

Manipulation then constitutes a sort of sneaky influence, pulling people's strings and prodding them behind their backs, without them being aware of it. It differs from *rational persuasion*, where someone tries to influence another by presenting information, reasons, and arguments that can then be considered by the interlocutor. It also differs from *coercion*, where options are taken away from the victim, making it impossible or at least much harder for him or her to choose otherwise. It differs from incentivizing, where options are changed materially and financially.

Sneakiness, we agree, is an integral part and characteristic of manipulation. And the means of exerting influence – through non-rational psychological mechanisms such as cognitive heuristics and biases – are key in establishing such sneakiness (Bovens 2009). However, while central to manipulation, this aspect is not the only relevant one and by focusing too much on it, the literature risks missing out on other crucial aspects.[5] In fact, an all-too-strong focus on manipulation's means raises two significant problems.

4.2 Why Means-Based Conceptualizations of Manipulation Are Problematic

The first is a conceptual problem: focusing exclusively on manipulation's means categorizes too many cases as manipulation, including some that are clearly not. If manipulation is thought to consist in influencing someone through (quasi-)automated psychological processes and implicit biases (the means), the problem arises that such non-rational or less-than-reflective processes and influences are often inevitable and will be steering people's decisions anyway. Yawning, for example, is highly contagious. My involuntarily yawning makes it more probable that whoever sees me starts yawning as well. While this is a highly automated process that might occur without people noticing it (i.e., that their behaviors are 'mimicking' each other) and that bypasses people's rational capacities, it doesn't make sense, in our view, to claim that I am manipulating others into yawning.

Or take body language, which is known to influence others (e.g., subtle cues like crossed arms that signal rejection or the mirroring of behavior when we like a person). When this happens unintentionally, we believe it fails to amount to manipulation. Manipulation is not the same as merely exerting non-rational influence. Let us call this first problem the *Catch All Problem*: understanding manipulation merely[6] in terms of its means and thus as 'non-rational influence' is too broad and over-inclusive.

This also applies to the set-up of online environments. Any website visited or app used will inevitably exert *some* influence on their users (where to click, etc.). But that does not make all of these environments manipulative. What might be lacking is a specific intention on part of the designer to exert a particular influence on another. Manipulation, in other words, requires an agent (component 1) who wants to steer another agent (component 2) in a specific direction (component 4).

The second problem is a normative one. Authors who merely or exclusively focus on the means of manipulation will tend to locate the ethical worry in that aspect as well. Bypassing a person's rational faculties by making use of more automated psychological processes and implicit biases will be considered wrong because it does not respect the person as a rational and autonomous decision-maker (for a discussion of this in the case of nudging, see Hausman and Welch 2010; Engelen and Nys 2020).

The problem here is that we do – and in our view also should – respect many non-rational decisions or choices. By focusing on the means, these authors focus too much on manipulation's effects upon manipulees, that is, its *victims*.[7] If someone sneakily influences another into doing something, then that action, the worry goes, is no longer fully her own but partly attributable to the manipulator. The victim was then somehow tricked into doing something, the wool was pulled over her eyes.

The problem here lies in the assumption that, without the manipulative interference, the manipulee's actions would otherwise have been (more) rational, reflexive, autonomous, conscious, deliberate, and so on. But this counterfactual is misleading in many, if not most, cases. Given the ubiquity and inevitability of non-rational influences on behavior (see the earlier *Catch All Problem*), we have reason to believe that the quality of decision-making processes in the absence of deliberate manipulation will likely be very similar (in relevant aspects) to the quality of those process in the presence of manipulation. Hence, it is not obvious at all how (and often simply not true that) the means used here degrade the quality of the decisions made and actions performed.

As a consequence, this approach raises the requirements for morally unproblematic decision-making significantly. If the normative worry indeed lies with manipulation actually rendering decisions less rational, less reflective, and less autonomous, then there should be something amiss with *all* decisions properly characterized as non-rational. If the problem with you manipulating me into buying something lies in the decision-making processes at play, that problem will not disappear when you stop manipulating me while the same processes remain at play (as they do in most circumstances).

Critics of manipulation who focus on the means thereby raise the threshold for 'ordinary' decisions. Let us call this second problem the *Raising the Bar Problem*. The requirement that non-rational factors should be absent for decisions and decision-makers to be respected is unrealistic and overly demanding and thus raises the normative bar (for what it means to respect decisions) to implausible heights.

In sum, understanding manipulation by focusing only on its means is both conceptually (*Catch All Problem*) and normatively problematic (*Raising the Bar Problem*). These problems are related, as they both bear on non-rational factors and processes being at play not only in cases of manipulation but in most cases, including those we deem morally unproblematic. To avoid these problems, one should steer clear from accounts of manipulation that ignore, downplay, or underestimate the intentions of manipulators (such as Manne 2014; Noggle 2018; Susser, Roessler, and Nissenbaum 2019). There is no manipulation without intention (conceptual claim) and the worry of manipulation at least partially resides in the manipulator's intention (normative claim). In the next section, we analyze more fully how to understand (the role of) those intentions.

4.3 Ends-Based Conceptualizations of Manipulation

Instead of defending a full-fledged account of manipulation, we want to make four conceptual points on how we approach manipulation here.

First, whatever one has to say about the specific kind of means involved in manipulation this shouldn't detract from the other three components. In our view, there is no manipulation without a manipulator whose underlying intention is to influence manipulees towards her own ends. Manipulation is a form of 'social influence'[8] and has a '*mens rea*' condition: a 'mental component necessary for something to count as an act of manipulation' (Baron 2014, 100). This effectively avoids the *Catch All* and the *Raising the Bar Problem* as it excludes cases where non-rational factors influence behavior unintentionally (as in cases of yawning or body language). In addition, it captures the idea that there is something distinctively problematic about intentionally using knowledge about non-rational influences for steering people in some direction. If I *deliberately* use body language to get you to do something, this is different, both conceptually speaking (it does count as manipulation) and normatively speaking (it is worrisome in a way that unintentional body language is not).

Second, when assessing whether COCAs are manipulative or not, we will consider them as *tools* of manipulation. As such, we are not assessing whether machines and automated systems, artificially intelligent or not, can meet requirements of agency or intentionality (as, e.g., Klenk or Pepp et al. do in their respective chapters in this volume). Important to us is that COCAs – or any other means involved in manipulation for that matter, such as the psychological tricks grandma is pulling when emotionally blackmailing us into visiting her more often – are deliberately put to use as influencing tools. The crux of our claim is that there are intentions *behind* COCAs and that those are crucial in assessing COCA's manipulativeness, which does not imply that COCAs are intentional systems doing the manipulating themselves.

We thus approach COCAs and the question whether they count as manipulative as one would approach some offline or analogue commercial influence, such as billboard advertising. Instead of analyzing whether advertisements themselves are doing the manipulation (and count as manipulative *agents*), we think it suffices to say that these are purposefully designed and can thus count as *manipulative tools* in the hands of commercial agents.[9] This approach, in our view, in no way implies that COCAs cannot be manipulative (just like advertisements can be manipulative) or that there is nothing wrong with them (just like advertisements can be worrying). Like advertising, the design of COCAs is a prototypical case where commercial agents aim to get consumers and users (their target) to do something that benefits those agents, and where they are investing a lot of time, money, and resources in perfecting the tools to achieve exactly that end.[10]

Third, while manipulators can have a lot of different ends and purposes in mind (commercial or not, self-interested or not, benevolent or not), these typically remain hidden in manipulation. Manipulators tend to conceal what and how they try to achieve their ends. Manipulation, it is often said, operates 'in the dark' and influences people 'behind their backs'. In fact, it can be argued that this is also what makes it morally problematic (see Bovens 2009 and Hansen and Jespersen 2013 for a discussion on the transparency and manipulativeness of specific kinds of nudging techniques). While it is in principle possible that a manipulator's ends and means are known to manipulees, this is not how the manipulator envisages it. As a manipulation expert, you might be able to see the tricks that Facebook and YouTube designers are pulling and what they are hoping to achieve, but that transparency and openness are not part of the design. Quite the contrary, manipulative designs aim to be non-transparent, hiding the underlying ends and means from users' sight.

Fourth, our aim here is not to definitively settle which conceptualization of manipulation, if any, is the correct one. This will depend crucially on what one wants this definition to do. A first desideratum is 'accuracy': a definition of manipulation should accurately include only those cases that are plausibly called manipulative. Our earlier Catch All Problem basically argues that means-based understandings of manipulation are inaccurate as they are over-inclusive.

A second desideratum is normative and provides the basis for the rest of our chapter: a definition of manipulation should help clarify whether and why cases of manipulation are morally objectionable. In our view, a proper investigation of the moral worries that COCAs raise requires attention to their underlying intentions, purposes, and ends. What sets manipulation apart from ordinary instances of non-rational decision-making is where it is coming from (the actual manipulators) and where it is heading (the specific purposes manipulators have in mind). While there may be nothing wrong with choosing a marriage partner (and being chosen as such, of course . . .) on the basis of hunches and the kind of *je-ne-sais-quoi* that characterizes the laws of attraction, it is another thing altogether to be deliberately steered in choosing such a partner by someone who intentionally 'plays' those hunches in a way that suits their purposes. Similarly, there is a normative difference between a clumsily designed pub or (web)shop with no specific intention in mind and one that is deliberately and cleverly designed by a whole team of clever marketeers with a lot of resources and psychological knowledge at their disposal.

4.4 Why COCAs Are Tools of Manipulation

Having defined what COCAs are and identified what the main elements of manipulation are, we can now argue that COCAs are manipulative, as they embody those key elements.

The key element we have stressed – manipulation's intentional aspect or the 'mens rea condition'– is definitely fulfilled when it comes to COCAs. They qualify as tools of manipulation since they are purposefully and deliberately designed by commercial agents with the specific goal of monetization. Google even makes this quite explicit: '[t]o maximize your revenue, consider multiple monetization models for your app' (quoted in Sax 2021). Other companies like Facebook or Spotify might be less open about it but are in the same business of making money through the careful design of their online choice architectures.

There are other elements of manipulation as well. COCAs exert non-rational influence on users and typically do so sneakily, behind people's backs. In fact, obfuscating both its means and ends is often part and parcel of COCA's purposeful design. While there is something paradoxical in trying to prove the hiddenness of something, let alone its deliberate hiddenness, this arguably does occur. While COCA designers non-rationally influence users with specific ends in mind, the who, how, and why of this process are not (meant to be) obvious to the latter. This further substantiates the claim that this is a form of manipulation, as non-manipulative influences – think of (rational) persuasion, coercion, or reliance on incentives – are much more out in the open, with targets being aware of what is happening (and this awareness is *needed* for the influence to work).

With COCAs, this sneakiness and hiddenness is achieved by the design's attempt to make it all about the users, their experiences, and their goals. When you are in COCAs, you feel it is about you (and not about them): you are receiving personal recommendations (Netflix' 'because you watched' section), you are shown things that you might want to buy, etc. COCAs typically give us, as users, the impression that they are doing us a service. This helps 'obscure' the underlying ends of COCAs: to commercially benefit those designing and running the COCAs.

5 COCAs and Personal Autonomy: Do Ends Align?

Having established that COCAs often manipulate their users, let us address what is plausibly regarded as the main normative worry, namely that they undermine users' personal autonomy.

5.1 Conceptualizing Autonomy and Heteronomy

Normative worries about manipulation are typically spelled out in terms of personal autonomy: 'Perhaps the most common account of the wrongness of manipulation claims that it violates, undermines, or is otherwise antithetical to the target's personal autonomy' (Noggle 2018). Manipulation undermines autonomy exactly *because* rational decision-making capacities are bypassed, perverted, or precluded. But remember that both intentional and non-intentional cases may be similar in this respect, which means that this approach fails to capture the distinctive worry about *intentional* manipulation.

A first response here is to argue that intentional manipulation violates not so much autonomy but some other value, such as dignity or freedom in the republican sense of 'non-domination'. The distinctive worry with manipulation – compared to non-rational factors influencing our decisions unintentionally – arguably lies in the fact that manipulators dominate others and exploit their vulnerabilities at will, exercising worrisome levels of power, oppression, and subordination over them (Noggle 2018; Schmidt and Engelen 2020).

A second response argues that personal autonomy *is* at stake but should not be characterized in terms of rationality. Your beliefs, desires, and decisions are autonomous when they are 'properly yours' and 'speak for you' or in slightly different terms, when you identify with them, even if you are not (fully) rational. How do COCAs affect *that* property? Sure, spending a lot of time on YouTube or Facebook, where algorithms determine what you see, is bound to have an impact on your beliefs, desires, and decisions, but the question to ask is whether the latter still qualify as sufficiently your own. Smartly designed COCAs arguably undermine autonomy when their users can no longer be conceived as the 'authors' of their own beliefs, desires, and decisions.

Note that the commercial aspect seems to be an attenuating factor here. Commercial transactions are usually seen as quite innocuous exchanges where the preferences of buyers and sellers neatly align. Nobody is forcing or tricking anyone. Remember Adam Smith's famous quote about the market's invisible hand:

> Give me that which I want, and you shall have this which you want. . . . It is not from the benevolence of the butcher, the brewer or the baker that we expect our dinner, but from their regard to their own interest.
> (Smith 1986, I.ii.2)

On the condition that this kind of self-regard is indeed autonomous – each party is merely pursuing their own preferences and goals – the market is autonomy respecting *par excellence*. After all, it simply provides parties with opportunities and incentives to mutually benefit each other (Sugden 2018). COCAs then are one of the many ways in which markets bring together buyers and sellers, who 'use' each other to get what they want (and they are assumed to know what they want). In short, whenever customers visit butchers, brewers, or bakers, their ends align. And should they not align, then each party can always opt out of the transaction. This, arguably, is a reassuring thought for those who worry about market mechanisms undermining autonomy.

There are a couple of big 'ifs' here, and the one relevant to our purposes is that consumers often do not have fixed preferences and that smartly designed pubs and (web)shops can nudge them into buying more and more expensive (or unhealthy, or whatever other property that is not in their interest) items. As such, the alignment of ends between buyers and sellers may well be only skin-deep. Sure, they each make their 'own' choices but in what way and on what basis? What makes preferences and decisions for A over B really and properly 'their own'? These considerations are relevant for assessing how COCAs affect someone's personal autonomy.

To answer them, we need to conceptualize autonomy. Here, again, rather than defending one specific conception of autonomy (the literature is rife with these and doing this requires much more space than we have here), we will focus on autonomy's core idea, namely having some form of control over our actions, desires, and beliefs. We will treat 'personal autonomy' here as a placeholder: we will take one influential family of autonomy theories and consider how COCAs impact the autonomy of their users. We will show that someone's autonomy could, but should not, be negatively affected by such choice architecture and that, in fact, the autonomy-preserving, or even promoting, effect is not enough to get COCAs off the normative hook.

This last point will bring us back to our emphasis on the underlying purposes of COCAs.[11]

Important structural and historical accounts of personal autonomy focus on the notion of *identification* (Frankfurt 1988) or on the broader notion of *non-alienation* (Christman 2001).[12] A preference, choice, or decision is then autonomous, that is, is a person's own, if they identify with it or if they do not regard it as alien to themselves. Consequently, a person acts on a heteronomous or non-autonomous preference (or another psychological element) if she does not identify with it or is alienated from it. Prototypical examples in the literature involve cases of addiction, compulsive disorders, and phobias. Take Harry Frankfurt's 'unwilling addict' who is rendered a 'helpless bystander to the forces that move him' (Frankfurt 1988, 21). He is alienated from his own desire (to take drugs), does not want to be moved by it, and would like to distance himself from it. Similarly, a person who frantically tries to avoid stepping on the cracks in the sidewalk could wish to be cured of that compulsive behavior. But the examples are not limited to near pathological cases: we all act heteronomously from time to time when we are moved by a desire that we do not want to be effective (e.g., to act out of jealousy or spite).

5.2 COCAs Exploiting Heteronomy

Now, can COCAs be said to undermine a person's autonomy in this way? Do they induce or promote heteronomy? We claim that they can, namely by deliberately playing into desires that users of these COCAs do not want to be moved by. Online gambling sites, for example, arguably do so by making it harder for users to withstand the pull of unwanted and alien desires and urges. Here we see the interplay between the design(ers) of the platform and the psychology of its users. By triggering and shaping certain desires, making certain kinds of options more salient or kinds of decisions more tempting, they push or pull at least several of their users into a direction they do not want to be moved in. If it takes effort and willpower to maintain one's autonomy and remain in control, then offering a push in a certain direction can mean the proverbial straw that breaks the camel's back.

Presumably, Oscar Wilde once said: 'I can resist anything, except temptation'. And it's funny because it's true. To be tempted by something is to find our power of resistance weakened. Hence, temptation threatens our autonomy, or more accurately, puts it to the test. While this is true, the real problem shows itself only when including the manipulator's intentions. Any online design, purposeful or not, can steer people in directions and ways they do not identify with, but sometimes the design *willfully targets* those who are prone to heteronomy. While, for example, any unwilling addict can fall off the wagon, an additional worry arises when someone deliberately pushes her off and thereby exploits someone's vulnerabilities to achieve her own ends.[13]

As a real-life example of COCAs engaging in such practices, think of Facebook's targeting of young teenagers. Allegedly, Facebook's algorithm could predict when teens would be feeling down or sad by examining and tracking their online behavior. In 2017, Facebook experimented with 'manipulating' these emotional states by tweaking the news feed they were exposed to (see also Wildman, Rietdijk, and Archer, in this volume). This could enable Facebook to target teenagers with specific ads that would 'help' them overcome their sadness or insecurities. For example, it could 'target moments in which young users are interested in "looking good and body confidence" or "working out and losing weight"' (Machkovech 2017). Here, the users are led to 'give in' to motivating factors they would rather resist.

What matters here, and this is why this is a case of 'exploiting heteronomy', is that groups like young teenagers are in some way *vulnerable* and that vulnerability is exploited by COCA designers. While there may not be anything inherently wrong with making others feel sad (*Watership Down*, anyone?) or even inadvertently bringing about someone's heteronomy (e.g., by offering a drink to someone you don't know is a former alcoholic), there is at least reason to object to cases where *someone intentionally attempts to exploit another's heteronomy*.[14]

This is clearly and exactly what COCAs like Facebook or online gambling sites are able to do. First, they can search out and target a specific group they know to be vulnerable in specific ways. Second, they can trigger and tap into those vulnerabilities and exploit their targets' weaknesses to serve their own ends. Facebook's founding president admits this much when stating that Facebook has always been out to make you 'consume as much of your time and conscious attention as possible' by 'exploiting a vulnerability in human psychology' (Solon 2017).

Two remarks are in place here. First, companies will claim that they do not intentionally try to make people heteronomous. Their operating notion is that of predictability: they offer stimuli that predictably get the desired results, treating the users' minds as black boxes. A desire to gamble, to lose weight, or to impress peers, is not, in and of itself, heteronomy inducing. To argue that it is, one would need a substantive account of autonomy and argue that companies undermine autonomy because they lead us to make the wrong choices (i.e., choices that, on the basis of their content, jar with autonomy like the choice for being a 'contented slave' or a 'submissive housewife').[15]

How to respond to companies claiming that their goal is not to induce heteronomy but simply to sell stuff or make money in some other way? This may be true, and they may be succeeding in their goal by designing COCAs that facilitate their users to achieve their own goals, that is, buy the things they want or watch some free videos. Yet, even if the ends of companies and users arguably align, there is something wrong here as the whole process predictably undermines the autonomy of some of their users, which is a foreseeable wrong. The fact that it is unintended does not make up for the fact that it has this detrimental impact on some people's autonomy. In addition, targeting groups that are especially vulnerable amounts to willfully committing that wrong to a specific audience. What is problematic here is that companies pursue their goals by foreseeably or intentionally causing the heteronomy of some of their users. It reveals a blatant and worrisome attitude of indifference or carelessness on the part of companies concerning the autonomy of their users (see also Jongepier and Wieland, in this volume, for an argument why this reveals an inappropriate attitude when it comes to political micro-targeting).[16]

Second, many theorists have rightfully questioned whether the autonomous–heteronomous divide is so clear-cut. In Frankfurt (1999, 99), there is the phenomenon of ambivalence: a person being divided over the question whether she wants to be moved by some desire or motive, or not. Such cases of what he calls the absence of 'wholeheartedness' are different and should not be conflated with those of heteronomy. Perhaps instances of heteronomy, as exemplified in the literature by nigh-pathological examples of addiction, phobias, and compulsion, are pretty exceptional. Perhaps even in Facebook's vulnerable teenagers example, we should say that these youngsters are not really alienated from the desire to 'look good' and 'work out'; perhaps they do identify to a certain extent with what they are pushed to pursue. As such, we should acknowledge that COCAs may *not* result in or aggravate heteronomy or alienation but rather respect personal autonomy.

5.3 COCAs 'Respecting' the Perimeters of Personal Autonomy

Some influential and more recent accounts have indeed argued that autonomy is not an all-or-nothing affair. Persons can be autonomous with regard to a motivating aspect to greater or lesser degrees. Laura Ekstrom (2005), for example, talks about how desires and beliefs can be *integrated* to various degrees. Nomy Arpaly and Timothy Schroeder argue that we should consider how deep certain motivational elements are embedded and supported in a person's psychological constitution (Arpaly and Schroeder 1999). The point is often that heteronomy-as-alienation does not do the trick: sometimes the 'alien' desire still speaks for the agent in spite of her rejection or repudiation of it. Arpaly and Schroeder discuss an example of a person who frequently 'succumbs' to his urge for shoplifting (a desire he does not identify with), but whose proclivity for petty theft is still very much his own, because it is sufficiently rooted – integrated – in his personality.

This is another way of describing and theorizing what we have elsewhere dubbed the 'perimeters of autonomy' (Engelen and Nys 2020).[17] We are autonomous across a *range* of options, actions, and motivating factors. Vegetarians, for example, may want to resist the desire to eat meat, but their commitment still leaves open a wide array of possible food choices that remain in line with their autonomy (and even if they hate the taste of broccoli, plenty of options are still on the table). The perimeters of our autonomy are constituted by things we care about (cf. Frankfurt 1988) and those 'volitional commitments' are typically quite broad. Hence, our (specific, first-order) preferences may change and be subject to all kinds of influencing factors – including purposefully designed and manipulative COCAs – without necessarily leading us to cross the boundaries of our autonomy.

When it comes to vegetarians, for example, clever marketeers can influence them by buying shelf space at eye level in supermarkets or making their products more salient in their web shop. However, the perimeters of their clients' autonomy will limit their wiggle room, as vegetarians will never be manipulated into buying meat. When successful, such marketeers can be said to *respect* autonomy because their efforts have not resulted in a person overstepping the boundaries of her autonomous decision-making. If Netflix successfully nudges us into binge watching its brand-new series, then claiming that we really do *not* want to give in to that temptation sounds like a rather poor excuse. It sounds more plausible to claim that such 'giving in' actually reveals something about ourselves and what we care about. What Netflix does, we would argue, is exploiting this bandwidth of our autonomy.

Notice how this can be used to cast doubt on the idea that COCAs violate autonomy or induce heteronomy. In quite some cases, COCAs work within the perimeters of users' autonomy. You might actually endorse a certain ideal of physical beauty and COCA designers who know this can use it to get you to buy clothes, gym subscriptions, or beauty products. This only violates your autonomy if this set of motivating elements were completely alien to you and you'd rather be relieved from them. (And even if *you* regard it as alien, then others, who know you well, could still claim that it does speak for you, despite your avowed reservations.)

If personal autonomy has perimeters, one can understand how an influence may be an instance of manipulation while still respecting autonomy. Choice architects can play into the space that is allowed by these perimeters and tap into the autonomy of the individual to monetize their products. Think of a person who enjoys running and who believes it to be very healthy; she wants this, and she values it. Now, at some point she will need running shoes, and knowing what she likes and what she values will enable choice architects to push the product – *their* product – that fits the bill. This, we believe, is not a *violation* of autonomy but should be understood as an *exploitation* thereof (Engelen and Nys 2020). Whatever is morally wrong about such an influence, we believe, has less to do with its effect on the autonomy of manipulees (they remain autonomous) than with the intentions of the manipulators instead.

Note that we do not claim that *all* COCAs *always* respect users' perimeters of autonomy. There is the case of pushing people outside these boundaries we discussed earlier (e.g., getting the struggling vegetarian to buy that hamburger), and even if each interaction in COCAs typically changes preferences only ever so slightly, the long-term effect can still be quite dramatic. The recommendation systems and algorithms of YouTube, Facebook, Twitter, and other online platforms can gradually draw people into traps of conspiracy theories and extremist ideologies. This can cause people to end up way beyond their initial perimeters of autonomy (and thus have their initial self's autonomy violated).

5.4 COCAs Promoting Autonomy

Understanding autonomy along the lines of identification and non-alienation, there is one possibility left to discuss: COCAs not so much violating or respecting autonomy but actually promoting it. Because we have already dealt with this in the context of nudging (Nys and Engelen 2017), we want to be brief here. Clever product placement in supermarkets can actually dissuade aspiring vegetarians from buying meat, and the gamification in activity trackers, health and fitness apps can encourage people who are planning to adopt a healthier lifestyle to go for that extra run or that extra mile.

COCAs can thus be designed to effectively promote users achieving their own ends and may even do so by helping them overcome temptation, procrastination, or weakened wills. As much as COCAs can stack the deck against autonomy, they can also facilitate users achieving their goals and making them act upon the desires they identify with, thus promoting their personal autonomy. If consumers realize that procrastination and akrasia can inhibit them from realizing their own ends, companies can and will supply products and services that meet the resulting demand for (online) tools to fight these autonomy-thwarting impediments.

So, up until now, we have described three possible scenarios of users engaging with COCAs. The result in terms of their autonomy could either be negative, neutral, or positive, and it is possible to give clear examples of each. One conclusion could be that the moral wrongfulness of manipulation hinges on these outcomes. Should COCAs undermine autonomy that should be regarded a prima facie wrong. If they respect it, they are morally permissible. And if they promote autonomy, that counts in favor of them and makes them commendable (pro tanto).

But that conclusion would be too quick. In each of these three scenarios, the wrongness eludes us if we leave out the underlying intentions, or rather the *relationship between the means–end structure of the manipulator and that of the user*. Even in the neutral and positive cases, the manipulators are only instrumentally using the predictable relation between the agent and her goals. The design that is used is not put in place to secure, or guarantee, or foster an autonomous relationship between the user and her ends; it *merely exploits* it. This explains why it can be strategically helpful for companies to 'play into' the autonomy of the user simply because it is more effective. It just works better. While the purposefulness of the design can be autonomy impairing, preserving, or promoting, it also importantly reveals a deeper attitude of *indifference*.

Recall how companies can try to parry the criticism that they intentionally seek to increase and exploit users' heteronomy. They could say that they have no such intention. In fact, they would rather have their users autonomously decide to gamble, buy beauty products, or vote conservative. Moreover, who are *we* to judge and question their motives?

One answer to this is to stress that the design speaks louder than words. If, for example, many people would deem their autonomy negatively affected by fatigue, then designing-for-fatigue leaves little room for questioning these revealed intentions. Also, the absence of a motive to negatively affect someone's autonomy may coexist with an attitude of *indifference* toward it.

Note how our claim that COCA designers are problematically indifferent to the users' autonomy is not incompatible with our focus on their intentions (i.e., the commercial interests they aim to serve). Let us take a closer look at Marcia Baron's 'mens rea' condition that we invoked earlier:

> The *mens rea* of manipulation can be a combination of intent and recklessness: the aim of getting the other to do what one wants, together with recklessness in the way one goes about reaching that goal is insufficiently concerned about the other qua agent. The recklessness amounts to a disregard for whether one is treating the other with respect.
>
> (Baron 2014, 103–4)

Precisely because COCA designers are primarily out to make money, they will be drawn to manipulating users without sufficient regard or respect for the latter's autonomy. If they would show such sufficient regard and actually care about users' autonomy (and thus not let commercial interests be their primary aim in designing their algorithms), their design would be less sneaky, less deceptive, and morally less problematic.

5.5 COCA Designers' Indifference Toward Users' Autonomy

This attitude of indifference or carelessness about users' autonomy is a common thread in all three scenarios. While it is obvious in cases where COCA designers are willing to violate users' autonomy to promote their own commercial ends (Section 5.2), the same attitude underlies the neutral and even the positive set of cases discussed earlier (Sections 5.3 and 5.4). COCAs respecting or even promoting personal autonomy is either *accidental* or *strategic*. Even when they happen to promote it, companies do not really care about our autonomy. Or better: they only care to the extent that this benefits them (if, for example, they can discover and tap into our autonomous pursuits). If the basis on which COCAs operate is indeed predictability, then we should acknowledge that people are predictable in their autonomous as well as non-autonomous behavior. So, COCAs can and will work both ways. But autonomy-respecting or autonomy-promoting manipulation is far less conspicuous as it does not trigger the negative experience of alienation. So strategically speaking, it might be better for COCA designers to stay on the safe side.

In sum, companies who pay COCA designers to promote their commercial ends display a worrisome attitude of indifference or carelessness toward the autonomy of their customers. Companies typically do not show sufficient regard for the ends users have and for their capacity to set and pursue those ends themselves. Or rather, if our analysis is correct, then the remaining worry is that the respect they show in not undermining the personal autonomy of their users, even sometimes promoting it, is only *superficial*. If one then still wants to argue that the online manipulation that COCAs commit is morally problematic because it does not properly respect people's autonomy, one will need to dig deeper *and* – again, if we are on the right track – one will need to take into account the intentions of the manipulator. A shift of perspective toward a concept of *moral autonomy* along the lines of Immanuel Kant's readily comes to mind because, on his account, it is not *what* people do (in our case, leaving other people's personal autonomy intact or even promoting it) but *why* they do it (in our case, out of a concern for their autonomy or just in order to make money) that determines the moral worth of an action. But that would require another chapter.

6 Conclusion

Let us briefly summarize our main line of argument. As we are making more decisions than ever in COCAs, we need to assess whether and why there is something morally wrong about the way they are designed and implemented. We have focused specifically on the worry that COCAs are tools of manipulation that arguably undermine or violate the personal autonomy of users. An adequate concept of manipulation, we argued, should include the intentions of the manipulator for both conceptual and normative reasons. This has enabled us to claim that COCAs can indeed be manipulative and that their wrongness can indeed be cashed out in terms of violating personal autonomy.

However, we have also argued that COCAs can respect and even promote personal autonomy. Focusing on the underlying intentions of the manipulating COCA designers, however, revealed that even in these cases, where their 'concern' for autonomy is only secondary, strategic, or instrumental, we have argued that there is something morally problematic about companies' underlying attitudes of carelessness and indifference. Even in those cases where COCAs do not violate personal autonomy, normative issues surrounding COCAs' manipulative potential remain prevalent.

Notes

1. Of course, in developed countries, almost everyone is an internet user. In 2019, 97 % of the Dutch population older than 12 years had access to the internet, and nine out of ten of these users were online on a daily basis (CBS 2020). In fact, 93.5 % of those between 12 and 18 and 98.4 % of those between 18 and 21 uses the internet daily, primarily for social media (Nederlands Jeugdinstituut 2021). The general trend is that these numbers continue to rise and that younger users systematically spend more time online (Kemp 2020).
2. About three out of ten Google users, for example, click on the very first search result and eight out of ten never make it past the first ten results (Southern 2020).
3. Next to other chapters in this collection (see the introductory chapter 1 by Jongepier and Klenk, this volume), the literature on how digital choice environments can nudge, influence, or otherwise engage with user behavior is growing (Benartzi and Lehrer 2015; Jameson et al. 2013; Schneider, Weinmann, and vom Brocke 2018; Weinmann, Schneider, and vom Brocke 2016; Yeung 2017).
4. Marijn Sax explicitly discusses Google and Apple's 'monetization models' (Sax 2021, 31).
5. In line with Jongepier and Klenk's terminology chapter 2 of this volume (Jongepier and Klenk, chapter 1 2022a), one could call sneakiness an "aggravating factor" instead of a necessary condition of manipulation: perhaps not all manipulation is sneaky, but it often is and when it is, it makes things worse than they already are.
6. The devil is in the details. Many would say that they do not *merely* focus on the means. While they can argue that focus on cases where these 'sneaky means' are intentionally used by another, the meaning and importance of this added intention are hardly ever developed or scrutinized more fully.
7. We will argue that these effects in terms of the victim as a decision-maker are actually quite minimal if not non-existent.
8. This social aspect implies a narrow conception of manipulation here, where there is someone who is a victim to manipulation. This narrow conception is the focus of most of the psychological work on manipulation (e.g., Simon 1996). As such, we ignore the broader understanding of manipulation as 'skillfully operating something', which is at play in the following cases: a researcher manipulating data statistically or a writer manipulating a pencil.
9. We diverge here from Allen Wood (2014, 27), who argues that manipulation can be unintentional and occur without manipulators but still claims that, for example, advertising can manipulate. Our approach, which is to require intentionality on behalf of manipulators (as for example, Klemp 2011 does as well) and to claim that something can be manipulative if it is a tool in the hands of such manipulators, is closer to everyday language and intuitions about manipulation (Baron 2003, fn 11). Our claim that COCAs are manipulative is then structurally similar to the uncontroversial claim that 'laws are coercive', where the latter should be understood as 'laws are tools in the hands of agents, i.e. the state, who are engaged in coercive activities and use laws for their specific purposes'.
10. Cass Sunstein (2016, 213) also claims that motives (in our words, intentions) matter, ethically speaking: 'the manipulator's motives become more self-interested or venal, and as efforts to bypass people's deliberative capacities becomes more successful, the ethical objections to manipulation become very forceful'.
11. Because we use autonomy as a placeholder, other families of autonomy theories, like relational and substantive accounts, could be fitted in as well. We just want to note that, on these alternative accounts, it isn't *obvious* either that COCAs simply diminish personal autonomy. What, for example, is COCA's impact on a person's self-trust, self-respect, and self-esteem? Is that always negative? Sufficiently negative? Never positive?

12. Suppose that non-alienation is only a *necessary* condition for autonomy. We suppose that consensus about this is sturdier than about the *sufficient* conditions for autonomy.
13. The Bible warns us about the devil's work that is involved here: 'And *lead* us not into temptation, but deliver us from evil'.
14. Advertising has always been about enticing customers to buy products, but that was always 'one size fits all'. Other issues in this volume address the issue of personalization in online manipulation (e.g., Jongepier and Wieland 2022; Miotto and Chen 2022).
15. For a famous substantive account of autonomy, see Marina Oshana (2006).
16. According to Michael Klenk (2021), '[w]e have a case of manipulation if and only if the manipulator does not care whether his or her means of influence reveals eventually existing reasons to the manipulatee'.
17. In our earlier paper (2020), we make a distinction between autonomy proper, that is, the ability to *set* ends, and autocracy, that is, the ability to *translate* these ends into action. A failure to distinguish between these interpretations is a source of confusion in the debate on autonomy. In this entire fifth section we talk about autonomy-as-autocracy. In terms of getting what you 'really, autonomously want' (related to your deepest cares and concerns), manipulation can either thwart, respect, or promote that ability.

7 References

Arpaly, Nomy, and Timothy Schroeder. 1999. "Praise, Blame and the Whole Self." *Philosophical Studies* 93 (2): 161–88.

Baron, Marcia. 2003. "Manipulativeness." *Proceedings and Addresses of the American Philosophical Association* 77 (2): 37. doi:10.2307/3219740.

Baron, Marcia. 2014. "The Mens Rea and Moral Status of Manipulation." In Coons and Weber 2014, 98–109.

Benartzi, Shlomo, and Jonah Lehrer. 2015. *The Smarter Screen: Surprising Ways to Influence and Improve Online Behavior*. New York, NY: Penguin Books.

Bovens, Luc. 2009. "The Ethics of Nudge." In *Preference Change: Approaches from Philosophy, Economics and Psychology*, edited by Till Grüne-Yanoff and Sven O. Hansson, 207–19. Dordrecht: Springer.

CBS. 2020. "453 Thousand Dutch People without Home Internet Access in 2019." Accessed August 23, 2021. www.cbs.nl/en-gb/news/2020/14/453-thousand-dutch-people-without-home-internet-access-in-2019.

CBS. 2021. "11 Percent More Online Purchases in First Half of 2020." www.cbs.nl/en-gb/news/2021/02/11-percent-more-online-purchases-in-first-half-of-2020.

Christman, John. 2001. "Liberalism, Autonomy, and Self-Transformation." *Social Theory and Practice* 27 (2): 185–206. https://doi.org/10.5840/soctheorpract20012729

Coons, Christian, and Michael Weber. 2014a. "Manipulation: Investigating the Core Concept and its Moral Status." In Coons and Weber 2014b, 1–16.

Coons, Christian, and Michael Weber, eds. 2014b. *Manipulation: Theory and Practice*. Oxford: Oxford University Press.

Ekstrom, L. 2005. "Autonomy and Personal Integration." In *Personal Autonomy: New Essays on Personal Autonomy and Its Role in Contemporary Moral Philosophy*, edited by James S. Taylor, 143–61. Cambridge: Cambridge University Press.

Engelen, Bart, and Thomas Nys. 2020. "Nudging and Autonomy: Analyzing and Alleviating the Worries." *Review of Philosophy and Psychology* 11 (1): 137–56. doi:10.1007/s13164-019-00450-z.

Engelen, Bart, and Andreas T. Schmidt. 2020. "Digital Nudging: Opportunities and Threats." *The Habtic Standard* (3). https://www.thehabticstandard.com/articles/choosing-to-digitally-nudge.

Frankfurt, Harry G. 1988. *The Importance of What We Care About: Philosophical Essays*. Cambridge: Cambridge University Press.

Frankfurt, Harry G. 1999. "The Faintest Passion." In *Necessity, Volition, and Love*, edited by Harry G. Frankfurt. Cambridge: Cambridge University Press.

Gilbert, Ben. 2018. "How Facebook Makes Money from Your Data." *Business Insider*. www.businessinsider.com/how-facebook-makes-money-according-to-mark-zuckerberg-2018-4.

Globenewswire. 2020. "Global Online Content Consumption Doubles in 2020, Research Shows." https://www.globenewswire.com/news-release/2020/09/23/2097872/0/en/Global-Online-Content-Consumption-Doubles-in-2020-Research-Shows.html.

Hansen, Pelle G., and Andreas M. Jespersen. 2013. "Nudge and the Manipulation of Choice: A Framework for the Responsible Use of the Nudge Approach to Behaviour Change in Public Policy." *European Journal of Risk Regulation* 4 (1): 3–28. doi:10.1017/S1867299X00002762.

Hausman, Daniel M., and Brynn Welch. 2010. "Debate: To Nudge or Not to Nudge." *Journal of Political Philosophy* 18 (1): 123–36.

Jameson, Anthony, Bettina Berendt, Silvia Gabrielli, Federica Cena, Cristina Gena, Fabiana Vernero, and Katharina Reinecke. 2013. "Choice Architecture for Human-Computer Interaction." *Foundations and Trends in Human – Computer Interaction* 7 (1–2): 1–235. doi:10.1561/1100000028.

Jongepier, Fleur, and Michael Klenk. 2022a. "Online Manipulation: Charting the Field." In *The Philosophy of Online Manipulation*, edited by Jongepier, F. and Klenk, M., 15–48. New York: Routledge.

Jongepier, Fleur, and Michael Klenk, eds. 2022b. *The Philosophy of Online Manipulation*. New York, NY: Routledge.

Jongepier, Fleur, and Michael Klenk. 2022c. "Introduction." In Jongepier and Klenk in 2022b.

Jongepier, Fleur, and Jan Willem Wieland. 2022. "Microtargeting People as a Mere Means." In *The Philosophy of Online Manipulation*, edited by Jongepier, F. and Klenk, M. (pp. 156–179). New York: Routledge.

Kemp, Simon. 2020. "Digital Trends 2020: Every Single Stat You Need to Know about the Internet." https://thenextweb.com/growth-quarters/2020/01/30/digital-trends-2020-every-single-stat-you-need-to-know-about-the-internet/.

Klemp, Nathaniel. 2011. "When Rhetoric Turns Manipulative: Disentangling Persuasion and Manipulation." In *Manipulating Democracy: Democratic Theory, Political Psychology, and Mass Media*, edited by John M. Parrish and Wayne Le Cheminant, 59–86. New York, NY: Routledge.

Klenk, Michael. 2022. "Manipulation, Injustice, and Technology." In *The Philosophy of Online Manipulation*, edited by Jongepier, F. and Klenk, M., 108–132. New York: Routledge.

Klenk, Michael. 2021. "Manipulation (Online): Sometimes Hidden, Always Careless." *Review of Social Economy*. 80: 1, 85–105. doi:10.1080/00346764.2021.1894350.

Machkovech, Sam. 2017. "Report: Facebook Helped Advertisers Target Teens Who Feel 'Worthless'." https://arstechnica.com/information-technology/2017/05/facebook-helped-advertisers-target-teens-who-feel-worthless/.

Manne, Kate. 2014. "Non-Machiavellian Manipulation and the Opacity of Motive." In Coons and Weber 2014, 221–46.

Miotto, Lucas, and Jiahong Chen. 2022. "Manipulation, Real-time Profiling, and their Wrongs." In *The Philosophy of Online Manipulation*, edited by Jongepier, F. and Klenk, M., 392–409. New York: Routledge.

Nederlands Jeugdinstituut. 2021. "Cijfers over mediagebruik." https://www.nji.nl/cijfers/mediagebruik.

Noggle, Robert. 2018. "The Ethics of Manipulation." In *Stanford Encyclopedia of Philosophy: Summer 2018*, edited by Edward N. Zalta. Milton Park, UK: Routledge.

Nys, Thomas R. V., and Bart Engelen. 2017. "Judging Nudging: Answering the Manipulation Objection." *Political Studies* 65 (1): 199–214. doi:10.1177/0032321716629487.

Oshana, Marina. 2006. *Personal Autonomy in Society*. Hampshire: Ashgate.

Pepp, Jessica, Rachel Sterken, Matthew McKeever, and Eliot Michaelson. 2022. "Manipulative Machines." In *The Philosophy of Online Manipulation*, edited by Jongepier, F. and Klenk, M., 91–107. New York: Routledge.

Sax, Marijn. 2021. "Between Empowerment and Manipulation: The Ethics and Regulation of For-Profit Health Apps." PhD thesis, University of Amsterdam.

Schmidt, Andreas T., and Bart Engelen. 2020. "The Ethics of Nudging: An Overview." *Philosophy Compass* 15 (4),e12658. https://doi-org.tilburguniversity.idm.oclc.org/10.1111/phc3.12658

Schneider, Christoph, Markus Weinmann, and Jan vom Brocke. 2018. "Digital Nudging." *Communications of the ACM* 61 (7): 67–73. doi:10.1145/3213765.

Simon, G. K. 1996. *In Sheep's Clothing: Understanding and Dealing with Manipulative People*. Little Rock, AR: A. J. Christopher & Co.

Smith, Adam. 1986. *An Inquiry into the Nature and Causes of the Wealth of Nations*. Düsseldorf: Verlag Wirtschaft und Finanzen.

Solon, Olivia. 2017. "Ex-Facebook President Sean Parker: Site Made To Exploit Human 'Vulnerability'." *The Guardian*, September 11. Accessed August 23, 2021. https://www.theguardian.com/technology/2017/nov/09/facebook-sean-parker-vulnerability-brain-psychology.

Southern, Matt. 2020. "Over 25% of People Click the First Google Search Result." https://www.searchenginejournal.com/google-first-page-clicks/374516.

Sugden, Robert. 2018. *The Community of Advantage: A Behavioural Economist's Defence of the Market*. Oxford: Oxford University Press.

Sunstein, Cass R. 2016. *The Ethics of Influence: Government in the Age of Behavioral Science*. Cambridge: Cambridge University Press.

Susser, Daniel, Beate Roessler, and Helen Nissenbaum. 2019. "Technology, Autonomy, and Manipulation." *Internet Policy Review* 8 (2): 1–22. doi:10.14763/2019.2.1410.

Thaler, Richard H., and Cass R. Sunstein. 2008. *Nudge: Improving Decisions About Health, Wealth, and Happiness*. New Haven, CT: Yale University Press.

Weinmann, Markus, Christoph Schneider, and Jan vom Brocke. 2016. "Digital Nudging." *Business an Information Systems Engineering* 58 (6): 433–36. doi:10.1007/s12599-016-0453-1.

Wildman, Nathan, Natascha Rietdijk, and Alfred Archer. 2022. "Online Affective Manipulation." In *The Philosophy of Online Manipulation*, edited by Jongepier, F. and Klenk, M., 311–326. New York: Routledge.

Wood, Allen W. 2014. "Coercion, Manipulation, Exploitation." In Coons and Weber 2014, 17–50.

Yeung, Karen. 2017. "Hypernudge: Big Data as a Mode of Regulation by Design." *Information, Communication & Society* 20 (1): 118–36. doi: 10.1080/1369118X.2016.1186713.

8 Microtargeting people as a mere means[1]

Fleur Jongepier and Jan Willem Wieland

1 Introduction

Is it morally problematic to manipulate people online in order to get, say, their attention, their data, their money, or even their political vote? If so, what makes it wrong exactly? When confronted with practices involving (apparent) online manipulation, common reactions include pointing out how such practices threaten or violate personal autonomy (Nys and Engelen in this volume; Keeling and Burr in this volume; Susser, Roessler, and Nissenbaum 2019; Williams 2018). When it comes to political microtargeting in particular, that is, tailoring online political messages to specific individuals, scholars further stress the potentially growing information asymmetry between citizens and political actors and the risks posed for "voter manipulation" or how one's autonomy in the political domain is affected and the threats posed to democracy as a result (Barocas 2012; Zuiderveen Borgesius et al. 2018).

Common responses to manipulative practices typically share two features: they are broadly consequentialist and risk-based. They are consequentialist in the sense that the primary focus is on the consequences for affected individuals, be it in their capacity as consumers or voters. They are risk-based for the simple reason that there is still little research on and evidence of the actual effectiveness of online manipulation. In fact, there has been a serious debate about whether (political) microtargeting is even possible. Scepticism about the effectiveness of microtargeting strategies has tempered somewhat, as there is now evidence illustrating the causal effectiveness of political ads for which microtargeting techniques were used (Zarouali et al. 2020), which we will turn to shortly. Scholars however continue to stress the "crucial importance" for further research in this domain. Further research is indeed important, if only for the reason that it is unclear how experimental settings in which causal effects were found relate to the more dramatic claims that political microtargeting is to be blamed for the outcome of the Brexit referendum in the UK and Trump's victory in the US election. And thus it makes sense to take a risk- or threat-based approach.

DOI: 10.4324/9781003205425-10

However, there are potential downsides to focusing on consequences for people's (political) autonomy in terms of the risks posed. Firstly, it makes one's moral evaluation of the agents responsible for such manipulative practices hostage to whether bad consequences have indeed resulted and whether the risks posed are real. But just how influential *are* political microtargeting strategies really, in the actual world? How much of a role did they play in the Brexit and US election events? We don't know, and perhaps there's not even a way to (ever) tell (cf. Frederik and Martijn 2019). On a rather radical extrapolation of a consequentialist approach to online manipulation, it tells us that if there aren't any bad consequences (e.g., hardly any consumer buys differently or hardly any voter votes differently after being microtargeted with ads), no wrongs have been committed. That may not be likely in practice, but it is suboptimal in terms of principle.

A second disadvantage is that common responses are, as we might put it, "victim-centered". By focusing on consequences, one principally focuses on the people affected by digital misconduct. One's interest in the actual perpetrators of online misconduct only come in, as it were, *indirectly* through and after an examination of how individuals (or democracies) are affected or likely to be affected. An "agent-centered" view, by contrast, which takes the manipulating agent as its starting point may be preferable or at least equally valid.

These downsides, we believe, motivate looking for other ways of morally evaluating manipulating agents. We should add that the moral analysis offered here is not meant to replace consequentialist approaches but rather expand on them. We too believe that political microtargeting is (morally) worrisome because of the risks for increasing divisiveness among voters, voter discrimination, political chilling effects, and how it steers toward single-issue politics (Barocas 2012). The overall aim is to strengthen the overall case that political microtargeting is a morally suspect practice; if not for consequentialist reasons, then for principled reasons of how citizens should (not) be treated.

In this chapter, we develop an approach which argues that corporations or political agents involved in online manipulative practices act wrongfully because they use persons purely instrumentally or as "a mere means". This approach, we believe, first of all allows for stronger moral resources to criticize the agents involved in such practice, even if those practices turn out not to make much of a difference to the (political) choices people make. Second, the alternative approach revisits the debate about the (in)significance of (online) consent (see, e.g., McDonald and Cranor 2008) and provides new theoretically grounded support for calls that stronger action be taken against corporations that insufficiently care about acquiring proper consent. Third, the approach fits with widely held views that part of what's wrong about agents engaging in online manipulation is that they "instrumentalize" or "objectify" people, treat them as objects and fail to respect them as persons.

As a colleague of ours once remarked, discussing personalized ads: "it's not that I worry about buying the shoes I do or don't want, it's their indifferent *attitude* towards me that bothers me". We think this colleague is onto something – the challenge we take up in this chapter is to try and explain what it is.

2 Political microtargeting

In this chapter, we will take political microtargeting as our central case. We focus on microtargeting as it is, we believe, one of the most acute forms of manipulative influence online, and we choose to focus on political microtargeting because it's distinctly worrying in threatening our people's democratic agency and also, as we'll see, because caring about consent in political contexts raises distinct questions and introduces special requirements for political actors.

Political microtargeting is a technique that involves "collecting and analyzing people's personal data to send them tailored political messages" (Zarouali et al. 2020). The technique enables a political party to "identify the individual voters which it is most likely to convince" and "match its message to the specific interests and vulnerabilities of these voters" (Zuiderveen Borgesius et al. 2018). The overall goal is thus to persuade potential voters to vote for the given party.

It should be mentioned that the technique, though often discussed because of the threats to democracy and autonomy, *might* also have positive upshots, such as increasing voter participation and reaching uninterested citizens who are otherwise much harder to reach, by "reaching out to them in ways that are personally relevant" (Zarouali et al. 2020). Furthermore, political microtargeting, if utilized in a certain way, may help increase citizen's political knowledge and could diversify political campaigns (Zuiderveen Borgesius et al. 2018).

Political microtargeting works by tailoring a message (political ad) to fit an individual's interests or personality trait through the use of psychometric profiling models. Zarouali et al. conducted two studies on personality-based political microtargeting that we will discuss in some detail in order to better understand the strategy of microtargeting.[2]

In the two studies, the researchers focus on the personality traits of extraversion and introversion; traits known to be an important factor in affecting political outcomes. They take extraverts to have "an energetic approach to the social and material world" and as usually being "upbeat, energetic, active, talkative, and assertive, while introverts are rather reserved or even shy". The researchers draw on the popular self-congruity theory in psychology according to which people are said to prefer stimuli (ads, in this case) that are in line with their own self-concept. Interestingly, whereas personality traits in psychology are often measured through self-report questionnaires, the authors – like most big tech companies – used users' "digital footprints" (online text) instead, following scholars who have previously shown that personality traits can be "assessed based on language on social media" and that "computer algorithms can sometimes be as (or even more) accurate than humans in predicting these traits" (Zarouali et al. 2020, 5). Specifically, the authors used the technique of "automated personality profiling" which starts from the hypothesis that a person's characteristics can be inferred on the basis of their writing style and is built on existing studies that show how user-generated content can be "automatically collected for different personality types and how machine learning techniques can be trained on this data to build classification systems that can automatically identify the personality type of social media users with a fairly good degree of accuracy" (Zarouali et al. 2020, 6).[3]

After having digitally identified the extraverts and introverts among their participants, participants were subsequently shown (fake) political ads, either congruent or incongruent. In the first study, they showed (Dutch) participants a political ad promoting a progressive, left-wing green party in the Netherlands and gave extraverts a version with "stronger, confident and dominant language consisting of assertions and commands". In the second study, participants were shown a message advertising a right-wing liberal party. They found that, in general, participants were "more persuaded when they receive a political ad containing a text that is tailored or framed based on data relating to their psychological make-up" and that they "reacted differently to affect-based political ads based on their psychometric profile". The first study revealed that "the extravert-framed political ad was significantly more effective in increasing their attitude toward the political party than the incongruent ads".

In the second study, introverts for instance responds more strongly to fear-based political ads ("the safety of our country is at stake"), whereas extraverts respond "better" to enthusiasm-based ads ("our country is safer than ever"). The authors conclude, with some important caveats, that there's causal evidence for the idea that people are more easily persuadable – indeed manipulable – when they are confronted with ads that fit their personalities. In light of the fact that large, global companies have, since 2016, "added personality-congruent targeting to their toolboxes and offer these services to any political actor willing to pay", and that there are all kinds of possibilities to make such profiling even more accurate, such that Facebook who could "start to offer the information on the basis of their WhatsApp data" – this conclusion is cause for concern.

The reason for focusing on political microtargeting is that this is considered by many to be an especially worrisome practice if effective, but also because even if such techniques turn out to make *little to no difference* to people's voting behaviour, there is reason to be worried about such ways of treating citizens all the same, thus making a non-consequentialist analysis worth considering. A (perhaps unintended) reading of the Zarouali et al.'s study is that it's epistemically and morally problematic to infer people's personalities on the basis of their digital traces and to use those (probably inaccurate) profiles to attempt to steer them politically, even if it doesn't work.

Central to our proposed moral lens to examine digital malpractices is that the moral wrong of such practices is due to the fact that they involve using people – internet users – as mere means. The language here is typically associated with Kantianism, but our proposal can be taken on board without heavy Kantian luggage.[4] As Miranda Fricker puts it:

> Since this captures such a common ethical idea about what it is to treat fellow human beings as full human beings, I think we can lift this bit of Kant's terminology without dragging the rest of his considerable philosophical apparatus along with it.
>
> (2007, 134)

A good starting point to think why microtargeting is morally problematic is to compare it to why making a dishonest promise would be problematic. In Kant's classic case, I want to get your money (for new shoes, say) and make a false promise to you, namely that I will pay it back soon while I don't really intend to do so. What's wrong with this? As Kant suggested: I want "to make use of another human being *merely as a means*" in the sense that: "he whom I want to use for my purposes by such a promise cannot possibly agree to my way of behaving toward him, and so himself contain the end of this action" (Kant, *Groundwork*, 4:429–3; Kant 1785, 4:429–3). According to Pauline Kleingeld (2020) we should interpret this passage as saying that I use you as a mere means when I don't really care about your consent to be used.

Despite many worries about moral analyses that hinge on the classic notion of consent – which we are sympathetic to and which we will discuss later on – the reason for exploring the moral wrongs involved in microtargeting though a "consent lens" is that caring about consent, at its core, is about respecting another's capacity to make their own choices. The thought can, and has been, expressed in scholarly Kantian language, but the underlying thought is really quite simple: before we do anything with, or to, another person, we need to carefully check with them how they feel about it and adapt or cease our course of action if they are, or we have reason to think they might be, nonplussed. Even if caring about consent is no sufficient threshold to make some action or way of treating others acceptable, which it isn't, it is a minimal threshold that ought to be met.[5]

In the following, we will assume that someone acts wrongly if they use another as a mere means. Our aim is to show, drawing on Kleingeld, that not caring about someone's consent is a way of treating them as a mere means. We argue that microtargeting of the sort discussed earlier constitutes a form of not caring about consent and thus microtargeting constitutes a moral wrong. We might not want to draw the strong conclusion, however, that *all* possible forms microtargeting are morally wrong – and will return to this point at the end of the chapter.

3 Using internet users

Before asking when an agent uses another person *as a mere means*, let's see when someone uses another in the first place. This appears fairly straightforward in, for example, vaccine trials where people are used quite literally. But how should we see this in our cases? How does making a false promise to, or microtargeting, someone amount to *using* that person?

First of all, someone has to be to some extent aware that one is using another as a means. For instance, if you decide to sit down on what you take to be a bean bag, not being aware that it is actually a person, then this does not count as using that person as a means. Hence, it's not about whether the agent *in fact* does something to someone, but whether she knows she does so and assigns an instrumental role to other people.

This criterion raises some tricky questions about culpable ignorance that may be especially relevant in discussions about online manipulation where agents might not know that they are using people but *should* have known. Imagine, for instance, that some political party asks an advertiser to attract more voters and offers the latter a bonus for all extra voters. Unbeknownst to the party, the advertiser uses microtargeting. If the party is indifferent or self-deceived about how they get the extra votes – if they have the wrong *mindset*, so to speak – then their ignorance should not, we think, let them off the hook.[6] They can still be using voters as a mere means.

Next, one might think that someone uses another only if they assign an instrumental role to the other person in the service of the agent's own ends and that the other person's ends don't count. For example, if your doctor gives you a vaccine against COVID-19, she acts in the interest of *your* ends (namely, protection against the virus). In such a case, one may wonder: does it make sense to say that you are used at all?

Here, our position is that people (as well as corporations or political parties) may act for various ends at once, and the ends may be a mix of self-interested and other-directed goals. Consider Google. They help you with finding sites on the internet and show the results that are most relevant to you. Clearly, though, this is not their only goal, as they are not a charity organization. They want to also sell personalized ads and protect their monopoly position. Similarly, Facebook says that they are concerned with connecting people (cf. Mark Zuckerberg in 2012: "Helping a billion people connect is amazing, humbling and by far the thing I am most proud of in my life") and world peace (Jones 2009), but of course they are also interested in making a profit. In all these cases, if the corporations are at least acting partly concerned about their own ends, they are using others.[7]

There is another concern. Suppose I make a false promise to get your money, but not to spend it on something for myself, but on something for you. For example, I invest it and then donate everything to a charity organization you care about and want to support as much as possible. For the sake of the argument, let's assume that I really act on your behalf and that I have no further concerns. In such a case, it seems I am still using you (moreover, I am using you as a mere means).

Such paternalist cases suggest that the lying or manipulating agent's specific ends don't really matter. In the following, we will therefore assume that an agent uses another person if they are aware of using that person (or are culpably ignorant of this), regardless of the particular ends they are trying the achieve.

4 Kleingeld on using others as a mere means

The subsequent question is: when does an agent not only use another person but also do this in a morally problematic way? There are countless situations in which we use others where this isn't necessarily problematic, indeed where this is often an enjoyable aspect of social life. One might use another in order to make a cake, paint the walls, have sex, build a treehouse, write a paper, test vaccines, get elected, sell products, and so on. There's nothing wrong with this in principle. These activities become troublesome when someone uses another only, or merely, as a means. The question is how to specify the "merely" clause.

A first, simple proposal would be to work it out along the lines of lack of transparency: someone uses another person as a mere means if and only if that person keeps the other *in the dark* about what they are doing with, or to, the other person. This account seems to apply to our political microtargeting example: the microtargeter does not inform people that they get to see ads that specifically relate to their personality types. However, this account is too narrow. For instance, if people are used against their will in vaccine trials, they seem to be used as a mere means, even when they know full well what is going on.

Another simple account would be that an agent uses another person as a mere means if and only if the other person *protests* to being used by them, and yet the person goes ahead and uses them anyway. But again, this account is not broad enough. If someone makes a false promise to you, then you won't protest. After all, you didn't know that the other person wanted to deceive you and did not intend to keep their promise. Still, you were used as a mere means all the same.

Clearly, we need a more sophisticated account that is neither too broad nor too narrow. The notion of consent is at the heart of many recent accounts. One popular proposal is that an agent uses another person as a mere means if and only if the other person does not give consent to being used (cf. Kleingeld 2020, 392–93).[8] If, say, someone hasn't given consent to participate in vaccine trails, they are used as a mere means on this view.

There are many worries about the notion of consent, especially in the digital domain (e.g., Richards and Hartzog 2019). As is well known, many people as a matter of fact consent to being tracked, targeted, having their privacy violated, and so on, not because they enjoy being tracked and targeted but rather because they want to navigate the web and annoying pop-ups stand in the way and reading privacy policies takes way too long (McDonald and Cranor 2008). People "often do not read disclosed information, do not understand it when they read it, and do not use it even if they understand it" (Ben-Shahar and Schneider 2011, 665). Internet users are often pressured or coerced into giving consent, hence online "consent" should normally be placed in quotes (Williams 2018). The consent people give online is often spurious and not indicative of what they actually want to agree to. The phenomenon of spurious consent, though, should not be a reason to throw away the baby with the bathwater by saying that consent itself is not valuable. Indeed, many scholars discuss the weaknesses of consent in order to formulate ways in which proper consent may be guaranteed.

Aside from the problem of spurious consent, there's another problem to specifying the "merely" clause in terms of consent, though. Imagine a scenario in which some evil tech corporation, call it Oxbridge Analytica, asks their users to consent to using microtargeting strategies that involve constructing profiles of people's moral principles (cf. Griffin 2021), in order to get a fascist political leader elected.[9] Now suppose that, after careful deliberation, one particular user – call him Frank – genuinely consents because he truly believes that any such means are okay, so long as they serve the right cause and he thinks the cause is a good one. We might think Frank should not have consented, but even if an individual's consent seems unwise or even immoral, their person's consent may still be genuine, that is, not spurious.

It would follow from a consent-based account that, because genuine consent was given (it was informed, voluntary, etc., however unwise), Frank wasn't used as a mere means. But that does not seem plausible. Imagine that Oxbridge Analytica does not *care* in any way whether anyone consented to their way of using them. They don't care whether Frank consented and others didn't. We can even imagine that Oxbridge Analytica actually assumes that no one consents, and were they to hear about Frank's consent, the CEO would laugh with contempt.[10] This brings out the fact that everyone, including Frank, was used as a mere means, and that the uncaring attitude is what plays the key role in determining this.

An account in which evil or indifferent mindsets takes central stage is the recent approach by Kleingeld. What we need, Kleingeld suggests, is not something the manipulee does (e.g., consent or protest) – so-called victim-centered account of consent – but rather something the manipulator does, or so-called agent-based account of consent (Kleingeld 2020, 404). It is true that Frank gave his genuine consent, but, according to the alternative approach, we should rather look at whether Oxbridge Analytica gave any weight to it.

Kleingeld's agent-based consent account is as follows: an agent uses another person as a mere means if the manipulating agent does not make her use of the other person conditional on their genuine consent.[11] In other words, in order to use someone without using them merely as a means involves being open and ready to change or discontinue one's course of action depending on how the other person feels about it. Making your conduct conditional in this way, Kleingeld explains, means that one regards the consent of the people you use as a "limiting condition" (Kleingeld 2020, 400–401). You use them *only if* they agree. More precisely, their consent "is to function as a general rider on one's practical reasoning" (ibid.). Oxbridge Analytica might, for example, have reasoned as follows: "If I get people's permission, I'll go ahead. But, if they don't consent, I will either try to motivate them in another way to vote for the fascist political leader, or else give up the campaign". Given that Oxbridge Analytica does not reason like this – they do not even pay attention to what people think of it – they do not make their conduct conditional on their consent. They do not have what we referred to as having the right *mindset* vis-à-vis (online) people.

Kleingeld introduces a further requirement, namely that the agent has to restrict her use of others "as a matter of moral principle" (Kleingeld 2020, 405–6). The reason for this addition is the possibility that an agent might care about consent but for the wrong reasons, for instance because they just feel like it or because they just want to show how powerful they are, say, or because caring about consent gives the twenty-first-century tech corporation a competitive edge.

In real life, you might well restrict your use of others partly because to some extent you care about what they think of it but also because you want to feel good about yourself. Or you might also do it to avoid legal sanctions or from a partial concern for your reputation. It's likely that many corporations and political parties have such concerns, and it is not strange to think that it should be permitted for them to act in some way.[12] Caring about consent for moral reasons, even when that's not your only reason, should do. Indeed, as we'll see, this is already a highly demanding moral norm for big tech corporations and microtargeting political parties to meet.

5 Object of consent

Consent to what exactly? Kleingeld suggests that one should make one's conduct conditional on people's genuine consent to be "used, in a particular manner, as a means to the agent's end" (Kleingeld 2020, 398).

We already suggested (in Section 3) that the agent's ends might not be relevant. There can be cases where you try to help me with achieving my own ends, but where I still wouldn't agree with the manner in which you do this. Similarly, there can be cases where I can endorse your ends but not your means to them. For example, I might well endorse your aim to get more money or votes but not that you steal these from me. Therefore, we assume that you should make your conduct conditional on the person's consent to be used by you (regardless of the exact end).

Even so, if you should be interested in my consent to be used by you, then what should this consent be about exactly? After all, one's use of others can be described on different levels. In Kant's case, I am using your trust (to borrow your money) but also your ignorance about my plans (to deceive you) as well as your money (to spend it myself). In principle, we think that all these descriptions of how you are using me can be relevant. Especially in the digital domain, it is important to consider the various possible interpretations of the object of consent.[13]

First, consider the thought that in the case of the lying promise to return the money, the consent should concern the deception. In a way, that's a strange object of consent. Why would you consent to being deceived? Moreover, as soon as you know about my plans, my deceit is likely to fail. You are not going to lend me – or indeed give away – the money if you know that I am not going to pay it back. (Even in cases where you see why I would want the money (and want to support me), you would want me to be honest with you.)

Rather, it's more plausible that the agent should make their conduct conditional on the other person's genuine consent to *lending their money*. False promising, after all, is not so different from stealing, where one also uses another's money without being concerned about their permission for doing this. In the case of stealing, one simply doesn't ask the other person anything. In the case of false promising, one does ask the other if they can use the other's money, but then the other person is tricked into consenting (by being lied to about returning the money), and in this way, as we will explain later, the agent is still not *genuinely interested* in the other's permission.

Next, consider the microtargeting case. The political party is using your time and attention (to get advertised), but also your personality and your ignorance about microtargeting (to manipulate you), and your vote (to gain power). Again, if you were to give consent in order not to be used as a mere means, what should your consent be about exactly?

First, the party should be interested in your consent to using manipulative tactics, that is, whether you give them a permission to target your character and subsequently present personalized ads. For example, they could flag a clear warning before showing the ad (e.g., "the following ad is selected on the basis of your online interests and/or targets your personality") similar to warnings to alcohol ads, for example. However, as soon as you see or hear the warning, the microtargeting is likely to fail. If you know that they are going to use certain language only to target your introvert or extravert character (and to induce misleading beliefs about safety), it might be that you won't be triggered by the ad or much less so. And, again, it sounds strange to say that internet users should be asked whether they want to be manipulated.

More plausibly, then, a political party should be interested in whether an internet user-cum-citizen is okay with giving their vote to them. Importantly, what matters here is not what potential voters think of it and what they would consent to (as the victim-centred approach has it). Some people might think that microtargeting is fine, and some might agree with giving their vote to the given party (recall Frank). But, following the agent-focused approach, that's all less relevant. Instead, what matters is whether the party *genuinely cares about* their answer to the question "can we have your vote?" As we'll see, genuine caring is hard work.[14]

Similarly, in the false promising and stealing examples, what ultimately matters is not whether or not you consent to giving your money away. Instead, what matters is whether the relevant agent is genuinely interested in their answer (as opposed to, for example, circumventing it by lying to you or by secretly taking it). In the following, we will work this out in some more detail.

6 Caring adequately about consent

Parties or corporations might think that microtargeting people is fine so long as the latter have clicked on some cookie consent button, or they show you personalized ads only if you have accepted the terms and conditions. But virtually all participants in this debate consider such "cookie consent" inadequate. Passively waiting for a mere click – knowing that typically people don't read any of the terms and conditions – does not suffice. As Kleingeld puts it: "apparent consent could in fact be spurious, for example if it is the result of deceit, misinformation, misunderstanding or manipulation" (Kleingeld 2020, 404; cf. O'Neill 1989, 106–12).

At this point, though, a circularity worry looms. We want to say: microtargeting (a common example of online manipulation) is wrong if the party uses voters as a mere means. But now, given that consent is typically spurious if it is the result of manipulation, it seems we also have to say: the party uses voters as a mere means if it manipulates them. So it seems the "merely" clause is being explained in terms of microtargeting (or manipulation) and vice versa. To break this circle, we want to offer an analysis of *why forms of microtargeting are incompatible with adequate care*. If we have that, we can say: microtargeting is wrong when – and because – the party doesn't adequately care about the voter's consent. The moral weight, so to speak, falls on the not caring part rather than the manipulation part. And so the next step is to explain the caring part.

Kleingeld, as we saw, explains adequate care in terms of practical reasoning (motivated in the right way). But one might wonder whether adequate care is a mere mental affair.[15] Arguably, more is required than only running through certain pieces of reasoning. Suppose Oxbridge Analytica reasons as follows "if I get their permission, I'll go ahead and use their personal implicit bias data, but not if they object", but then forgets to ask them anything, fails to properly listen to their answer, doesn't hire any personnel to make sure proper permissions are asked for and received. Hence, adequate care also requires corresponding actions on behalf of the agent.[16] What are those?

Inadequately caring (or not caring) about someone's consent can take different forms, ranging from mere omissions (indifference) to active strategies to circumvent another person's opinion and possible dissent:

a) not paying attention and even noticing this person;
b) ignoring and not listening to her;
c) not informing her and keeping her in the dark;
d) not checking if she has understood it properly;
e) not asking her for her permission to go ahead;
f) obscuring her opportunity to protest by making this opportunity less salient or more difficult;
g) pressuring her, obscuring her opportunity not to be used;
h) forcing her to consent to be used;
i) undermining her actual protest by silencing her or deflating her credibility;
j) tricking her into consenting; and
k) tricking her into being used in the given way.[17]

It's instructive to briefly contrast this account to Christine Korsgaard's. According to Korsgaard (1996: 139), an agent uses another person as a mere means if and only if she *prevents* the latter from choosing whether or not to contribute to her end. This account has difficulty with cases where a third party – not the agent who subsequently uses the victim – prevents the victim from choosing whether or not to contribute (cf. Kerstein 2013, 74). Our account resolves such problem cases. There are many distinct ways in which one can fail to care enough about the consent of the people one uses. Preventing someone from choosing whether or not to contribute to what you want to achieve is one way. But not asking for any permission, for example, is another way.

We might not want to make the strong claim that an agent cares adequately about another person's consent to be used only if each and every one of these requirements are fulfilled. Consider one of Kerstein's cases where some hikers are lost in the mountains and are following another person to find the way back (Kerstein 2013, 63). The hikers don't ask the person for a permission to follow her. Even so, we might not want to say that they don't care about her permission (or use her as a mere means for that matter). The same goes for throwing surprise parties (cf. Kerstein 2013, 75) and other activities where not informing another person of your (true) plans – and thus, inevitably, not asking them for their consent – is part of a perfectly respectable (though perhaps not particularly enjoyable, depending on whether one enjoys surprises) plan.

In the light of such cases, we are open to a contextual approach according to which the requirements we listed do not kick in when, in some context, it is sufficiently likely or obvious that the person used would consent. It is plausible to assume for instance that the hiker would agree to being followed by the two lost hikers. In that particular context, then, the aforementioned requirements need not kick in. But if we tweak the example only so slightly we can already see just how fickle this assumption about its being "clear" whether someone would consent is. For if our hiker is a woman being followed by two men, she might be well adviced to make other assumptions, and the men should not just trail on but explicitly ask her if she's OK with being followed. (The more natural question is of course simply to *ask her the way*. Only a hiker philosopher of a particular bent will ask her if she would *consent to being followed*.)[18]

So, despite the fact that contextualism about adequately caring about consent is important to emphasize, so as to avoid having to go through (a)–(k) every time one simply wants to have coffee with a friend, it is also important to realize that in many contexts, especially involving online consent, such assumptions cannot in fact be made. Indeed, making such assumptions about people's likely consent is precisely part of the problem. Genuine care also involves not making such hasty assumptions and involves not just the absence of malice and viciousness but most notably also the absence of negligence.[19] Hence, we believe in most contexts, meeting all requirements does make sense in spite of – or better, because of – setting an ambitious threshold.

7 Microtargeting and tricking

Next, we will argue that both false promising and microtargeting fail to satisfy the requirements for adequate care. If I make a promise to you, I do ask for your permission to use your money (so I satisfy aforementioned (e)), but I trick you into consenting to this by lying to you that I will return the money soon (and so fail to satisfy (j)). What about the microtargeting variety?

If a political party microtargets potential voters for instance by presenting messages that are more likely to appeal to their introvert or extravert characters, they do ask for permission to get their vote (so they satisfy (e)). Also, they do not in any obvious way force people to give their vote or obscure the opportunity to vote for alternative parties, so they may well even satisfy (g), though legitimate questions can be asked about the "soft" pressure that is exerted by being confronted with, say, political advertisements that are intended to instill fear, and so there is certainly room to argue that condition (g) isn't satisfied. Most strikingly, however, is the fact that political microtargeting strategies involve trying to *trick* people into giving away their vote and so fail to satisfy (k).

Let's see more closely why microtargeting amounts to some sort of tricking, and why the latter is incompatible with caring about the person's consent. Generally, tricking someone involves misleading another person (cf. Noggle 2018). For example, by making a false promise to you I trick you (into giving away your money) in the sense that I make you think, falsely, that I am going to pay back the money soon. In the microtargeting case, the party spreads disinformation and thus tricks potential voters in that they induce false beliefs about (say) safety in them, for it cannot be the case that, say, the country is both safer than ever *and* at risk.

One may wonder about the difference with traditional advertisement. Indeed, traditional ads (think of Axe Body Spray) may also induce misleading beliefs (i.e., that you will become irresistible if you use the spray) and subsequent desires to buy the given products. Even so, it is important to determine this on a case-by-case basis. Take, for example, billboards in public spaces. They are more intrusive than personalized ads, one might suggest, as we can hardly ignore them or object to seeing them. Even so, they need not trick us, that is, insofar as they need not spread disinformation or induce desires in us based on outright lies or misinformation. Hence, the problems should be kept distinct. Billboards are problematic to the extent that they force you to see the ad (though not necessarily to give your vote or money), and so the main problem is located in coercive territories. In contrast, certain ads are problematic to the extent they trick you into giving your vote or money.

Compared to outright lying, an important question is whether political microtargeting is also wrong if the message, tailored to fit people's personality, does not include obvious falsehoods. We want to argue it does, though the trickery takes a trickier form. One of the worrying aspects of political microtargeting is that different people get to see different messages. As Zuiderveen Borgesius et al. point out, "A party may highlight a different issue for each voter, so each voter sees a *different* one-issue party" which "could lead to a biased perception regarding the priorities of that party". The problem is not (necessarily) that the political party *lies* to voters, but rather the problem – call it the chameleon effect – is that the very same party says (or promises) different things to different people, without strictly speaking lying to them. Zuiderveen Borgesius et al. (2018, 88) offer the following example of this chameleon effect:

> [A] politician has a [digital] profile of Alice. The politician has information that suggests that Alice dislikes immigrants. The politician shows Alice personalised ads. Those ads say that the politician plans to curtail immigration. The politician has a profile of Bob that suggests that Bob has more progressive views. The ad targeted at Bob says the politician will fight the discrimination of immigrants in the job market. The ad does not mention the plan to limit immigration. Ads targeted at jobless people say that the politician will increase the amount of money people on welfare receive every month. To people whose profile suggests that they mainly care about paying less tax, the politician targets ads that say the politician will limit the maximum welfare period to one year.

A political party can be taken to trick voters by suggesting that, say, once coalitions get formed, they are going to give priority to curtailing immigration. But it might well be the case that the party drops this promise right away. Given that the technique of political microtargeting enables political parties to adapt themselves to what the voters want, it is no longer clear to voters what the party's agenda is and which issues they are going to stick to and prioritize. It is "a nearly perfect perversion of the political process".[20] The party does not lie in such a case, perhaps (though this may depend, to some extent, on one's definition of what constitutes a promise and how cheap they can come, or whether cheap promises aren't promises at all) but is disingenuous all the same, and disingenuity can be an effective form of trickery, as anyone who failed to tell the whole relevant truth, for strategic purposes, knows full well.

This also helps bring out why tricking people in this way is incompatible with caring about their consent (to giving their vote to the party) or, more generally, why our condition (k) is a key moral requirement. One might legitimately wonder, after all, whether *caring* about or being interested in getting someone's consent might come cheap. Obviously, Facebook cares about your consent because clearly they want to get it.[21] At this point, to explain the relevant (lack) of care in contexts involving political microtargeting, it is necessary to disambiguate different conceptions of caring or being interested in (which we are here using interchangeably). On a thin conception, a political party caring about people' consent might just mean whether the political party *wants* it or not; whether it is in some way profitable or instrumental for them. Call this the "consumer" conception of caring about consent for political microtargeting. Notice that even on a consumer conception, one might meet the requirements on the list in the previous section. The only thing is that one's motivation for meeting them is wholly instrumental.

Alternatively, there is a *political* conception of when one genuinely or adequately cares about getting someone's voter. This is the conception we have in mind. On this conception, in order for a political party to care, in the right way, about getting a person's consent to giving their vote involves wanting to get their vote *as a political choice* (and not, say, a commercial choice or a choice out of habit or indeed a choice out of trickery). In other words, for a political party to be genuinely interested in a person's vote is to be interested in them in their capacity as democratic agents, as citizen with the ability to make up their own minds. This is not in principle incompatible with politically microtargeting them. In fact, doing so and in so doing trying to communicate with them as democratic citizens with specific values and concerns can be an expression of caring about their vote in this way. It is not, however, the typical case.

On the political conception, if you are *genuinely* interested in people's consent to giving their vote, then you would, for example, advertise information about the standpoints, achievements, and priorities of your party.

You would say that safety is your top priority. But you would not spread conflicting, and so misleading, messages about safety, be disingenuous about your priorities, or promise things you are hardly committed to. Doing so obstructs people's ability to decide whether they are okay with voting for you. In doing so, you are not really interested in what voters think and don't respect their ability to think and decide for themselves.[22] Rather, you are merely interested in getting their vote. You are using them as a mere means.

We should like to point out that our account differs from the one by Nys and Engelen (in this volume). Nys and Engelen claim that an agent uses another person in a morally problematic way if the agent doesn't care enough about the other's *personal autonomy*. For example, companies fail to care in this latter sense if they induce desires in consumers to buy stuff that they don't really want - or even if they want it, the company couldn't care less. Caring about what people really want *sounds* very similar to caring about their consent to be used, yet the emphasis is distinct. Nys and Engelen's-type of caring is about finding out about what consumers really want, not just superficially might get hooked to, and taking that into account when you develop and offer them your goods. Our view is similar in that, in this context,[23] companies should care about whether consumers are genuinely okay with giving their money to them. Our specific contribution, here, is our proposal of what such care might amount to: our list (a)–(k).

8 Implications and responses to objections

Finally, we want to mention some clarifications and implications and consider a worry about our project as a whole.

First, a point of clarification. Our list of what adequate care for genuine consent comes down to should make clear that caring about the "Kantian" type of consent that is necessary in order not to use people as a mere means differs significantly from the type "online" consent that corporations "care" about. To care about people's consent in the relevant sense is, at bottom, to care about and respect someone's ability to think and decide for herself, and most forms of online consent hardly meet this norm. Many corporations fail to adequately inform persons by using hopelessly vague language in their privacy policies. Sax and Ausloos (2021) discuss the privacy policy of Epic Games (the developer of the online game Fortnite) as an example, mentioning that its "privacy policy uses vague language (using examples rather than clearly delineated definitions, often using the word 'generally' when explaining their data practices, leaving large grey zones) and often uses hypotheticals (such as 'we *may* receive')" and by mentioning overly general purposes such as "developing and improving services". Even worse are instances in which corporations use dark patterns[24] to get people to accept the terms and conditions or privacy policies. Gray et al. (2018) for instance give the example of a website where users, when registering for an account, are "given an option to accept the terms and conditions above a long list of small text" and "Hidden within this text is a small checkbox to opt out of the bank selling the user's information". Some websites thus deliberately try to hide or disguise relevant information. Many other examples of dark patterns are, or can be, used for undermining consent as well, such as "nagging" which involves distracting users by pop-ups or sudden audio notices, "preselection" or opt-in systems where an option (e.g., consent) is selected by default,[25] graying out one of the options that is actually still clickable "giving the user the false impression that the option is disabled", hiding tracking preferences in obscure locations, "sneaking" which is defined as "an attempt to hide, disguise, or delay the divulging of information that has relevance to the user" for instance by requiring a user "to consent to a privacy statement before they can unsubscribe from an email newsletter," and using confusing wording such as (multiple) double negatives.

All of these examples of a type of dark or 'evil design" which, as the authors point out, are often the result of "explicit, purposeful design intentions" illustrate a rather extreme form of not caring about people's consent. It's important to notice, however, in light of the concept of culpable ignorance briefly discussed earlier, that so-called anti-patterns which are "simply a result of poor design" can equally betray a lack of care. In the user design community the aphorism of Hanlon's Razor is often shared, which is to "never attribute to malice that which is adequately explained by stupidity" (Gray et al. 2018).[26] This aphorism is relevant in moral discussions of online manipulation as well, given that there's a tendency to focus on malicious corporations and evil intentions. Looking at what adequate care really comes down to, though, we should be at least equally concerned about stupidity, ignorance, naiveté, and negligence.

The implication of this chapter (and to simultaneously respond to a possible worry) is in any case certainly not a defence of the notice-and-consent paradigm, in fact we agree with critics that the ways of securing informed consent is currently "fundamentally inadequate" (e.g., Barocas et al. 2014). As Barocas and Nissenbaum also mention though, the point of the critique is not that consent can "play no possible role in relation to behavioral targeting" but rather that "the surrounding context as currently holds, unlike in the medical arena, does not properly support a meaningful role for it".

Notice, also, that if what matters is that corporations and political parties genuinely care about consent in the ways suggested previously, one can go in either of two ways. The first is simply that stronger measures must be taken such that agents start caring more about consent, meaning at the very least that the people they track and target should be given the real opportunity to give or withhold genuine consent. Such measures may also include making changes at the design level such as disrupting user experience by introducing friction precisely *with an eye on* enabling people to engage in critical reflection rather than the opposite (Terpstra et al. 2019). Call this approach the *optimistic* approach.

The second route is the *pessimistic* approach: if genuine care about getting consent from persons in their capacity as democratic agents is what should be protected, then perhaps we should plead for the implementation of a ban on microtargeting practices (cf. Zarouali et al. 2020). This is a more radical route, obviously, but one that the basic argument in this chapter provides possible motivation for. Our general argument about what adequately caring about consent involves does not rule out microtargeting as an acceptable way of using people; it is not impossible, in principle, to engage in microtargeting techniques whilst genuinely caring about (and getting) people's consent. In Section 2, we mentioned some possible opportunities of microtargeting. For example, it may not only be used to convince people to vote on some specific party but also to activate them to vote regardless of party or to reach them in a language that appeals to them. Even in such cases where the microtargeting benefits the manipulee rather than the manipulator, we would say that many, if not all, of our requirements apply.

However, in practice, it is highly unlikely that microtargeting will be implemented in such a way in the near or far future. Given the unhappy marriage between the strategy of a) sending targeted political advertisements to people based on, say, their personality on the one hand and b) actually caring about consent – that is, about people's ability to make their own choices – advocating a ban does not seem like such an exaggerated measure. Truly caring about consent may well require it.

Before we end the chapter, we want to discuss a final objection that takes issue with the project as a whole. We already mentioned existing criticism of consent paradigms. Our reply at bottom was that such criticism is correct but that we should not be throwing away the baby with the bathwater. There is another, deeper, objection though, which concedes that consent as such should perhaps not be discarded altogether but claims that to focus on consent is really to misanalyze the problem – why microtargeting is morally problematic. The thought is this: certain ways of treating people are wrong *as such*, that is, regardless of people's consent. Thus, when considering what lies at the core of what is morally problematic about (political) microtargeting, saying that it's the fact that certain agents fail to (sufficiently) care about consent is just a very roundabout and indirect way of analyzing the issue. It's just wrong, and talking about consent is a distraction.

This deeper worry ties in with concerns about the broader consent paradigm more generally, which is that it is sometimes problematic or plain wrong just to *ask* for consent. Asking, say, the first woman who comes into a meeting *whether she consents* to making you some coffee constitutes a wrong.[27] The fact that you genuinely cared to know her answer and thus to ask her explicitly, and in clear language, does nothing to change that and in fact only makes it worse. The same seems to go for asking someone for consent to sell their organs, adopt their child, or an inappropriate request for sex. In general, asking someone whether they consent to something – as well as in effect forcing them *reflect* on certain issues in the first place, or forcing them to say no about something that should never have been on the table – is by no means always the mark of morally acceptable behaviour and can be precisely the opposite. This also has application in the online world: it may well be undesirable in and of itself to ask people whether they consent to being tracked across the internet, have cookies stories on their computers, have psychometric profiles made, and so on. Asking someone whether they're okay with being treated immorally doesn't make it okay, even if they end up consenting in non-spurious ways.

So what are we to say in response to what we might call the "deeper worry" to our view? A first thing to say is that we agree, as argued in the section on the object of consent, that it's problematic to ask people whether they're okay with being microtargeted just as it's problematic to ask people if they would consent to being lied to. We can now see *why* this is problematic: not just for conceptual and pragmatic reasons (because it sounds strange and no one would end up consenting) but also for moral reasons.[28] Also in the online context, the very asking of questions can add to the already high cognitive load of internet users, as many have pointed out before us.

This may leave much of the deeper worry intact, though. The deeper worry may also boil down to a methodological disagreement about opting either for a constructivist or broadly realist paradigm. To clarify: the consent paradigm is not only a broadly liberal paradigm, anchored solidly in (respect for) individual's choices, values, and preferences but also a constructivist one in the sense that our moral analysis of why microtargeting is wrong is arrived at through considering the microtargeter's mindset (i.e., not caring about consent). On a moral realist paradigm, certain conduct may be immoral regardless of the agents' mindset. The thought that, say, microtargeting is "just wrong", regardless of consent asked or gotten, is likely to emanate from this broader realist paradigm.[29]

One option for constructivists would be to admit that adequate care about consent does not always involve asking something (i.e., our requirement (e)). In Section 6, we suggested a contextual approach according to which our requirements kick in only when, in some context, it is sufficiently unlikely or unclear whether the person used would consent. At this point, we may want to expand this and say that requirement (e) ("Someone cares adequately about another person's consent only if they ask them for her permission to go ahead") does not kick in when, in some context, it is sufficiently likely that the person used would dissent. This strategy would keep the constructivism intact – it still depends on proper mindsets – though much greater weight is now placed on the (un)likeliness of dissent and trusting the agent's ability to make a proper assessment of this. We've already mentioned the risks of making assumptions about (un)likely dissent in online contexts.

The advantage of taking a realist perspective instead is that one doesn't have to explain the wrongness of microtargeting in such a roundabout way via the importance of (caring about) consent. One doesn't have to say that microtargeting is bad because consent is good and proper consent wasn't asked for. It's just not how you treat people – period. But the realist approach comes with a tricky methodological challenge of having to explain *why* microtargeting is problematic as such or constitutes immoral treatment, and to do so without in the end – not even through some backdoor – relying on the wrong of failing to respect what people would (not) agree to, as that would lead to a collapse back into the individualist, constructivist paradigm.[30]

We won't be able to settle this tricky methodological dispute here, if only because the two authors of this chapter are not undivided about how to respond to it. What we instead hope to have achieved is to bring out both the methodological advantages and disadvantages of the liberal-constructivist background that the caring-about-consent approach hinges on, and to clarify that if one is persuaded by the deeper worry, then one does have a much more direct way of saying that political microtargeting is wrong, but one has to tackle some tricky questions about justification and specification of relevant (realist) norms that do not get their justification from what, broadly speaking, goes on in microtargeting minds.

9 Conclusion

In this chapter, we have explored the view that someone uses another person in a morally problematic way when they do not genuinely or sufficiently care enough about their consent to be used in that way. In particular, we offered an analysis of the ambitious requirements of what adequate care might amount to. We argued that political parties should not, for non-consequentialist reasons, microtarget potential voters – especially if this involves tricking them – because such tricks are incompatible with genuinely caring about whether they are okay with giving their vote. We ended the chapter by responding to influential worries about consent and clarified the underlying methodological landscape.

Notes

1. We would like to thank the members of the online workshop series of this volume and members of iHub at Radboud University for their helpful comments and thoughts on the topic.
2. We focus on Zarouali et al. (2020) because it is a very recent and impactful study that explicitly addresses the earlier-mentioned question of effectiveness. Also, we chose to focus on political microtargeting in this chapter rather than microtargeting more generally, because of the extra worries it gives rise to, though much of this chapter can be read as dealing with microtargeting in general.
3. Methodologically, there's much to reflect on here, as important questions emerge about accuracy and limits of the information that algorithms can acquire about individuals' (true) personality on the basis of digital traces. Can true extraversion be inferred from people's online writing style? Who knows best whether a person is an extravert: the person herself or big data crunching algorithms?

What if the two come to different verdicts? As the authors rightly mention, they have to "be vigilant in claiming that text-data offers an undisputed window into someone true personality" (Zarouali et al. 2020: 20).

4. We do make some comments on what account of Kant's formula of humanity we consider plausible and systematically defensible. Even so, our primary aim is not to defend it against alternatives (or as an interpretation of Kant) but to use it and offer a plausible and interesting normative analysis of the case just discussed.
5. See, for example, J. Brison (2021) who forcefully shows that "consent is a very low bar". This goes for sex; it also goes for online manipulation. In this chapter we are limited to locating and securing the low bar first, even though subsequent bars are necessary.
6. We use the term "mindset" rather than particular mental states so that the account can be more naturally applied to group agents such as corporations, political parties, or governments.
7. Doctors, too, don't help only because they care about others but also partially because of other reasons (e.g., because they like their job or want to feel satisfied). If that's so, they do assign an instrumental role to the people they help (if not explicitly, implicitly), which, we take it, suffices for using others. Even so, we wouldn't say that patients are used as a mere means, since doctors generally deeply care about the consent of their patients.
8. To account for cases where people have no opportunity to consent (e.g. because they are unconscious and need immediate care), the proposal may also be read as: it's not reasonable for the agent to think that the person used would consent (cf. Kerstein 2013: 97, Kleingeld 2020: 292–3).
9. This example is based on Kleingeld's genocidal dictator case (Kleingeld 2020, 393), where an act-utilitarian consents to dangerous medical experiments.
10. Again adapted from Kleingeld.
11. We have simplified the account as stated in Kleingeld (2020, 398) in some ways.
12. Though, as with Kant's shopkeeper, one may still say that their conduct, albeit permitted, lacks full moral worth.
13. In the following, we distinguish between a "consumer" and "political" conception of what it might mean for a political party to be interested in getting someone's consent.
14. The upshot, which is especially relevant for the application to microtargeting, is that it makes the account more robust against "bad consent", as it does not make the wrongness microtargeting dependent on whether individuals are okay with it or not.
15. Alternatively, one might inflate the notion of the "mental" such that it necessarily includes for example, reliable dispositions to act in certain ways. We will not pursue this point here.
16. Apart from behavioural elements, it may also involve certain cognitive, conative, and emotional aspects (cf. Arpaly 2003; Wieland 2017) which we will set aside here.
17. This list is not meant to be exhaustive.
18. This quasi-humorous note actually raises interesting questions, such as: can one ask and get consent simply *through* asking world-directed questions? Is the consent-question about being followed *implicit* or somehow part of the question about which way to go? Alas, we do not have the space nor, more accurately, sufficiently articulate answers, to elaborate on them.
19. Negligence may well be understudied in the online manipulation debate. An excellent point of departure on the topic is Marcia Baron (2020). Kate Manne (2017) also makes a powerful case, in the context of discussing misogyny, for how focusing on individual monsters and explicit malice does not even amount

to doing half the job when it comes to doing something about misogyny – the problem lies in widespread implicit societal norms and practices. Debates about digital harm committed by corporations might have much to learn from such approaches in feminist ethics.

20. As expressed by Peter Swire, a legal scholar and Former Chief Counselor for Privacy in the Clinton Administration (cited in Barocas 2012). The full quote reads:

    > The nightmare scenario is that the databases create puppet masters . . . Every voter will get a tailored message based on detailed information about the voter. [This] means that the public debates lack content and the real election happens in the privacy of these mailings. The candidate knows everything about the voter, but the media and the public know nothing about what the candidate really believes. It is, in effect, a nearly perfect perversion of the political process.

21. Thanks to Jessica Pepp for pressing us on this.
22. Compare Klenk (2021) who defines manipulation in terms of a kind of "carelessness". Note that Klenk aims at offering a conceptual analysis of manipulation, not a normative analysis of what may be wrong with it.
23. In commercial contexts (that Nys and Engelen are mostly concerned with), caring about consent might not be important compared to the political case. To some extent, we propose to treat all such contexts alike: caring about people's consent is a minimal condition that should *always* be met when you want to use them.
24. Defined as "instances where designers use their knowledge of human behavior (e.g., psychology) and the desires of end users to implement deceptive functionality that is not in the user's best interest" (Gray et al. 2018).
25. Utz et al. (2019) note that under 1% of users would provide informed consent when opt-in systems were used.
26. https://en.wikipedia.org/wiki/Hanlons_razor
27. Thanks to Susan Brison for this example and helpful conversation on this point about deeper worries about consent-paradigms generally.
28. Recall we further argued that being genuinely interested in people's answer to the question of whether they consent typically requires *not* doing certain other things.
29. A good example of a similar constructivist-realist tension can be found in Nissenbaum (2011), where she criticizes the dominant online notice-and-consent paradigm and argues in favour of the idea of "contextual integrity" or the idea that there are context-specific "substantive norms" about how information ought to be collected and shared. She doesn't mention this explicitly, but plausibly such norms exist even if the majority of online users were to believe and act as if they don't.
30. Analogously, the challenge for a view like Nissenbaum's is to explain just why *those* norms are the relevant ones and what justifies them.

10 References

Arpaly, Nomy. 2003. *Unprincipled Virtue: An Inquiry into Moral Agency*. New York, NY: Oxford University Press.

Barocas, Solon. 2012. "The Price of Precision: Voter Microtargeting and Its Potential Harms to the Democratic Process." *Proceedings of the First Edition Workshop on Politics, Elections and Data*, 31–36.

Barocas, Solon, and Helen Nissenbaum. 2014. "Big Data's End Run around Anonymity and Consent." In *Privacy, Big Data, and the Public Good*, edited by Julia Lane, Victoria Stodden, Stefan Bender, and Helen Nissenbaum, 44–75. New York, NY: Cambridge University Press.

Baron, Marcia. 2020. "Negligence, Mens Rea, and What We Want the Element of Mens Rea to Provide." *Criminal Law, Philosophy* 14 (1): 69–89. doi:10.1007/s11572-019-09509-5.

Ben-Shahar, Omri. and Schneider, Carl E. 2011. "The Failure of Mandated Disclosure." *University of Pennsylvania Law Review* 159: 647–749.

Brison, Susan J. 2021. "What's Consent Got to Do with It?" *Social Philosophy Today*. doi:10.5840/socphiltoday202181983.

Frederik, Jesse, and Maurits Martijn. 2019. "The New Dot Com Bubble is Here: It's Called Online Advertising." Accessed September 15, 2021. https://thecorrespondent.com/100/the-new-dot-com-bubble-is-here-its-called-online-advertising/13228924500–22d5fd24.

Fricker, Miranda. 2007. *Epistemic Injustice. Power and the Ethics of Knowing*. Oxford: Oxford University Press.

Gray, Colin M., Yubo Kou, Bryan Battles, Joseph Hoggatt, and Austin L. Toombs. 2018. "The Dark (Patterns) Side of UX Design." In *CHI 2018: Proceedings of the 2018 CHI Conference on Human Factors in Computing Systems: April 21–26, 2018, Montréal, QC, Canada/sponsored by ACM SIGCHI*, edited by Regan Mandryk, Mark Hancock, Mark Perry, and Anna Cox, 1–14. New York, NY: The Association for Computing Machinery.

Griffin, Matthew. 2021. "Researchers Taught an AI to Successfully Detect People's Moral Principles." Accessed September 15, 2021. www.fanaticalfuturist.com/2021/01/researchers-taught-an-ai-to-successfully-detect-peoples-moral-principles/.

Jones, Sam. 2009. "Facebook Project Gives World Peace a Chance." *The Guardian*, October 28. Accessed September 15, 2021. www.theguardian.com/technology/2009/oct/28/facebook-world-peace-online-project.

Jongepier, Fleur, and Michael Klenk, eds. 2022. *The Philosophy of Online Manipulation*. New York, NY: Routledge.

Kant, Immanuel. 1998. *Groundwork of the Metaphysics of Morals*. Cambridge: Cambridge University Press.

Keeling, Geoff, and Christopher Burr. 2022. "Digital Manipulation and Mental Integrity." In *The Philosophy of Online Manipulation*, edited by Jongepier, F. and Klenk, M., 253–272. New York: Routledge.

Kerstein, Samuel J. 2013. *How to Treat Persons*. Oxford: Oxford University Press.

Kleingeld, Pauline. 2020. "How to Use Someone 'Merely as a Means'." *Kantian Review* 25 (3): 389–414. doi:10.1017/S1369415420000229.

Klenk, Michael. 2021. "Manipulation (Online): Sometimes Hidden, Always Careless." *Review of Social Economy*. 80: 1, 85–105. doi:10.1080/00346764.2021.1894350.

Korsgaard, Christine M. 1996. *Creating the Kingdom of Ends*. Cambridge: Cambridge University Press.

Manne, Kate. 2017. *Down Girl: The Logic of Misogyny*. Oxford: Oxford University Press.

McDonald, A. M., and L. F. Cranor. 2008. "The Cost of Reading Privacy Policies." *I/S: A Journal of Law and Policy for the Information Society* 4: 543.

Nissenbaum, Helen. 2011. "A Contextual Approach to Privacy Online." *Daedalus* 140: 32–48.

Noggle, Robert. 2018. "Manipulation, Salience, and Nudges." *Bioethics* 32: 164–70.

Nys, Thomas, and Bart Engelen. 2022. "Commercial Online Choice Architecture: When Roads Are Paved With Bad Intentions." In *The Philosophy of Online Manipulation*, edited by Jongepier, F. and Klenk, M., 135–155. New York: Routledge.

O'Neill, Onora. 1989. *Constructions of Reason: Explorations of Kant's Practical Philosophy*. Cambridge: Cambridge University Press.

Richards, Neil M., and Woodrow Hartzog. 2019. "The Pathologies of Digital Consent." *Washington University Law Review* 96: 1461–1502.

Sax, Marijn and Ausloos, Jef. 2021. "Getting Under Your Skin(s): A Legal-Ethical Exploration of Fortnite's Transformation Into a Content Delivery Platform and Its Manipulative Potential." Interactive *Entertainment Law Review* 4: 3–26.

Susser, Daniel, Beate Roessler, and Helen Nissenbaum. 2019. "Online Manipulation: Hidden Influences in A Digital World." *Georgetown Law Technology Review* 4 (1): 1–45.

Terpstra, Arnout, Alexander P. Schouten, Alwin de Rooij, and Ronald E. Leenes. 2019. "Improving Privacy Choice Through Design: How Designing for Reflection Could Support Privacy Self-Management." *FirstMonday* 24 (7): 1–13. doi:10.5210/fm.v24i7.9358.

Utz, Christine, Martin Degeling, Sascha Fahl, Florian Schaub, and Thorsten Holz. 2019. "(Un)informed Consent: Studying GDPR Consent Notices in the Field." https://arxiv.org/pdf/1909.02638.

Wieland, Jan Willem. 2017. "Responsibility for Strategic Ignorance." *Synthese* 194: 4477–97.

Williams, James. 2018. *Stand out of Our Light: Freedom and Resistance in the Attention Economy*. Cambridge: Cambridge University Press.

Zarouali, Brahim, Tom Dobber, Guy de Pauw, and Claes de Vreese. 2020. "Using a Personality-Profiling Algorithm to Investigate Political Microtargeting: Assessing the Persuasion Effects of Personality-Tailored Ads on Social Media." *Communication Research* 1–26. doi:10.1177/0093650220961965.

Zuiderveen Borgesius, Frederik J., Judith Möller, Sanne Kruikemeier, Ronan Ó Fathaigh, Kristina Irion, Tom Dobber, Balazs Bodo, and Claes de Vreese. 2018. "Online Political Microtargeting: Promises and Threats for Democracy." *Utrecht Law Review* 14: 82–96. doi:10.18352/ulr.420.

9 Manipulation as digital invasion

A neo-republican approach

Marianna Capasso

1 Introduction

Political actors in the public sphere often manipulate others: they provide incentives and other means to purposely influence and alter individuals' behaviours and beliefs. In general, manipulation is deemed to be a kind of intentional disruption or imposition in the expected functioning of individuals' decision-making processes. However, there is no consensus on the definition of manipulation (Sunstein 2016; Coons and Weber 2014b). At the same time, technology ethicists have raised concern about the possible manipulative nature of new emerging digital technologies, since the pervasive and interconnected nature of such systems can undermine users' autonomy and their capacity to make free and meaningful choices in certain cases (Klenk and Hancock 2019; Burr, Cristianini, and Ladyman 2018; Burr and Floridi 2020a, 2020b).

The general aim of this chapter is to contribute to the creation of a more systematic interaction between the fields of philosophy of technology and political philosophy. Moreover, its specific goal is to give an original contribution to the issue of manipulation in relation to digital nudging. To do that, this chapter relies on a critical analysis of neo-republican political philosophy. Contemporary theorists, such as Philip Pettit, Quentin Skinner, Maurizio Viroli and others, have developed a civic republican (or neo-republican) political theory that, further implementing insights from republicans, individuates the salient nature of political freedom in the absence of domination or alien control. Recently, some scholars have used neo-republican political theory as a general framework to argue that automated profiling (Gräf 2017), systems of mass surveillance and Big Data Analytics (Smith 2020; Hoye and Monaghan 2018; van der Sloot 2018), and algorithms (Danaher 2019) are all domination-facilitating tools. All those approaches share the idea that such technological systems facilitate the introduction of a pervasive and implicit master in an internet user's life, which can monitor their acts and impact on their privacy protection and autonomy. Pettit himself is pessimistic about the dominance of openly partisan and unreliable

DOI: 10.4324/9781003205425-11

corporations and media organizations (Pettit 2019), which facilitate online relationships in which everyone "wears the ring of Gyges" (Pettit 2004).

In most cases, this literature is constrained by its almost exclusive focus on systems' negative impact on privacy and security. Instead, my proposal is to further extend neo-republican political conceptions to show how they can also provide the other side of the same coin: original conceptual clarifications for the discourse on digital nudging and manipulation. The reflection on the use of digital nudging has sparked much controversy, and criticisms often identify transparency as the most important criterion to distinguish nudging from manipulation, raising ethical concerns on the use of non-transparent digital nudges. In this chapter, by contrast, I try to individuate criteria to distinguish nudging from manipulation and to assess the degree to which digital nudges can be deemed to be wrongful manipulative – and, thus, dominating – technological influences or, conversely, part of a democratic net of control and protection.

The neo-republican political theory may offer a promising account of the conception of manipulation in digital contexts for several reasons. After all, neo-republicans predominantly focus on the mere power to manipulate as a possible risk of domination. Hence, their frameworks can better address the normative issue of manipulation in the digital domain, where actual or potential behaviour steering practices by technological systems, private and market-driven groups or institutions may affect society without being held adequately accountable for their power. Moreover, neo-republicans individuate specific criteria to assess when manipulation – as a kind of interference – is problematic and amounts to domination or not. In this sense, neo-republicanism can provide new tools for conceptual clarification and normative justification for possible practices of digital manipulation, clarifying when a digital practice can lead to a loss of freedom or what kind of digital social relations and influences can infringe upon individuals' meaningful choices.

The chapter is structured as follows. In the second section, I outline Pettit's notion of 'freedom as non-domination' and explain how manipulation is a kind of interference. Also, I distinguish the conceptual definition (as an activity) of manipulation from its normative status (as an invasion). In Section 3, I provide examples and critical evaluations of a specific technological influence: digital nudges. After having introduced digital nudges, I propose an evaluative framework to assess when and to what extent digital nudges can be classified as dominating manipulative interferences (invasions) (Section 4). Finally, I discuss in Section 5 the sense in which freedom in the digital sphere requires not the absence of 'manipulation' as interference but rather the absence of alien control on such activity and the presence of a democratic net of protection against the latter. The chapter concludes by raising some open issues and suggesting avenues for future research.

2 Freedom as non-domination: a sketch

The core of neo-republican theory advanced by Pettit is the ideal of freedom as non-domination. Pettit defines domination as follows: someone, A, is dominated *as long as* another agent or agency B (1) has a power of interfering (2) that is arbitrary or not itself controlled by A (3) in a certain choice that A is in position to make (Pettit 1997, 52, 2012, 50). This conception of freedom differs from traditional liberalism, for instance, Isaiah Berlin's account of negative freedom, according to which an agent is free if there is no interference from others, which means that his or her freedom of choice between chosen as well unchosen options remains intact (Pettit 2011, 704).

By contrast, freedom as non-domination is freedom of agents, not of options (Pettit 2003). An agent can be subject to domination at any time, even in those cases where there is no actual interference from others, where interference is understood as an intentional or quasi-intentional intervention by one party in the choice of another (Pettit 2008, 110). The paradigmatic neo-republican example is the relation between the slave and the master. The master can be benevolent and might not actually interfere with the slave but nonetheless remains in a position and standing to do so and to exercise on the slave the constant threat of being interfered with.

Neo-republicanism allows for two main theses. The first is there can be *domination without interference*, as in the master–slave example. The second is that there can also be *interference without domination*. This happens when interference is non-arbitrary (Pettit 1997), controlled (Pettit 2012) or non-alien (Pettit 2008).[1] 'Non-arbitrary' or 'non-alien' are the terms that Pettit uses to indicate the legitimacy of certain kinds of interference but without a moralized intent (Pettit 2008, 117). In his recent work, Pettit prefers to talk about domination as "exposure to another's uncontrolled power of interference" instead of arbitrariness (Pettit 2012, 50–58).[2] A lack of freedom is not about interfering into a set of options but rather derives from *uncontrolled* interference, that is, "interference that is uncontrolled by the person on the receiving end" (Pettit 2012, 58).

However, something more is needed to characterize interferences as dominating interventions: the absence of control or arbitrariness. This clarification on interference may have profound implications for the assessment and use of the conception of manipulation. Manipulation is not domination, as some scholars have sustained (Wood 2014; Grüne-Yanoff 2012) but is one of the possible kind of interferences individuated by Pettit in his taxonomy. Manipulation is indeed an interference that has an impact on the cognitive capacities of individuals and involves what Pettit calls "misrepresentation": it changes how the options are presented to the agent according to his or her perceptions. Specifically, manipulation affects the proper understanding of options, leading to the creation of 'distorted' options for the decision-making processes of the manipulated (Pettit 2012, 54).

Manipulation falls along a *continuum* and adopts a wide range of behaviours: it can be either an extreme intervention that uses hypnosis, brainwashing or intimidation (radical manipulation) (Pettit 2008, 110–11) or an intervention that takes a non-rational form; that is, it appeals to people's emotions, desires and beliefs. Moreover, it can even take a rational and deliberative form, in the rigging of the actual or expected consequences and outcomes of people's actions or in the relevant intrusion in people's values-metric with rhetoric (Pettit 1996, 578–79, 2012, 56). And above all, manipulation is not deceptive about its means and intentions: it does not imply stating falsities or purposely misinforming. In summary, manipulation, as a kind of misrepresentation, leads to "forming your will in the dark" (Pettit 2012, 54).

Manipulation as a practice is not necessary for realizing domination: being an interference, it reduces freedom but does not eradicate it. Nonetheless, it can be one source of subjection, if accompanied by the loss of control on the part of the agents. Pettit uses a specific term to define the wrongful – and, thus, uncontrolled – interference: invasion.[3]

Manipulation, understood as a practice, can be defined as a direct, non-contingent and non-deceptive misrepresentation that affects the manipulated agents' cognitive capabilities in understanding a set of options and leaves them unsure about the means (*how*) and intentions (*why*) of such misrepresentation.

Conversely, manipulation as an invasion is one of the possible realizations of *alien control* or *domination with interference*. The latter results in being dependent on the will of another that negatively intervenes and subverts the agent's deliberative choice, and that does not leave to the agent the ability to respond to and counter-control the interference. Manipulation is not domination as such but a peculiar form of domination that occurs in combination with a specific kind of uncontrolled interference (uncontrolled manipulation) (Pettit 2008, 110–11).

Under this account, manipulation is an invasion since it leads to a complete displacement of individuals' will. 'Will' should be understood not in a metaphysical or ethical sense but as political: a social free will, which allows individuals to be in the position to make free and meaningful choices according to their interests and preferences (Pettit 2012, 36–38, 49). This displacement implies that A's authorship over a decision-making process is transferred as a whole to B. Indeed, B subverts A's cognitive resources in identifying relevant valued options, options that do matter in the social sphere. As a misrepresentation, the invasive manipulation leaves agents unsure that such interference in their choice has been put in place and unsure about its methods and B's intentions behind it. Nonetheless, what connects such covertness to a loss of freedom is the fact that the misrepresentation is uncontrolled or unjustified by the part of A, that is, A is not located into a net of protection that makes covertness unacceptable, or at least suitably difficult or costly, and/or easy to detect and to contest.

This account of manipulation distinguishes two different accounts of manipulation: conceptual and normative (the latter based on a neo-republican approach). They are answers to the following questions: (I) what is manipulation? (II) what makes manipulation problematic? The conceptual account is descriptive and helps to individuate a set of activities without connecting them to moral commitments or to a specific normative theory of justice. In this, I follow other scholars in recognizing that the analysis of the normative status of a practice should be preceded by a prior conceptual definition of such practice (Coons and Weber 2014a; Wood 2014; Whitfield 2020).

As a matter of fact, the conceptual definition of manipulation shows how such practice is prima facie wrongful: it fails to respect the integrity of our cognitive capacities, leading to a series of acts whose nature consists in misrepresenting a state of affairs. However, this only means that manipulation stands in need of normative justification, without providing one. What makes it incompatible with freedom and gives it a moral or political valence depends on the normative theory through which we look at the concept. The neo-republican normative account proposed here is one of the possible attempts to fill this gap.

Second, this account provides a clear-cut distinction between deception and manipulation. In manipulative acts there is no need to employ deceitful communication. To be effective, manipulators can simply use correct arguments, or abundance of information and rhetoric, or work on an agenda to push the manipulated agents towards their preferences. As some scholars noted, this is what makes manipulation indistinguishable from persuasion and difficult to reveal and challenge from an objective basis (Whitfield 2020).

Approaches to manipulation that define it as an influence that does not engage or appeal to individuals' rational capacities for deliberation and reflection are misleading (Sunstein 2016; Blumenthal-Barby 2012). On the contrary, manipulators often use and employ an adequate knowledge of individuals' cognitive mechanisms and perceptions as means to ensure that manipulated agents make decisions and take actions they prefer. The use of rational claims can be manipulative (Klenk 2020; Gorin 2014; Barnhill 2014). The reduction of individuals' deliberative capacities is not necessarily achieved with the adoption of falsities or reason-bypassing means but rather by winnowing down options without notifying them about the *ratio* behind such intervention and thus by misrepresenting a state of affairs.

Third, the normative account of manipulation defines it as an interference that not only tries to reduce and shift the authorship of decision-making processes but also to subvert it while obscuring such intention. This is what I mentioned as "displacement" of manipulated agents. In neo-republicanism, one of the aims is to promote "non-manipulability" of institutions and norms, which means that they should promote public ends and be resistant "to being deployed on arbitrary, perhaps sectional, basis"

(Pettit 1997, 172). Pettit warns against "false positives", which are sectional misrepresentations that pretend to be initiatives supported by public reason (Pettit 2000).

Therefore, to avoid sectional and partisan advantages that violate the functioning of public decision-making processes, institutions should promote the normative ideal of deliberative democracy. This is based on the creation of common good and standards that are recognized as fair and relevant by all social actors (Pettit 2019). The public decision-making processes should respect interests and ideas, "under an efficacious form of control that you share equally with others in imposing" (Pettit 2012, 178). Thus, this account of manipulation is political rather than ethical: it warns against socially powerful citizens or groups and institutions and points out that there is a need for adequate forms of institutional design, starting from tracking and accountability relationships.

3 Digital nudging

The term "digital nudging" refers to the "use of user-interface design elements to guide people's behaviour in digital choice environments" (Weinmann, Schneider, and vom Brocke 2016). It is based on the work of Thaler and Sunstein (2008) that advocates a libertarian and paternalistic choice architecture. "A nudge . . . is any aspect of the choice architecture that alters people's behaviour in a predictable way without forbidding any options or significantly changing their economic incentive" (Thaler and Sunstein 2008, 6).

Digital nudges allow for a greater versatility and opportunities for choice architects due to the much more dynamic and automated character of the digital environment (Meske et al. 2019). As a matter of fact, Big Data nudges have been defined as a special kind of nudge: *hypernudges*, since they can shape people's choice context and collect their data in more efficacious, targeted and interconnected modalities (Yeung 2017).

As mentioned, neo-republican interference is a term broadly enough to cover any activity that intentionally intervenes in choice (Pettit 2012, 50). Digital nudges as activities arguably have an interfering nature, since they are direct interventions embodied in user-interfaces or websites (choice architecture) by designers (choice architects) that seek to influence users' choice.[4] Moreover, digital nudges rely on the use of psychological mechanisms, such as *framing*, which implies an alteration of the (perceived) presentation of the environment, or *priming*, which aims to elicit intentions by using statements or images that steer towards a specific action before a decision is taken (Mirsch, Lehrer, and Jung 2017) and many others.

Therefore, digital nudges in certain cases may arguably lead to forms of manipulation: subjective interferences that change how a set of options presents itself according to the cognitive perceptions of users and leave the nudged unsure about the *ratio* of such change. Namely, they may lead to misrepresentations that leave the nudged unsure about the means (how) and intentions (why) behind them. This is what scholars called the 'transparency' of a nudge (Hansen and Jespersen 2013).

Some scholars, relying on republicanism, worry that nudges can help governments or corporations to dominate individuals because they lack transparency (Grüne-Yanoff 2012; Hausman and Welch 2010). Similarly, some identify transparency as the most important criterion to distinguish nudging from manipulation, raising ethical concerns on the use of non-transparent digital nudges (Hansen and Jespersen 2013; Caraban et al. 2019). Digital nudges have been defined as manipulative when they affect the un-reflective cognitive abilities of individuals and are non-transparent (Heilmann 2014). When these digital nudges are overt and identifiable and allow for the users' consent and general awareness, then they are ethically justifiable (Meske and Amojo 2020).

However, the problematic aspect of digital nudges should not be reduced only to transparency. Nudges' ability to interfere and their possible lack of transparency can be enough to subject people to domination, as other scholars have argued. Nonetheless, their manipulative character is neither a sufficient nor a necessary condition to describe these technological influences as forms of domination *per se*. Indeed, digital nudges can be designed either to be sources of invasion on users and society at large, implicating a significant alien interference in relevant valued choices or to be vehicles for reflection and freedom. The key element that allows to differentiate between the two results is not the fact that an interference – such as manipulation – can take place, but that such interference can be accompanied by a denial of users' power and control or, conversely, by a recognition and promotion of that same power. Freedom in the social and political sense does not require the absence of "manipulation", understood as an activity whose effects and reasons are likely to be unrecognized by the individual manipulated agent but rather the absence of alien control on such activity and the presence of a systematic net of protection against the latter.[5]

Not all digital manipulation amounts to forms of domination. Digital manipulation is a form of *domination with interference* as long as it intervenes on choices that are significant in social life and is neither suitably justified and transparent nor under a democratic form of control. There may be cases in which target acts in digital nudging are relevant choices in social life that have been selected and evaluated by an alien values-metric. This alien values-metric is such if, not checked and controlled, alters the set of options before agents and leads to the creation of different evaluative profiles, which introduce changes "that do matter: changes that affect the probabilities of various valued or disvalued consequences" (Pettit 2008, 122). Digital nudges may run the risk of radically misrepresenting a set of options, with the result that the original options are no longer available for agents. If not controlled, this could be a feature that might make some specific nudges ethically problematic and controversial.

4 Manipulation as digital invasion: examples and critical evaluations

Digital nudges may range from desirable interventions to questionable and even radical interventions. Thus, what matters is establishing a solid evaluative framework to assess when and to what extent digital nudges involve a denial or deprivation of users' freedom and undermine their social and political relationships.

According to the proposed framework based on neo-republican political philosophy, to be classified as wrongful manipulative interferences (invasions) and thus dominating, digital nudges should fall within at least one of those cases: a) nudges do not track and do not conform to the agent's interests (inherently hostile); b) nudges subvert relevant valued options for the agent in distorted ones; c) the agent is exposed to uncontrolled misrepresentation; d) nudges do not leave the possibility to check and counter-control their interferences (displacement).

In the first case, when digital nudges do not track and conform to users' general values and metrics, users are exposed to radical manipulation, which undermines their overall ability to choose and imposes a goal or result in contrast with their interest and ends. Examples comprise the promotion of bias, discrimination or fraud against the self-interest of users (Letzler et al. 2017).

In the second case the manipulative nature of digital nudges lies in the fact that they may be interventions in valued and relevant options in the set of options before agents. On a neo-republican understanding, the free person is not someone who avoids interventions or burdens but rather someone who is systematically protected and empowered against interventions in the choices that are deemed to be significant in social life (Pettit 1996). It is thus necessary to define which choices or which domains of choice should be protected in the social sphere.

Digital nudges shape users' behaviours and beliefs that may or may not be conducive to various social values. For example, due to the nature of their feedback, digital nudges can drive self-reinforcing biases and lead to the creation of filter bubbles and echo chambers (Bozdag and van den Hoven 2015; Pariser 2011).

Relevant value options might refer to the specific values-metrics of a group in society, whose own interests and peculiarities need to be meaningfully taken into consideration. There are cases in which digital nudges exacerbate side effects in vulnerable groups, such as persons with eating disorders (Levinson, Fewell, and Brosof 2017) or may increase addictions rather than reduce them. The latter is known as the "backfire effect", which triggers users to adopt the opposite target behaviour (Stibe and Cugelman 2016). Thus, in the design of digital nudges, a focus on contextual sensitivity (Pettit 1997, 53) should be predominant.

Another important theme is the fact that establishing which options should be understood as valuable may be controversial. For example, the permissibility of nudges may vary considerably in terms of which values they support – general social values or values tailored for nudged agents – or of which domain they interfere with (Alfano and Robichaud 2018). In recent years, policy decisions have given citizens the choice to opt out rather than opt in for organ donation (i.e., consent to donate is presumed) (Shepherd, O'Carroll, and Ferguson 2014) and have thus increased the number of organ donors. The same has been realized for increasing the participation in corporate pension schemes (Beshears et al. 2017).

In one passage, Pettit explicitly wonders whether nudges could amount to manipulation. According to him there is no general answer, but the specific case of default rules for organ donation does not amount to manipulation, since it provides more information on "the *correctness* of the message conveyed" and does not constitute a distortion of valued options (Pettit 2012, 56n32, my italics.; See also Pettit 2014, 242).[6]

However, one may argue that not all opt-out and other psychological mechanisms are free from concerns on their acceptability. For example, one of the psychological mechanisms used in digital nudges is the application of social norms, that is, standards that constrain and guide a group (Mirsch, Lehrer, and Jung 2017). Amazon nudges us to buy further products based on what other customers bought. Social norms – or even credible and apparent norms – emerge from social interactions and networks and can even change the evaluative and normative sense of rightness. Indeed, it may happen that a powerful group in society has an additional share of influence over collective decisions and on certain norms for arriving at a social choice.[7]

In neo-republicanism, there is a prior 'baseline' to which any effects of interferences by groups or institutions must be understood, and this underpins a set of basic liberties that may vary "across differences in culture and technology" (Lovett and Pettit 2018). These liberties are the ones identified by law, such as freedom of speech, association, employment, and others, but this does not imply that they should be necessarily restricted and resistant to discussion or expansion.

New digital interactions may require a discussion and a justification drawn from this prior baseline due to the unprecedented and risky possibilities they entail. Therefore, even the set of liberties should be subject to an ongoing reassessment, considering the present-day conditions and technologies. The current debate on the introduction of the right to mental integrity to protect the individual from "many different forms of manipulation, that the mind encounters on a daily basis . . . in reaction to new challenges and technologies" could be an example (Michalowski 2020, 411).

There may be cases in which the presentation of options by default rules or other means may impose a burdened or distorted option on what informed people would have chosen in counterfactual scenarios. A default rule in a domain like politics may endanger the self-government or other norms of the political body. Thus, to avert manipulation we can ask which tools we have at our disposal to evaluate nudges – such as balancing, proportionality, reasonableness, or others – and if there are *ex ante* or *ex post* measures that make the choice of (digital) nudges open to participation and discussion (Cassese 2016).

Finally, a misrepresentation can be deceptive or manipulative, according to Pettit (Pettit 2012, 54). The latter can involve true statements in the sense that it does not imply deceitful communication but nonetheless can give misleading impressions, for example in the relevant omission or abundance of information. Moreover, we can distinguish between negligent or innocent misrepresentations from fraudulent ones. In common law, for example, to be fraudulent, a misrepresentation should be accompanied by recklessness to the truth of its statements: a state of mind that deliberately and unjustifiably takes an action while disregarding the associated risks. In criminal law, some scholars call it a kind of "culpable carelessness".[8] But also negligence for risk-taking can equally be a kind of culpable lack of care.

For example, US college students are often unaware of the fact that Google or Facebook personalization algorithms track their data and filter and prioritize and "nudge" contents accordingly, in ways that may not be recognized by them (Powers 2017). A Facebook experiment intentionally changed many users' new feeds but omitted to inform users about it (Kramer, Guillory, and Hancock 2014). Finally, the recent COVID-19 pandemic has shown how an infodemic – understood as an overabundance of information online during a pandemic – may include deliberate attempts to undermine the public health response and promote alternative agendas of individuals or groups (World Health Organization 2020).[9]

Uncontrolled misrepresentation may involve the use of "false positives", that is, partisan misrepresentations that pretend to be supported in the name of the common good, as already mentioned. These partisan misrepresentations can be translated in the digital domain as interventions that pretend to empower certain common and recognizable interests for shaping governments or institutions' decisions, while promoting objectives and goals of sectional and partisan providers. In the literature in philosophy of technology, for example, there has been a growing concern on the predominant impact that market-driven systems, such as private big tech corporations like Google or Amazon, may have on shaping public agendas and research.[10] Moreover, an uncontrolled misrepresentation can be supported by a "culpable carelessness" attitude, which without justification disregards or neglects the potential worrisome effects associated with an action. These actions in turn can expose the others to the risks of suffering foreseen harmful consequences that could have been avoided.

5 A net of protection and empowerment

The account of manipulation as invasion that I propose in this chapter groups together a series of practices in the digital domain in which users are not fully aware that they are compromised in their actions. The risk is that users may accept the worldview or misrepresentation of choice environments that market-driven tech corporations can sustain, internalize it, and do not see what is arbitrary about it. What I define as the risk of "covertness" associated to manipulation may imply different levels. Beyond the failure to adequately inform users or the use of dark patterns or hidden agenda by corporations,[11] such covertness may extend to the unequal distribution of social powers in which members of a group tend to reproduce a norm that do not adequately rely on rules, regulations or procedures that are in line with democratic standards and protect individuals' rights and interests.

"Being in the dark" (Pettit 2012, 54) can be prima facie related to the unawareness of the intention or means behind an influence. Of course, big tech corporations are moved by the motive of profit and users have some growing intuition and awareness that their data and actions in the digital domain are placed and shaped in such a market environment.[12] However, "being in the dark" may also refer to the fact that users can unthinkingly – often in a manner that is habitual – reproduce in their actions a social norm that pretends to endorse an equal social status for all individuals while exploiting a partisan advantage of some over others and undermining the collective ability to safely rely on the law. As already mentioned, manipulation as invasion affects social free will, which allows individuals to choose meaningfully in line with their interests: in doing so, it brings about an unequal distribution of power and knowledge of whose implications the manipulee can be not completely aware.[13]

However, the problematic aspect of digital nudges should not be reduced merely to transparency and awareness. Digital nudges often lack transparency and do not reveal and exhibit to people the reasons and procedures behind their interactions with them. To be invasive and thus morally problematic, digital nudges should deny not merely the full or adequate knowledge of their means and supposed aims but even a status to manipulated agents: a position which allows them to be recognized and to see, uncover, and even contest nudges.[14] The lack of transparency can exacerbate and also be a symptom of another more dangerous risk: the failure to respect the status of users as citizens and thus sources of the norms that govern them.

Indeed, the further step introduced by neo-republicanism extends the scope of freedom, making it a robust and normatively justified status (Pettit 2003). Perspectives that reduce neo-republicanism to liberalism, arguing that in both approaches the right to individual freedom and privacy is predominant over instances for public and political protection (Stahl 2016), overlook a fundamental feature of Pettit's framework. Indeed, with the term "status" Pettit does not merely imply acts or strict formalizations of rights, but relationships of power: the individuation of right forms of relational balance of power, where one can have the possibility to be heard and authorized by the others (Pettit 1996).

A principle advanced by Thaler and Sunstein to prevent manipulation via nudges is Rawls's publicity principle, according to which public institutions or groups cannot adopt policies that they would not be able or willing to defend publicly (Thaler and Sunstein 2008, 244–45). However, as scholars have pointed out, this principle is ineffective in digital contexts, since monitoring and interactions often take place without citizens' consent. Also, institutions and public or private agencies openly defend their behaviours without any concern on the possible consequences of their acts (Yeung 2015, 462).

This is where neo-republicanism may turn out to be helpful since it focuses on the power to manipulate rather than the acts of manipulation themselves. It sheds light on the fact that the absence of manipulative acts or the awareness that such acts has been put in place[15] are not sufficient to guarantee freedom. A benevolent manipulator remains someone who has the power to manipulate. There can be unfreedom even in those cases where actual or possible practices of manipulation are publicly communicated through transparent means and people are aware of those practices.[16]

Therefore, in the case of digital nudges, the regulatory challenge consists not only in the implementation of awareness by the part of users and of transparency about means and intentions by the part of private providers but also in providing public tools and means of empowerment, communication, and contestation. Any kind of interference should be made not only transparent but also explainable and justifiable: it should be subject to public protection, debate, and contestation, especially in all those cases where groups in society have a power to interfere with relevant valued options, that is, options that are significant in the social sphere. When Pettit analyzes domination, he is interested in the social relation of power between individuals and the kind of choices that can have more weight and significance in the social arena. Some choices and some relationships are more important than others for our freedom, and neo-republicanism helps to differentiate normatively different kinds of influences and social standings.[17]

The last criterion adopted in this chapter to assess when digital nudges are dominating interferences (invasions) is the one related to the "displacement" of individuals or the lack of checking and counter-control. A displacement does not merely imply an intervention into users' choices but an *uncontrolled* intervention by those whose set of options is affected.

What makes digital manipulation morally problematic is not the fact that it can interfere with the set of options of individuals or that it is non-transparent. Rather, what makes that digital manipulation lead to a loss of freedom is the fact that is democratically uncontrolled: it has an impact on options that do matter in social and political reality, without being sufficiently or adequately justified by the part of groups or powers that should be held accountable for their actions. Opaque digital nudging by private big tech corporations is often a sign that such social actors do not care much about a democratically controlled system that can oversee and warn against their actions. Political and social freedom does not just concern the absence of interferences such as manipulation or the doors that are open to individuals but also requires that no doorkeeper has the power to close or conceal a door without significant costs (Pettit 2011, 709). In this sense, the development of a systematic net of protection serves to make unacceptable, or at least suitably difficult or costly, this kind of uncontrolled digital manipulation.

Such a systematic net is brought about by a "cultural, legal and political matrix of protection and empowerment" (Pettit 2008, 104) and involves different tasks. For example, in the digital domain it can provide means to the public to hold the decisions and acts of private big tech corporations democratically accountable. Such a net of protection should raise questions about public accountability gaps, which, beyond the issues of information disclosing and visibility, affirm that we need modalities to make systems not only transparent, explainable, and understandable to the experts or the designers but also explainable and understandable to the users and audience at large (Pasquale 2015; Santoni de Sio and Mecacci 2021).

Moreover, in the digital context, over and above the manifest choice of a regulatory instrument that should be tailored to new systems' functionalities and overcome the limits of consent-based approaches, such a net may also require a regulatory overseeing body or group. This group could shape technological policies and foster public understandability and scrutiny. Individuating a mediator in the social environment is one of the modalities and solutions that a neo-republican perspective could provide, along with a preference for the notion of contestation over that of consent as the basis for political legitimacy (Pettit 1997, 202, 2012, 215–16).[18] The definition and construction of such a net are a work in progress (Pettit 2019) and can constitute a relevant alternative to our current approaches to digital choice architecture, which arguably have a predominant focus on individuals and neglect collective sociopolitical action.[19]

6 Conclusion

In this chapter, I showed how neo-republicanism can provide conceptual and normative tools to analyse and address the problem of manipulation in relation to digital nudges. This proposed shift to a neo-republican perspective can be a means to address collective and shared responsibility in relation to – and not in opposition to – individual freedom and agency. Indeed, with its emphasis on social and political relations, it may offer a promising account on the interconnection between digital choice architecture and human freedom. It should be noted, however, that this chapter neither addressed the issue of theorizing neo-republican forms of "control" that do not lead to a loss of freedom nor explored in detail the role that digital nudges may have in shaping and supporting a democratic net of protection and empowerment. Thus, future work consists in further implementing the proposed theoretical framework to understand the challenge of designing digital choice environments that avert forms of uncontrolled manipulation and promote the freedom of individuals and society.

Notes

1. On the frequently interchangeable use of the three terms in Pettit, see also Beckman and Rosenberg (2018).
2. Pettit states that the introduction of arbitrariness is not an evaluative justification (moral) but factual (Pettit 2012). On the exact understanding of the term "arbitrariness" there is a huge debate in the literature, which has generated ambiguity and different interpretations (see, for example, Arnold and Harris 2017). However, I am not going to explore in detail the issues related to the concept of arbitrariness and its procedural or substantive interpretations. On this point, see Gorin's chapter in this volume, where a "reason substantivism" is adopted.
3. (Pettit 2012, 46). Another kind of invasion is domination: the mere exposure to the power of another. Of course, Pettit's view is focused on the normative status of interference. This specific distinction between a conceptual definition and a normative status of manipulation is proposed in this chapter starting from and further developing Pettit's arguments in various works, among the others: Pettit (2012, 2008).
4. In Pettit's view, interference takes place only when there is an agent or a corporate that intentionally exerts it or has the capacity to do so (Pettit 1997, 52–53). This may raise questions about the nature of intentionality, the capacities of technological systems for intentionality and about the agency of corporates and groups that in this chapter I am not going to explore in detail. Nonetheless, I am focusing on the special emphasis that neo-republicanism places on the power to change and respond to possible sources of domination and interference in the wider environment, notwithstanding these are intentional, quasi-intentional, or not.
5. A similar suggestion was advanced for example by Schmidt (2018) and Schmidt and Engelen (2020), claiming that nudges to be acceptable should be suitably transparent and amenable to democratic control.
6. This specific nudge seems to be labelled by Pettit under the umbrella term of "persuasion", which makes the pros and cons of options more salient and does not infringe upon individuals' deliberative capacities (see Pettit 2015).
7. Some neo-republican scholars prefer to talk about "systemic domination" in such a case: a kind of domination that is not agent-relative, stemming from the epistemic or material resources of a group. Conversely, it is mediated through a set of social norms and practices (Laborde 2010; Gädeke 2020).
8. Where carelessness is defined as "a suitably clear demonstration of the defendant's insufficient concern for the interests of others" (Stark 2016, 9). Under the term of "culpable carelessness", Stark (2016) wanted to analyze two terms that have been individuated in the Standard Account of Anglo-American criminal law and doctrine: "awareness-based culpability (recklessness) and inadvertence-based culpability (negligence) for unjustified risk-taking" (Stark 2016, 6).
9. In recent years, scholars have noticed that social networks are a space for targeted and polarized political propaganda, as the case of the Cambridge Analytical scandal and US political elections have demonstrated (Howard et al. 2018; Milano, Taddeo, and Floridi 2020).
10. On this point, see Sharon (2016, 2021).
11. See Jongepier and Wieland's chapter in this volume.
12. This point is highlighted also in Grill's chapter in the volume.
13. According to Sandven, for example, when social norms bring about an unequal distribution of status and credibility, they ground epistemic injustice, making individuals unable to exercise "responsive" control, the kind of control that

people should have after having experienced an interference (Sandven 2020; Schmidt 2018).

14. Reckless actors are culpable when they are "unmoved" by beliefs that show how they can be "insufficiently motivated by the interests of others", see Stark (2016, 122).
15. Awareness is not enough:

> (alien control) will remain true if B becomes aware of the invigilation and virtual control exercised by A and can do nothing about it . . . Apart from living under the control that goes with being invigilated, B will suffer the inhibition that goes with being consciously invigilated.
>
> (Pettit 2008, 113)

16. A condition for a system to be considered under adequate civic control lies in the fact that it is *unconditioned*, which means that "people have an influence on government that is not conditioned on the willingness of government, or of any third party, to play along" (Pettit 2012, 80).
17. On the contrary, this is a limitation of Foucaultian approaches, which hold that any reconfiguration of power relations may in principle amount to domination (Shapiro 2012; Hoye and Monaghan 2018).
18. Contestation is provided by open assemblies, critical media, watchdog bodies, tribunals, independent ombudsman and courts through which contestations can be heard and appealed. They allow a "pre-contestation, for transparency in the decisions contested, and post-contestation, for impartiality in resolving the charges raised" (Pettit 2012, 215; Farrell 2020, 871).
19. See also Schmidt and Engelen (2020).

7 References

Alfano, Mark, and Philip Robichaud. 2018. "Nudges and Other Moral Technologies in the Context of Power: Assigning and Accepting Responsibility." In *The Palgrave Handbook of Philosophy and Public Policy*, edited by D. Boonin, 235–48. Cham: Palgrave Macmillan.

Arnold, Samuel, and John R. Harris. 2017. "What is arbitrary power?" *Journal of Political Power* 10 (1): 55–70. doi: 10.1080/2158379X.2017.1287473

Barnhill, Anne. 2014. "What is Manipulation?" In Coons and Weber 2014b, 51–72.

Beckman, L., and J. H. Rosenberg. 2018. "Freedom as Non-domination and Democratic Inclusion." *Res Publica* (24): 181–98.

Beshears, J., S. Bernatzi, R. Mason, and K. Milkman. 2017. *How Do Consumers Respond When Default Options Push the Envelope?* https://papers.ssrn.com/sol3/papers.cfm?abstract_id=3050562.

Blumenthal-Barby, J. S. 2012. "Between Reason and Coercion: Ethically Permissible Influence in Health Care and Health Policy Contexts." *Kennedy Institute of Ethics Journal* 22 (4): 345–66.

Bozdag, Engin, and Jeroen van den Hoven. 2015. "Breaking the Filter Bubble: Democracy and Design." *Ethics and Information Technology* 17 (4): 249–65.

Burr, Christopher, Nello Cristianini, and James Ladyman. 2018. "An Analysis of the Interaction Between Intelligent Software Agents and Human Users." *Minds and Machines* 28 (4): 735–74. doi:10.1007/s11023-018-9479-0.

Burr, Christopher, and Luciano Floridi, eds. 2020a. *Ethics of Digital Well-Being: A Multidisciplinary Perspective*. Cham: Springer.

Burr, Christopher, and Luciano Floridi. 2020b. "The Ethics of Digital Well-Being: A Multidisciplinary Perspective." In Burr and Floridi 2020, 1–29.

Caraban, A., Evagelos Karapanos, Pedro Campos, and Daniel Gonçalves. 2019. "23 Ways to Nudge: A Review of Technology-Mediated Nudging in Human-Computer Interaction." In *Proceedings of CHI Conference on Human Factors in Computing Systems (CHI'19)*, 1–15. New York, NY: Association for Computing Machinery.

Cassese, S. 2016. "Exploring the Legitimacy of Nudging." In *Choice Architecture in Democracies: Exploring the Legitimacy of Nudging*, edited by Alexandra Kemmerer, Christoph Möllers, Gerhard Wagner, and Maximilian Steinbeis, 241–46. Baden-Baden: Nomos.

Coons, Christian, and Michael Weber. 2014a. "Manipulation: Investigating the Core Concept and its Moral Status." In Coons and Weber 2014, 1–16.

Coons, Christian, and Michael Weber, eds. 2014b. *Manipulation: Theory and Practice.* Oxford: Oxford University Press.

Danaher, John. 2019. "The Ethics of Algorithmic Outsourcing in Everyday Life." In *Algorithmic Regulation*, edited by Karen Yeung and Martin Lodge. Oxford: Oxford University Press, 98–117.

Farrell, Liam. 2020. "The Politics of Non-domination: Populism, Contestation and Neo-republican Democracy." *Philosophy and Social Criticism* 46 (7): 858–77.

Gädeke, D. 2020. "Does a Mugger Dominate? Episodic Power and the Structural Dimension of Domination." *Journal of Political Philosophy* 28: 199–221.

Gorin, Moti. 2014. "Towards a Theory of Interpersonal Manipulation." In Coons and Weber 2014b, 73–97.

Gräf, E. 2017. "When Automated Profiling Threatens Freedom: A Neo-Republican Account." *European Data Protection Law Journal* 4: 1–11.

Grüne-Yanoff, Till. 2012. "Old Wine in New Casks: Libertarian Paternalism Still Violates Liberal Principles." *Social Choice and Welfare* 38 (4): 635–45.

Hansen, P. G., and A. M. Jespersen. 2013. "Nudge and the Manipulation of Choice: A Framework for the Responsible Use of the Nudge Approach to Behaviour Change in Public Policy." *European Journal of Risk Regulation* 4 (1): 3–28. doi:10.1017/S1867299X00002762

Hausman, D. M., and B. Welch. 2010. "Debate: To Nudge or Not to Nudge." *Journal of Political Philosophy* 18 (1): 123–36.

Heilmann, C. 2014. "Success Conditions for Nudges: A Methodological Critique of Libertarian Paternalism." *European Journal for Philosophy of Science* 4 (1): 75–94. doi:10.1007/s13194-013-0076-z

Howard, Philip N., B. Ganesh, D. Liotsiou, J. Kelly, and C. François. 2018. *The IRA, Social Media and Political Polarization in the United States, 2012–2018.* Oxford: Project on Computational Propaganda, 47 p.

Hoye, J. M., and J. Monaghan. 2018. "Surveillance, Freedom and the Republic." *European Journal of Political Theory* 17 (3): 343–63.

Klenk, Michael. 2020. "Digital Well-Being and Manipulation Online." In *Ethics of Digital Well-Being: A Multidisciplinary Perspective*, edited by C. Burr and L. Floridi, 81–100. Cham: Springer. doi: 10.1007/978-3-030-50585-1_4.

Klenk, Michael, and Jeff Hancock. 2019. "Autonomy and Online Manipulation." *Internet Policy Review*, 1: 1–11. Accessed February 28, 2020. https://policyreview.info/articles/news/autonomy-and-online-manipulation/1431.

Kramer, A. D. I., J. E. Guillory, and Jeff Hancock. 2014. "Experimental Evidence of Massive-scale Emotional Contagion Through Social Networks." *Proceedings of the National Academy of Sciences* 111: 8788–90.

Laborde, Cécile. 2010. "Republicanism and Global Justice: A Sketch". *European Journal of Political Theory* 9 (1): 48–69.

Letzler, Robert, Ryan Sandler, Ania Jaroszewicz, Isaac Knowles, and Luke M. Olson. 2017. "Knowing When to Quit: Default Choices, Demographics and Fraud." *The Economic Journal* 127 (607): 2617–40.

Levinson, Cheri A., Laura Fewell, and Leigh C. Brosof. 2017. "My Fitness Pal Calorie Tracker Usage in the Eating Disorders." *Eating Behaviors* 27: 14–16. doi:10.1016/j.eatbeh.2017.08.003.

Lovett, Frank, and Philip Pettit. 2018. "Preserving Republican Freedom: A Reply to Simpson." *Philosophy and Public Affairs* 46 (4): 363–83.

Meske, C., and I. Amojo. 2020. "Ethical Guidelines for the Construction of Digital Nudges." In *53rd Hawaii International Conference on Systems Sciences (HICSS)*, 3928–37. Maui, Hawaii: HICSS.

Meske, C., I. Amojo, A.-S. Poncette, and F. Balzer. 2019. "The Potential Role of Digital Nudging in the Digital Transformation of the Healthcare Industry." In *HCII 2019: Design, User Experience, and Usability. Application Domains*, edited by A. Marcus and W. Wang, 323–36. Cham: Springer.

Michalowski, S. 2020. "Critical Reflections on the Need for a Right to Mental Self-Determination." In *The Cambridge Handbook of New Human Rights*, edited by A. von Arnauld, K. von der Decken, and M. Susi, 404–12. Cambridge: Cambridge University Press.

Milano, S., M. Taddeo, and Luciano Floridi. 2020. "Recommender systems and their ethical challenges." *AI & Society* 35: 957–67.

Mirsch, T., C. Lehrer, and R. Jung. 2017. "Digital Nudging: Altering User Behavior in Digital Environments." In *Proceedings der 13. Internationalen Tagung Wirtschaftsinformatik (WI 2017)*, edited by Leimeister J. M. and W. Brenner, 634–48. St Gallen, Switzerland: WI 2017.

Pariser, Eli. 2011. *The Filter Bubble: What the Internet Is Hiding from You*. New York, NY: Penguin Books.

Pasquale, Frank. 2015. *The Black Box Society. The Secret Algorithms that Control Money and Information*. Cambridge, MA: Harvard University Press.

Pettit, Philip. 1996. "Freedom as Anti-power." *Ethics* 106: 576–604.

Pettit, Philip. 1997. *Republicanism: A Theory of Freedom and Government*. New York: Oxford University Press. http://search.ebscohost.com/login.aspx?direct=true&scope=site&db=nlebk&AN=1204675.

Pettit, Philip. 2000. "Republican Liberty and its Constitutional Significance." *Australian Journal of Legal Philosophy* 25 (2): 237–56.

Pettit, Philip. 2003. "Agency Freedom and Options Freedom." *Journal of Theoretical Politics* 15 (4): 387–403.

Pettit, Philip. 2004. "Trust, Reliance and the Internet." *Analyse und Kritik* 26: 108–21.

Pettit, Philip. 2008. "Republican Liberty: Three Axioms, Four Theorems." In *Republicanism and Political Theory*, edited by C. Laborde and J. Maynor, 102–30. Oxford: Blackwell.

Pettit, Philip. 2011. "The Instability of Freedom as Noninterference: The Case of Isaiah Berlin." *Ethics* 121: 693–716.

Pettit, Philip. 2012. *On the People's Terms: A Republican Theory and Model of Democracy*. The Seeley Lectures. Cambridge: Cambridge University Press.
Pettit, Philip. 2014. *Just Freedom: A Moral Compass for a Complex World*. New York, NY: W.W. Norton & Company.
Pettit, Philip. 2015. "Freedom: Psychological, Ethical, and Political." *Critical Review of International Social and Political Philosophy* 18 (4): 375–89.
Pettit, Philip. 2019. "The General Will, the Common Good, and a Democracy of Standards." In *Republicanism and the Future of Democracy*, edited by Yiftah Elazar and Genevieve Rousseliere, 13–40. Cambridge: Cambridge University Press.
Powers, E. 2017. "My News Feed is Filtered?" *Digital Journalism* 5 (10): 1315–35.
Sandven, H. 2020. "Systemic Domination, Social Institutions and the Coalition Problem." *Politics, Philosophy & Economics* 19 (4): 382–402.
Santoni de Sio, Filippo, and Gulio Mecacci. 2021. "Four Responsibility Gaps with Autonomous Systems: Why they Matter and How to Address Them." *Philosophy and Technology* 34: 1057–84.
Schmidt, A. T. 2018. "Domination without Inequality? Mutual Domination, Republicanism, and Gun Control." *Philosophy and Public Affairs* 46 (2): 175–206.
Schmidt, A. T., and B. Engelen. 2020. "The Ethics of Nudging: An Overview." *Philosophy Compass* 15 (4): e12658. https://doi.org/10.1111/phc3.12658.
Shapiro, Ian. 2012. "On Non-domination." *University of Toronto Law Journal* 62 (3): 293–336.
Sharon, Tamar. 2016. "The Googlization of Health Research: From Disruptive Innovation to Disruptive Ethics." *Personalized Medicine* 13 (6): 563–74.
Sharon, Tamar. 2021. "From Hostile Worlds to Multiple Spheres: Towards a Normative Pragmatics of Justice for the Googlization of Health." *Medicine Health Care and Philosophy* 24 (3): 315–27. doi:10.1007/s11019-021-10006-7.
Shepherd, Lee, Ronan E. O'Carroll, and Eamonn Ferguson. 2014. "An International Comparison of Deceased and Living Organ Donation/Transplant Rates in Opt-in and Opt-out Systems: A Panel Study." *BMC Medicine* 12 (1): 131. doi:10.1186/s12916-014-0131-4.
Smith, P. T. 2020. "A Neo-Republican Theory of Just State Surveillance." *Moral Philosophy and Politics* 7 (1): 49–71.
Stahl, T. 2016. "Indiscriminate Mass Surveillance and the Public Sphere." *Ethics and Information Technology* 18: 33–39.
Stark, Findlay. 2016. *Culpable Carelessness: Recklessness and Negligence in the Criminal Law*. Cambridge: Cambridge University Press.
Stibe, A., and B. Cugelman. 2016. "Persuasive Backfiring: When Behavior Change Interventions Trigger Unintended Negative Outcomes." *Lecture Notes in Computer Science*, 65–77.
Sunstein, Cass R. 2016. *The Ethics of Influence: Government in the Age of Behavioral Science*. Cambridge: Cambridge University Press.
Thaler, Richard H., and Cass R. Sunstein. 2008. *Nudge: Improving Decisions About Health, Wealth, and Happiness*. New Haven, CT: Yale University Press.
van der Sloot, B. 2018. "A New Approach to the Right to Privacy, or How the European Court of Human Rights Embraced the Non-domination Principle." *Computer Law & Security Review* 34: 539–49.
Weinmann, Markus, Christoph Schneider, and Jan vom Brocke. 2016. "Digital Nudging." *Business & Information Systems Engineering: BISE: The International Journal of Wirtschaftsinformatik* 58 (6): 433–36.

Whitfield, G. 2020. "On the Concept of Political Manipulation." *European Journal of Political Theory* 1–25.

Wood, Allen W. 2014. "Coercion, Manipulation, Exploitation." In Coons and Weber 2014b, 17–50.

World Health Organization. 2020. "Managing the COVID-19 Infodemic: Promoting Healthy Behaviours and Mitigating the Harm from Misinformation and Disinformation." Accessed February 2021 from www.who.int/news-room/detail/23-09-2020-managing-the-covid-19-infodemic-promoting-healthy-behaviours-and-mitigating-the-harm-from-misinformation-and disinformation.

Yeung, Karen. 2015. "Design for the value of Regulation." In *Handbook of Ethics, Values, and Technological Design: Sources, Theory, Values and Application Domains*, edited by Jeroen van den Hoven, Pieter E. Vermaas, and Ibo van de Poel, 447–72. Dordrecht: Springer.

Yeung, Karen. 2017. "Hypernudge: Big Data as a Mode of Regulation by Design." *Information, Communication & Society* 20 (1): 118–36. doi: 10.1080/1369118X.2016.1186713

10 Gamification, Manipulation, and Domination[1]

Moti Gorin

1 Introduction

In this chapter, I will argue that a family of accounts of manipulation, though differing in their details, can explain why gamification is a form of interpersonal manipulation. The accounts I will describe are what I call *norm-based* accounts, as they all take the violation or misuse of norms to be central to manipulation. The normativity at issue is not or need not be moral normativity, though the moral status of manipulation ultimately will be determined by the way in which, and the purpose for which, the norms in question are violated or misused. Gamification will turn out to be a method of embedding game rules and the motivations generated by these rules within larger practices that are normatively independent of the game rules embedded within them. By embedding game rules within these larger normative structures and practices, gamifiers leave their players detached from the considerations that ought to govern their behavior within the larger practices. This is not always wrong, and though I cannot offer a comprehensive theory of the wrongness of manipulation, I will suggest that sometimes the arbitrariness of the relation between game rules and the wider norms that govern the practices within which these rules are embedded make gamification a form of domination. If this is correct, then the wrongness of gamification manipulation may be accounted for in terms of the unfreedom it promotes.

2 Gamification

For purposes of showing that gamification is manipulative it will not be necessary to settle debates about the precise nature of games or to distinguish between different kinds of games or game playing. Instead, I will offer what I hope is a fairly neutral and intuitive account of gamification that is friendly to a range of more detailed and elaborate accounts of games and gamification. The account I offer should be abstract enough to avoid commitment to contentious or otherwise unsettled claims regarding how best to understand games or gamification.

DOI: 10.4324/9781003205425-12

Whatever else is true of games, they always involve an end or a goal and a system of rules that constrains players' behavior in the achievement of that end or goal. Crucially, the goals and rules of games are *artificial* and they are *normatively independent*. They are artificial because they are stipulated by their creators (who may also be the players). They are normatively independent because they do not admit of other rules or goals or ends that have not been stipulated by their creators; a player cannot achieve the goal of the game, *qua* game, except by doing so in accordance with the rules that dictate how the goal is to be achieved in the game. This does not mean that game rules cannot change, even mid-game, or that rules or norms that exist outside games cannot figure in games. It does mean that when rules change or when non-game rules or norms are incorporated into a game – for example, when some driving game incorporates the rules (i.e., laws) of the road – these changes and rules/norms are subordinate to the rules of the game.

The system of rules need not be complex or explicit. For example, if we are on the beach and decide to play Who Can Skip a Stone the Most Times, we may realize that we are committed to the rule "you must use only stones that are found on this beach during the game" only when one of us pulls out from a backpack a collection of smooth, flat stones which were collected earlier from a river bottom. Or, we might have a debate about whether we should allow the use of river stones, which amounts to a debate about the rules of the game we are in the process of creating. And how we decide this question not only will determine whether use of river stones will from now on be part of the game – that is, consistent with the rules – but also perhaps whether the pre-stipulated use of the river stones violated the normative independence condition on what it is to be a game, that is, whether the skipping of the river stones was consistent with the rules of the game we were playing at the time the river stones were used.[2]

Games are systems of goals and rules that generate motivation on the part of players to achieve the goals of the game as constrained by the rules. The precise nature of this motivation and the role it plays within a player's broader motivational set differ depending on the motivational set of the player and the precise nature of the game. C. Thi Nguyen distinguishes between achievement play and striving play. Achievement players play for either the sake of winning or for the sake of gaining some desirable outcome of winning, such as financial gain, honor, the right to play another iteration of the game (as in tournaments) or against higher-ranked players, and so on. In short, achievement players play to achieve a win or some further good that is attached to the win. Striving play involves a different kind of motivation. Here, players adopt the goal of the game because they desire the experience of overcoming the obstacles to winning, or because they wish to test their mettle, or because working to overcome the constraints set by the rules sharpens some ability of theirs. For pure striving players, the primary role that adopting the goal of winning plays is to make striving possible, while for pure achievement players, the striving to win is merely a means to the end of achieving the win or the further goods attached to winning (Nguyen 2019).

Nguyen has many interesting things to say about the differences between achievement play and striving play and how the motivations that drive these kinds of play shape our agency and, consequently, the aesthetic qualities to which game playing can, on his view, give rise. For my purposes here, I wish only to focus on the similarities between achievement play and striving play. Whether one plays a game in order to win by overcoming or plays in order to overcome by aiming to win, or both, one's more specific motivations within the game itself are given by the rules of the game. If you and I are playing chess together, we both adopt the aim of winning the game, even if I purely am playing for the money while you purely are playing for the experience of overcoming my strategies to defeat you. This means that whatever role winning plays in our motivation to play (either as an end or as a means to an end), our play itself will be subject to the same standards of assessment, which are given by the game itself. A "good move" in chess is – perhaps among other things such as boldness or creativity – primarily a move that, according to the rules of chess, makes it more likely that the player who made that move will win the game, whatever her motivations. In this way, the system of rules and goals that constitute the game generates reasons for players to do this or that while playing. The powers granted to your queen by the rules of chess provide me a reason to take it with my bishop when you leave it exposed (though perhaps other features of the specific game of chess we are playing provide me stronger reasons to pass up this opportunity). Let's call the reasons generated by the rules of a game *game reasons*.

Gamification can be understood as the imposition of game reasons onto domains in which game reasons – that is, reasons generated by artificial, deliberately stipulated goals and rules – do not ordinarily apply.[3] The purpose of imposing game reasons in this way is to motivate behavior in the non-game domain by aligning game goals – and the motivations to which they give rise – with the desired non-game behavior. There are many examples of gamification. Here are just a few:

1. Allotting points to children for reading books over the summer.
2. Awarding "sobriety coins" to recovering alcoholics.
3. Quantifying and displaying certain forms of online platform engagement (likes, followers, consecutive days engaged, etc.).
4. Offering rewards (badges, discounts, etc.) for using a company's app (e.g., fast food apps, fitness apps).
5. Displaying a scoreboard at a place of work that tracks employees' completion of tasks.

In each of these examples, goals and rules are stipulated such that they generate game reasons that align with non-game reasons. For example, the alcoholic is motivated to earn a sobriety coin and thus abstains from drinking, which abstention is supported by – aligned with – welfare considerations. The alignment in question can be understood in terms of mutual satisfaction: by satisfying the demands of the game reason, "players" also satisfy the demands of the non-game reason and vice versa.

I will argue in the following that the alignment relation between game reasons and non-game reasons renders gamification manipulative and that the moral status of a particular instance of gamified behavior, including the extent to which the manipulation is an instance of domination, depends on certain features of the non-game reasons at play as well as on the capacities of the manipulated agents to respond to non-game reasons. It will turn out that allotting points to children for reading or to addicts for remaining sober is manipulative but generally not wrongful, while gamifying online engagement – the central target of this chapter – often is wrongful.

3 Manipulation

In my view, the most plausible accounts of manipulation are those Robert Noggle (2018a, 2018b) classifies as *trickery accounts* and, more specifically, the accounts defended by Noggle (1996), Anne Barnhill (2014), and myself (Gorin 2014a, 2014b). I prefer to call these accounts *norm-based accounts* and the difference is not merely terminological. Noggle describes trickery accounts as approaching manipulation by "t[ying] it conceptually to deception" (Noggle 2018b). I have argued elsewhere that manipulation need not involve deception and will not revisit those arguments here (Gorin 2014b).[4] For current purposes, addressing the question of whether manipulation is best understood as being "conceptually tied" to deception will take us too far afield and nothing I say here hangs on the answer to this question.

What the norm-based accounts of Noggle, Barnhill, and Gorin have in common is that they take the failure to conform with norms to be central to manipulation. More specifically, each of these accounts takes manipulators intentionally to be causing their manipulees to behave in ways that fall short of the ideals governing the relevant domain of behavior.[5] In an early and influential paper, Noggle puts it like this (Noggle 1996, 44):

> There are certain norms or ideals that govern beliefs, desires, and emotions. Manipulative action is the attempt to get someone's beliefs, desires, or emotions to violate these norms, to fall short of these ideals.

It is not always clear what the domain-specific norms or ideals are, exactly. It is perhaps easiest to discern these norms and ideals in the case of belief, though I cannot give a full account here. At a minimum, beliefs should conform to the evidence – the stronger the evidence or warrant, the more firmly epistemic agents should hold the belief. Plausibly, the ideal of belief is truth or knowledge and thus certain features of beliefs – the extent to which they are justified or their relation to the truth, perhaps – determine how close beliefs and the agents who hold them come to realizing this ideal. For example, a firm belief that Bill Gates and Dr. Anthony Fauci have conspired to put microchips in the COVID-19 vaccine in order to control our thoughts is the result of a violation of epistemic norms, at least when it is based on nothing but a Facebook post from some anonymous user.

The norms of emotion do not make references to truth or knowledge, at least not directly. Here, the ideal may make reference to a kind of fittingness relation which comprises of a more complex set of relations between the cause of the emotion and its intensity, duration, moral valence, and so on, as well as its relation to other mental states such as belief and desire. For example, absent some unusual backstory, deep and lasting grief is not a fitting emotion when it is a response to one's having temporarily misplaced one's car keys. The norms and ideal(s) of desire are more difficult to specify, in no small part due to Humean skepticism about the relation between reason and desire. But it should be clear enough what such ideas and norms might look like, if there are any. First, as Noggle points out, we might want desire to conform to the dictates of rationality such that we desire to do what we believe we have good reason to do (Noggle 1996, 45). Second, desires may, like emotion, be subject to norms of fittingness, where intensity, content, moral valence, and duration of desire provide metrics along which desires can be judged appropriate or not.

On Noggle's view, manipulators influence others by causing them to acquire mental states that do not conform to the norms that govern the relevant type of mental state. So, a manipulator may lead a manipulee to have beliefs that are false or unjustified, or emotions that do not "fit" their objects, or desires that do not conform to the agent's beliefs about what she has best reason to do. And of course, a manipulator ultimately may be less interested in the mental states themselves, aiming instead to bring about some outward action to which the faulty mental states will predictably lead.

Barnhill defends a view similar to that of Noggle but modified to make explicit the relation between the aims of the manipulator and the interests of the manipulee. On her view (Barnhill 2014, 72):

> [M]anipulation is directly influencing someone's beliefs, desires, or emotions such that she falls short of ideals for belief, desire, or emotion in ways typically not in her self-interest *or likely not in her self-interest in the present context.*

Manipulators bring about the manipulee's norm-violating mental states, but the norm violation must be such as to set back the interests of the manipulee in order for the influence to count as manipulative. The connection between the interests of the manipulee and the violation of the relevant norms is important because it allows us to distinguish between influence that is intuitively morally bad – because it makes the manipulee worse off – and influence that is not always intuitively bad, such as when the norm violation leads to an outcome that is good, or likely to be good, or typically good for the manipulee.

Consider a patient who is known to his doctor to be a stubborn curmudgeon. The doctor knows it is in the best interests of her patient to receive the COVID-19 vaccine and also knows he's unlikely to consent to it, given his hard-headed nature and general distrust of and antipathy to other people. When the patient arrives, the doctor is sure to spend some extra time with him, listening sympathetically to his complaints about his neighbors, his family, the nursing staff, the post office, barking dogs, kids these days, and so on. She also arranges, before his arrival, to have the office audio system play music she knows her patient loves – music is one of the few things that bring him joy and causes him to relax his otherwise grouchy temperament. Though he isn't thrilled to do so – he directs some ire against the CDC while getting the injection – the patient grants his informed consent.

Did the physician manipulate her patient? On Barnhill's account, this will depend on whether his acquisition of the mental states that led him to consent were 1) in violation of the norms of belief, desire, or emotion and 2) typically not in his self-interest or unlikely to be so in the present case. Establishing whether or not these conditions are met would require us to know much more about the patient and specifically whether his temperament typically leads him to acquire mental states and to do actions that are in his self-interest or not. It is not important that we answer this question here, so long it is clear enough how Barnhill's account could be applied to answer it.

On the account of manipulation I defend, A manipulates B if and only if A deliberately and non-coercively influences B to x and at least one of the following conditions is met (Gorin 2018, 238):

1. A believes that B lacks sufficient reason to x.
2. A believes that B has sufficient reason(s) to x, but A is not motivated by this reason(s).
3. A's influence of B is motivated by B's sufficient reason to x, but A leads B to x in light of some other reason.
4. A exploits means of influence that do not reliably track reasons.

It will be helpful to say something about the four conditions it sets out as well as the inclusion of deliberativeness as a necessary condition on manipulation.

It is mostly fine for my purposes here to take "deliberate" as being interchangeable with "intentional." But intentionality can be understood as existing on a scale, with some actions being less intentional and others being more intentional. A stressed and frustrated parent who has not completely lost self-control nevertheless may yell at his child and do so intentionally, but it does not follow from this that every word, intonation, or decibel has been selected with care or foresight. I understand "deliberate" to denote actions that are intentional to a higher degree, meaning the agent intends not only to do some action but to do it for a settled purpose where, moreover, some attention and consideration have been paid to the precise nature of the specific means chosen to achieve that purpose. One might intentionally put out a campfire, say, by dumping water on it, but one would do so deliberately, in the sense I am after, only if one considered things like how much water to use (using too much will make it difficult to start a fire tomorrow), whether to dump the water quickly or slowly, and whether to use water rather than sand or soil.

So, on my view of manipulation, manipulators will always be agents capable of deliberate – and not merely intentional – influence. When applied to the online context, this means that algorithms, bots, gamified platforms, platform design, and so on are only derivatively manipulative. The primary manipulators – in fact the only "true" manipulators – are the people who employ these artifacts to achieve their ends.

Consider a case of manipulation that is uncontroversially morally neutral: a chef's manipulation of a knife. At a rather abstract level, what a chef wants to do with a knife is to rearrange matter in space, mainly by making parts out of wholes or making smaller parts out of larger parts and by combining previously disparate parts with other parts, and so forth. To do this, the chef manipulates the knife. If a chef wears gloves to achieve a better grip or removes the factory handle from the knife and replaces it with an ergonomically superior handle, this will allow her to better manipulate the knife, that is, she will now have a degree of control over the knife that was previously lacking: she can be more deliberate in her cutting. In such a case, it would be quite odd to suggest that the glove or the new handle manipulates the knife. Instead, we should say that the chef has employed some artifacts that allow her better to manipulate the knife. My suggestion is that the same goes for gamification: it is an artifact that allows some people better to manipulate other people, but it is not itself a manipulator. Of course, there is much more to say about intentionality, agency, and manipulation, but the view I am laying out should be clear enough.

Now to the four conditions that, in addition to the deliberativeness and non-coercive conditions, are disjunctively necessary and sufficient for interpersonal manipulation. I will set them out again and give an example of each.

1. A believes that B lacks sufficient reason to x.

I want you to come to the Grateful Dead concert with me, but I know you have terrible taste in music and therefore hate the music of the Grateful Dead and cannot tolerate listening to it for any significant length of time without becoming miserable. I lie to you about how long the concert will last and only then do you agree to come with me to the concert.

2. A believes that B has sufficient reason(s) to x, but A is not motivated by this reason(s).

I want you to come to the Grateful Dead concert with me. You have never been to such a concert, and I sincerely believe you will have a terrific time and that this provides you with sufficient reason to come. But I do not in fact care, at all, whether you have a good time or not. I need a ride to the concert and this need of mine is what motivates me to invite you. I convince you to come with me by citing the sufficient reason I believe you have, namely that you will have a great time. In short, I have an ulterior motive in inviting you. You agree.

3. A's influence of B is motivated by B's sufficient reason to x, but A leads B to x in light of some other reason.

I want you to come to the Grateful Dead concert with me. You have never been to such a concert, and I sincerely believe you will have a terrific time and that this provides you with sufficient reason to come. If I did not think you would have a great time, I would not invite you. I really, truly think you will enjoy the concert, and this is what motivates me to invite you. But I know you will not go if I cite this reason. Perhaps, we generally disagree about what counts as a fun time, or perhaps you are a bit depressed and so will not be feeling motivated enough to go if I cite your enjoyment of the concert as a reason for you to go. So, rather than citing the good time I think you'll have at the concert as a reason to go, I remind you that your favorite burrito place is just next door to the concert venue. "Let's go the concert," I say, "and then pick up some of those burritos." I think these burritos are disgusting and the last time we went there we got food poisoning. So, I am definitely not motivated by the burritos, nor do I think you should be. But, you agree to come with me to the concert because you want the burritos.

4. A exploits means of influence that do not reliably track reasons.

I want you to come to the Grateful Dead concert with me. I may or may not believe you have sufficient reason to do so. I employ some nudging methods[6] to get you to agree or I cause you to feel some emotion that makes you more amenable to my influence. For example, I remind you of the gift I recently gave you in order to get you to feel guilty and indebted to me. I could exploit this guilt to make it less likely that you would come to the show with me, if that is what I wanted. Crucially, the means I adopt to influence you are neutral with respect to the direction that influence might take. They are attached not to whatever reasons you might have but instead to my will. This arbitrariness with respect to reasons will play a central role in the following discussion of gamification and domination.

This account of manipulation is a norm-based account in the following sense: following Tim Scanlon, I take a reason to x to be a consideration that counts in favor of x-ing (Scanlon 1998, 10). Reasons are, on this view, normative – they tell us what we should do. Reasons guide action, which means not only that we should do what we have reason to do but that we should do it because of or in light of the relevant reason, that is, because of or in light of the consideration that counts in favor of doing that thing. If I do what I have most reason to do because I have most reason to do it, that is better, from the point of view of practical reason, than if I do what I have most reason to do but do it for some other reason or for no reason at all. For example, if the strongest reason I have to consent to the COVID-19 vaccine is that it will protect my health and that of those around me and yet I consent only because, say, I wish to impress my co-workers with my "I Got the Vaccine!" sticker, then although I have done what I have good reason to do I have not done it for that reason.

When we influence others we often do so by giving them reasons. Sometimes we generate the reasons ourselves, as when we coerce someone or offer them an incentive, and sometimes, more often, we point to the reasons we believe they have, reasons that are independent of anything we have done. Reasons can be means of influence, with rational persuasion being the paradigmatic (but perhaps not the only[7]) example of influence via reasons. Ideally, interpersonal influence is reason-preserving in a way that is somewhat, though only loosely, akin to the way that valid arguments are truth-preserving. As every first-year logic student learns (sometimes with difficulty), an argument is valid when the truth of the premises guarantees the truth of the conclusion, that is, when assigning "true" to all the premises (whether they are actually true or not) and "false" to the conclusion is logically impossible. It is *logically* impossible, but not impossible in other ways, which is just to say that it is psychologically possible – probably we all do it quite frequently – to violate the laws of logic. When we do this, we violate the norms of theoretical reason. Sometimes, manipulative influence fails to preserve the normative force of reasons that should, ideally, transfer from the influencing agent to the target of her influence while sometimes such influence generates motivation that is lacking in rational support. Norm-violating accounts of manipulation are not merely accounts of how agents can fail to believe, feel, or desire in accordance with the ideals that govern these states. They are accounts of how one set of agents can cause another set of agents to fail in these ways.

What I want to suggest is that manipulators deliberately influence others in ways that run afoul of norms of practical reasons. They do this when, as in (1) and (3) earlier, they lead their manipulees to act for reasons the manipulator regards to be bad reasons, that is, for considerations that either do not speak in favor of the manipulee's doing that thing at all or for considerations that the manipulator regards to be merely motivating reasons (for the manipulee) but not normative reasons. Or when, as in (2) and (4), the manipulator is indifferent to the relation that obtains between the manipulee and whatever reasons the manipulator believes should guide the manipulee's actions.

4 Gamification, Again

I said earlier that gamification can be understood as the imposition of game reasons onto domains in which game reasons – that is, reasons generated by artificial, deliberately stipulated goals and rules – do not ordinarily exist.[8] So, when Twitter platform designers gamify online discourse by introducing a system of points (likes, shares, follower counts, engagement counts, etc.), what they do is introduce game reasons into a domain – human discourse – in which such reasons typically do not exist, at least not overtly.[9] Teachers who award points to students who read, designers of other online platforms or workplace managers who include "streak counts" or similar feedback mechanisms that mark the achievement of stipulated goals, and substance abuse treatment organizations that provide sobriety coins do the same. On the account of manipulation sketched earlier, *any* form of gamification is manipulative because the reasons given by the game are normatively independent of the reasons that govern the relevant domain. One may have good reasons to remain sober, or to read lots of books, or to engage in online communication, or to exercise every day, but when this is the case, then these good reasons should guide one's actions – this is just what it is to be a good reason.

At this point it is important to note that the view of manipulation set out earlier is morally neutral. The moral status of a particular instance of manipulation will depend on certain features of the case, namely the aims of the manipulator and the extent to which the manipulee(s) is reasons-responsive. Manipulation that aims at the good of a manipulee who is not responsive to reasons is not even pro tanto wrong. For example, a parent who gamifies her young child's meals in order to ensure a healthful diet does not need to cite some countervailing and more weighty consideration that trumps the reason she has to avoid manipulating her child – there is no such reason. This is because the beliefs, emotions, and desires of a young child are not subject to the norms applying to fully developed adults. On the other hand, manipulating reasons-responsive agents is pro tanto wrong, but sometimes, perhaps often, there will be countervailing considerations that render the manipulation all-things-considered permissible or obligatory. For example, gamifying a medication regimen so that a vulnerable but reasons-responsive patient takes her pills on schedule is pro tanto wrong but all-things-considered permissible. Though I will not argue for it here, I take it that the use of sobriety coins and points for reading typically are cases of morally permissible manipulation. Judgments as to the morality of any given case of manipulation will depend on the details of the case and sometimes it may not be easy to render a confident verdict. This is as it should be. Morality would be complicated even human beings were perfectly reasons-responsive and clearly reasons-responsiveness among humans comes in degrees, with no one approaching the ideal.

If I am right that gamification is always manipulative, then there is nothing conceptually special or distinctive about online gamification. Online gamifiers will make use of digitized forms of gamification but what they do fundamentally is no different than what offline gamifiers do as far as manipulation is concerned. It does not follow from this, however, that scholars and policy makers have no good reasons to distinguish between online and offline gamification. Just as nuclear weapons receive more attention than hand grenades do in virtue of their awesome power, there may be good reasons, grounded in differences in degree rather than in differences in kind, to pay special attention to online gamification. As things stand today, a small group of online gamifiers with narrow, mercenary motives and vast fortunes manipulate billions of people using artifacts that are, for the first time in history, figuratively if not (yet) literally physically attached to their users. The scope of the manipulation, its ubiquity, relentlessness, and the shabby, commercial ends to which it is so often put to use do give us strong reason to single it out for analysis, critique, and possibly regulation.

5 Freedom as Non-Domination

Up to this point I have argued that gamification is a form of manipulation. What does this tell us about the moral status of manipulation? Though I cannot provide a comprehensive answer to this question here, I will argue that manipulation generally and gamification specifically undermine the liberty of their targets. The argument draws upon the core idea of freedom as articulated in the republican tradition, that is, freedom as non-domination.

The republican tradition in political theory and philosophy offers a distinctive conception of political liberty. Rather than conceiving of liberty as non-interference or as the capacity to live autonomous or authentic lives – the former conception being "negative liberty" and the latter "positive liberty" on Isaiah Berlin's (1969) famous articulation of political liberty – republican theorists understand political liberty as non-domination. According to freedom as non-domination, a person is free to the extent that she is not subject to the arbitrary power of another. To be subject to such power is to be dominated. It does not matter whether this power is exercised in a manner that actually undermines the agent's will or causes her some other harms. What matters is the structural relation that obtains between the dominator and the dominated. The intuitive idea is that when one agent *can* interfere with another "on a whim," as it were, then the latter agent is not free. The classic example here is that of the benevolent slave master who treats his slave with a gentle hand or perhaps even leaves her alone entirely. Republicanism can explain why such a slave is still unfree while defenders of freedom-as-non-interference cannot.[10]

One of the debates within the literature on republicanism concerns the notion of arbitrariness. What does it mean to say that someone is subject to the arbitrary power or will of another? Some have argued that arbitrariness is best understood procedurally. On this view, roughly, domination can be avoided so long as the rules or procedures that govern the relations between people (or between people and their governments) are set out in ways that are publicly accessible, stable, and consistently applied. Such an arrangement allows for people to arrange their activities, expectations, and plans, without worry that at any moment others will disrupt their lives unpredictably in ways they cannot control (Lovett 2012). An alternative conception of arbitrary power/will requires that the rules, aims, and procedures be not merely consistently applied, publicly accessible, and stable but that their content be justified on substantive grounds. Defenders of the substantive account of arbitrariness point out, rightly in my view, that the stability, transparency, and consistent application of procedures or rules do not ensure that the procedures or rules can be justified in their content (Arnold and Harris 2017). For example, unjustly discriminatory laws could be passed, widely publicized, and consistently applied. Such laws would, indeed, allow everyone to take part in activities and form and implement life plans structured by procedurally impeccable rules or policies, such that no one need to worry about unexpected interferences in their lives. And yet unjust discrimination is still unjust, irrespective of whether or not it is consistently applied, publicly legislated, or expected.

Here is how Arnold and Harris, following Lovett, define substantive arbitrariness (Arnold and Harris 2017, 58):

> Substantivism: B's power over A is arbitrary insofar as it is not reliably constrained by effective rules, procedures, or goals that meet some further substantive requirement (or requirements) R (or R1. . . Rn).

Substantivism itself can take several forms, including interest substantivism, which is the account Arnold and Harris defend (Arnold and Harris 2017, 58):

> Interest Substantivism: B's power over A is arbitrary insofar as it is not reliably constrained by effective rules, procedures, etc., that force B to track A's interests in the exercise of that power.

The addition of the substantive requirement is meant to solve the problem that plagues pure procedural accounts of arbitrariness, described earlier. Interest substantivism holds that the relevant substantive content should be filled in by the interests of those to whom the procedures/rules apply. Here, it is not enough that the state or one's employer or whoever else may be exercising power apply rules or procedures consistently and transparently – these rules must not undermine the interests of those whose behavior they structure.

I wish to introduce another variant of substantivism, which can make clear the connection between manipulation and freedom as non-domination.

> Reasons Substantivism*: B's power over A is arbitrary insofar as it is not reliably constrained by effective rules, procedures, etc., that force B to track A's reasons in the exercise of that power.

Freedom as non-domination is ordinarily a theory of political freedom, that is, a theory of what makes people free or unfree in their relations to their government or to one another *qua* citizens.[11] Reasons substantivism*, though, may not be a good theory of political freedom. This is because political authorities, such as governments or their representatives, do not have a general duty to promote reasons-responsiveness, per se. Governments or their representatives do have duties to protect or promote our interests, though of course there is plenty of disagreement among political philosophers about what interests, or what kinds of interests, ought to be promoted by the state. The main point I want to make here is that one can have a reason to x, even a sufficient reason, without it thereby being the case that x-ing will promote one's interests. One might have good reason to believe, desire, or feel something without its being the case that it is in one's interests to believe, desire, or feel that thing. Thus, reasons substantivism* is meant to capture a broader range of types of arbitrariness, including those realized in domains that are not obviously political.

It may be objected here that insofar as domination is cashed out in terms of arbitrary *power*, it is not a useful frame through which to understand interpersonal manipulation. This is because the best account of power may be irreducibly political or weaker, that the term "power" connotes political relations and therefore muddies the waters when the phenomena that interest us are not, or need not be, political. Rather than responding to this objection with an account of power broad enough to capture political as well as non-political relations, I will opt instead for the following account of reasons substantivism which does not invoke power:

> Reasons Substantivism: B's influence of A is arbitrary insofar as it is not reliably constrained by effective rules, procedures, etc., that force B to track A's reasons in the exercise of that influence.

The parallels between domination understood as a violation of reasons substantivism, on the one hand, and the norm-based accounts of manipulation described earlier, on the other hand, should be emerging. Manipulators deliberately cause their manipulees to adopt mental states or to do actions in ways that are not reliably constrained by effective rules, procedures, and so on and that do not constrain the manipulators to track their manipulee's reasons in the exercise of their influence. On Noggle's account, the failure to track reasons is cashed out in terms of the violation of the norms that govern the relevant domain (belief, desire, emotion), while on Barnhill's view, these norms are not only violated but violated in ways that do not, or typically would not, be in the interests of the manipulee. And the same goes for my reasons-tracking view. In manipulation forms 1–4, either the manipulator is not constrained by his manipulee's reasons – as when he does not believe there are such reasons or when he is indifferent to their existence – or, when he is constrained by the manipulee's reasons (as in form 3), the means he selects are not effective rules or procedures that force him to track those reasons (as when the manipulator cites some bad reason).

Gamification, then, being a form of manipulation that works via the substitution of game reasons for non-game reasons (when the latter exist), qualifies as influence that violates reasons substantivism. Therefore, gamification is a form of domination, just as all manipulation is domination, that is, arbitrary influence.

At this point, it may be objected that the republican understanding of freedom as non-domination is a structural, relational account of freedom rather than one that requires any actual interference in the behavior of the agent whose freedom is undermined. That this account does not require actual interference is in fact the distinctive feature of republican freedom, the feature that sets it apart from its competitors. But to have been manipulated is to have been influenced in some way. Manipulees do not merely stand in a particular sort of relation to their manipulators; the latter actually influence the former. Because it is a form of actual interference, manipulation cannot be a form of domination. Thus, insofar as gamification or manipulation is said to involve domination, it must do so in a way that differs substantially from the kind of domination a slave suffers at the hands of a slave master or which citizens suffer at the hands of an unrestrained government. It is a mistake, then, to analyze manipulation or gamification through the lens of the republican conception of liberty.[12]

This objection gets one thing right: if manipulation is a form of interpersonal influence – which it is – then clearly it cannot be an example of domination conceived in purely relational terms. The best response to this objection, or in any case the response I wish to give here, is to distinguish between two kinds of domination. The first kind is relational or structural domination, which is the kind of domination at the center of the republican conception of political liberty. This kind of domination, crucially, requires no actual interference or interaction between dominator and dominated. The second kind of domination is interactive domination, where what the influenced agent does depends on what the influencing agent does, and not merely on what it is in the latter's power to do. In this way, interactive domination differs substantially from the sort of domination that forms the core of republican liberty, and thus what I have argued for here with respect to manipulation cannot directly tell us anything about political liberty, and vice versa. The feature of domination shared by those who are unfree in the republican sense, on the one hand, and those who are manipulated, on the other, is that both are subject to the arbitrary wills of others. The key difference is that in the case of the manipulated, the arbitrary wills to which they are subjected are not only potentially exercised, but actually exercised via the manipulative interaction.

6 Conclusion

I have argued that gamification is a form of manipulation, that by substituting game reasons for non-game reasons, gamification causes agents to behave in ways that violate norms of practical reason. I then went on to argue that this violation can be understood in terms of arbitrariness, where the manipulee's reasons do not constrain or guide the influence of the manipulator.[13] Finally, I suggested the arbitrariness of manipulative influence, the lack of alignment between the considerations that ought to govern one's behavior and the considerations that actually motivate one's actions when manipulated, makes manipulation a form of domination. I called this form of domination, which requires actual interference, *interactive domination* to distinguish it from the kind of structural or relational domination at the center of republican conceptions of political liberty.

I have thus far only very briefly gestured at the ethical dimension of gamification and manipulation.[14] This is due partly to considerations of space. The other constraint, which is no less real, is that I do not yet quite know what to think about the normative significance of interactive domination. Relational domination of the sort captured by the republican conception of political unfreedom is just that – an account of unfreedom in *political* relations. But it is not obvious to me that what is wrong with manipulation generally or gamification more specifically is that they undermine anyone's freedom, and it seems obvious that even if they do undermine freedom, the freedom at issue here is not limited to political freedom. There is also the question of whether or how to balance the moral badness of gamification and manipulation, when there is such badness, however characterized, with whatever benefits are brought about by the manipulation. On the face of it, it seems plausible that there would be an important moral difference between, say, gamifying online engagement in order to sell useless products, on the one hand, and doing so in order to improve the quality of civic discourse, on the other. In any case, exploring such matters will have to wait for another occasion.

Notes

1. I wish to thank the organizers and other participants of the workshop out of which this chapter and volume emerged for their careful attention and constructive criticism.
2. One can imagine how this debate might go. The player who used the river stones might say, "hey, we never agreed that we couldn't use river stones, and so my

use of these stones was not a violation of the independence condition." Another player might say, "well, we just took it for granted that everyone understood that only beach stones could be used. After all, we are on the beach now, and clearly the game would be worse if we allowed river stones, since only you have river stones, and river stones provide a clear advantage when it comes to skipping stones." If these people are philosophers, this debate might drag on for a long time, probably spoiling whatever fun they were having skipping stones.

3. Because games can themselves be gamified, this simple definition will be amended later.
4. But see Parmer, in this volume.
5. I'm using "behavior" in the broad sense here and intend it to cover things like outward actions, the adoption of intentional mental states like beliefs or desires/preferences or intentions, as well as mental state that can be but need not be intentional such as moods.
6. Maybe I provide you a menu of options that will likely trigger ordering effects, or perhaps I tell you "everyone is doing it" in an effort to trigger the bandwagon effect. There are plenty of ways to nudge. But see Levy (2019) for an argument that nudges do track reasons.
7. I might believe you have reason to experience the sublime and so I may direct your attention to some work of art, or some natural wonder, in order to influence you in a way that will lead you to have this experience. I leave it open whether this should count as an example of rational persuasion.
8. There is a caveat here: games themselves can be gamified. For example, some people engage in competitions to see who can solve some game in the shortest amount of time (Rubik's Cubes, video games, chess, etc.). There may even be gamified games that are themselves further gamified, such as a competition to see who can break the most game-completion speed records in some given amount of time. Maybe there are even gamified games that can be gamified. So, when I say that gamification can be understood as "the imposition of non-game reasons onto domains in which game reasons . . . do not ordinarily exist" what that really means is that the imposed game reasons are foreign to or independent of the reasons already found in that domain (whether these domain-internal reasons are game reasons or, as is in the cases that interest us here, non-game reasons).
9. For an excellent discussion of how Twitter gamifies discourse, see Nguyen (forthcoming).
10. Philip Pettit has done more than anyone else to articulate and defend the account of freedom as non-domination. For clear and early articulations and defenses of this account, see Pettit (1996, 1997).
11. I intend "citizen" here in the broad sense, which encompasses not only people with formal citizenship within some political structure (e.g., a country or nation or state) but also people to whom the authority of political organization applies (e.g., foreign nationals, undocumented immigrants, etc.).
12. I thank Marianna Capasso for pushing me on this point. For a more comprehensive account of republican freedom, see Capasso, in this volume.
13. See Klenk (2021a, 2021b).
14. I say more about the ethics of manipulation in Gorin (2018).

7 References

Arnold, S., and J. R. Harris. 2017. "What is Arbitrary Power?" *Journal of Political Power* 10 (1): 55–70.

Barnhill, Anne. 2014. "What is Manipulation?" In Coons and Weber 2014, 51–72.
Berlin, Isaiah. 1969. *Four Essays on Liberty*. Oxford: Oxford University Press.
Capasso, Marianna. 2022. "Manipulation as Digital Invasion: A Neo-republican Approach." In *The Philosophy of Online Manipulation*, edited by Jongepier, F. and Klenk, M., 180–198. New York: Routledge.
Coons, Christian, and Michael Weber, eds. 2014. *Manipulation: Theory and Practice*. Oxford: Oxford University Press.
Gorin, Moti. 2014a. "Do Manipulators Always Threaten Rationality?" *American Philosophical Quarterly* 51 (1): 51–61.
Gorin, Moti. 2014b. "Towards a Theory of Interpersonal Manipulation." In Coons and Weber 2014, 73–97.
Gorin, Moti. 2018. "Paternalistic Manipulation." In *The Routledge Handbook of the Philosophy of Paternalism*, edited by Jason Hanna and Kalle Grill, 236–47. New York, NY: Routledge.
Jongepier, Fleur, and Michael Klenk, eds. 2022. *The Philosophy of Online Manipulation*. New York, NY: Routledge.
Klenk, Michael. 2021a. "Interpersonal Manipulation." *SSRN Electronic Journal*. doi:10.2139/ssrn.3859178.
Klenk, Michael. 2021b. (Online) Manipulation: Sometimes Hidden, Always Careless, *Review of Social Economy*, 80: 1, 85–105. doi: 10.1080/00346764.2021.1894350.
Levy, Neil. 2019. "Nudge, Nudge, Wink, Wink: Nudging Is Giving Reasons." *Ergo* 6. doi:10.3998/ergo.12405314.0006.010.
Lovett, Frank. 2012. "What Counts as Arbitrary Power?" *Journal of Political Power* 5 (1): 137–52.
Nguyen, C. T. Forthcoming. "How Twitter Gamifies Communication." In *Applied Epistemology*, edited by Jennifer Lackey. Oxford: Oxford University Press.
Nguyen, C. T. 2019. "Games and the Art of Agency." *Philosophical Review* 128 (4): 423–62.
Noggle, Robert. 1996. "Manipulative Actions: A Conceptual and Moral Analysis." *American Philosophical Quarterly* 33 (1): 43–55.
Noggle, Robert. 2018a. "Manipulation, Salience, and Nudges." *Bioethics* 32 (3): 164–70.
Noggle, Robert. 2018b. "The Ethics of Manipulation." In *Stanford Encyclopedia of Philosophy: Summer 2018*, edited by Edward N. Zalta. Stanford: Stanford University.
Parmer, W. J. 2022. "Manipulative Design Through Gamification." *The Philosophy of Online Manipulation*, edited by Fleur Jongepier and Michael Klenk. York, NY: Routledge.
Pettit, Philip. 1996. "Freedom as Anti-power." *Ethics* 106: 576–604.
Pettit, Philip. 1997. *Republicanism: A Theory of Freedom and Government*. Place of publication not identified: OUP Premium.
Scanlon, Thomas M. 1998. *What We Owe to Each Other*. Cambridge, MA: Harvard University Press.

11 Manipulative Design Through Gamification

W. Jared Parmer

1 Introduction

If you are like me, you are beset on all sides by habits you wish you could break. You might wish you wrote before breakfast, with nothing but your first cup of coffee to get you going; instead, you lie in bed and scroll through social media until your stomach is growling and the golden hour has faded. You might wish you flossed three times a week, but just remembering to do so, never mind mustering the courage, proves too difficult. You might wish you ate chips only in the weekend but find yourself munching away at them on a Tuesday once the kids are down and you do not have to worry about setting a bad example.

So let me introduce you to *Habitica*. Habitica is an app in which you create a little avatar for yourself who starts as a lowly warrior with paltry gear and few skills. You can then input habits that you want to avoid – like eating junk food too often in a week – and habits you want to cultivate – like flossing three times a week. Performing a "Good Habit" gives you gold for equipment or points you can use to buff your character, making them stronger, smarter, and so forth; performing a "Bad Habit" harms your character, depleting their health or magic reserves. As your character becomes better, you can then "fight" increasingly more challenging "bosses" by, as before, performing Good Habits, avoiding Bad Habits, completing daily to-dos, etc.[1]

Habitica promises to "Gamify Your Life". By offering you a game-like system for cultivating habits you want and avoiding habits you do not, it *gamifies* personal development in the hope that you will be more motivated to cultivate better habits and have fun doing so.

"Gamification" is a way of talking about design that has taken the tech world by storm. In broad strokes, it refers to the implementation of *game-like* features in artifacts that are, strictly speaking, *not games*.[2] At the end of the day, for example, Habitica is not really a role-playing game: it is an app for time management with RPG trappings.

The evangelists for gamification are easy to find. Gabe Zichermann, for example, announces that "the revolution will be gamified", while Jane

DOI: 10.4324/9781003205425-13

McGonigal declares that "reality is broken", and games "can change the world", by which she means the world beyond games (McGonigal 2011).

But to assess these grand claims, we need a better understanding of just what, exactly, gamification *is*. Now, "gamification" and its cognates are basically terms of art, whose meanings their users are free to stipulate for their purposes. The previous gloss, however, is too imprecise to be useful for much of anything beyond a sexy ad copy (cf. Bogost 2014). More precision is needed.

Let me illustrate. In 1795, Napoleon Bonaparte announced a large monetary prize of 12,000 francs to the citizen who could best improve upon existing techniques for preserving food (such as drying, smoking, and pickling), so that his soldiers on the increasingly far-flung frontlines could get varied, nutrition-rich foods without paying the exorbitant prices that local merchants demanded. In 1810, Nicolas Appert's industrialized canning processes ultimately won that prize. For Zichermann, this is an example of gamification that long predates that term (Zichermann 2013, chapter 1).[3]

There is *some* sense in which Napoleon's competition is "game-like": there are, after all, plenty of games that are competitions for prize money. And if *that* is all it takes to gamify, it is easy to be seduced by the revolutionary potential of gamification, since industrialized processes that support continent-spanning war efforts promise to change the world in profound, enduring ways, and Napoleon's competition might very well have moved such innovation along.

But this notion of gamification is too broad to be of any use. For one thing, competitions arise for scientific breakthroughs, arms races, bank runs, debates, jobs, finding El Dorado, and everywhere else; and many of these competitions come with monetary prizes. At the same time, there are proper games that involve any number of things we find in the world beyond games, such as warfare, household management, state-building, railroading, farming, border control, air traffic control, and sidewalks (as in the classic childhood game, "Don't Step on the Cracks"). *Commonalities* between games and the non-game world are everywhere and easy to find; their existence does not give us any insight into any particular mode of design.[4]

So there is a challenge here to say exactly what we're talking about when we talk about gamification. But this question, at the end of the day, might not *matter*: why should we care how these semantic issues shake out?

My answer is that gamification, properly understood, is an opportunity to better understand and evaluate *manipulative design*. Here is a brief gloss on the phenomenon. In manipulation in general, a manipulator adopts a strategic stance vis-à-vis her victim: she wants her victim to do something and chooses the means of influence she takes to be most expedient to making that happen. In more familiar, interpersonal cases, these means are typically things like emoting, cajoling, peer pressure, or trickery. In manipulative design, the means is *design*: the manipulator designs an artifact in such a way that, when her victim uses it just as it is designed to be used, the victim is manipulated into doing something.

Manipulative design is interesting in a couple of ways. First, manipulative design enables, if you like, "manipulation at a distance". In ordinary cases, the manipulator typically has to play an active, guiding role in producing the desired outcome. She has to be ready to tune her tactics on the fly, appealing to now this or that argument, obscuring this or that bit of information, or taking this or that turn on a long guilt trip. In manipulative design, this need not be so: she can rely on her victim using the artifact as it has been designed to be used, at some remove from the manipulator's watchful eye.

Second, and for this reason, manipulative design can be accomplished by manipulators who are not, in a robust sense, unified agents. While such manipulators do have to get their act together enough that the artifacts they produce are designed for particular uses, their internal coordination need not rise to the level required for strategic supervision in the deployment of those artifacts. In particular, they need not exhibit the internal coordination we find in a skillful individual manipulator, who brings to bear a wide range of subtle emotional cues, argument, presentation, and so forth in service of her aim and in response to the developing interaction with her victim. Manipulative design is how corporations, academic disciplines, and democratic republics manipulate people (when they do).

Here, in brief, is the connection between gamification and manipulative design. As we will see in Sections 1 and 2, gamification is designing artifacts to induce the patterns of reasoning characteristic of striving play. So gamification is a species of a more generic mode of design, namely of designing artifacts to induce particular patterns of reasoning. As we will see in Section 3, this mode of design is manipulative when and because inducing such patterns of reasoning serves the designers' hidden purposes.

There are many ethical lenses through which we might evaluate manipulative design.[5] While many concerns are at stake, I am interested in the negative impact that manipulative design has on living a meaningful life.[6] So, in Section 4, I will use Duolingo as a case study to sketch an argument to the effect that, typically, when we use manipulatively designed artifacts, we are hindered in making our lives (more) meaningful. This will also enable me to sketch some principles of design ethics for supporting, rather than hindering, living (more) meaningful lives.

2 Striving Play in Games and Gamification

We often play games in a way that diverges from how we engage in ordinary life. For example, someone might try to score more points than her racquetball partner in order to beat him, or they might try to build a heavily research-focused civilization in order to be the first empire to send a colony into space (as in the *Civilization* videogame series). In these sorts of examples, the relationship marked by "in order to" is the familiar one between means and ends; and, within the context of the game itself, we can say that these *are* (some of) the means and ends. However, in the broader context that incorporates facts about the motivational psychology of the players themselves, the story we tell must change and in an illuminating way.

To see this, consider the perennial "good sport" who plays a game such as *Civilization* with well-defined rules and victory conditions, as many modern sports, boardgames, and videogames typically have. First, this person plays such a game for some broader purpose(s) – she plays to spend time with friends, have fun, or both. Second, the good sport genuinely *plays to win*, fair and square: she focuses her efforts on scoring the most points, or sending a colony to mars, and doing so in the ways the game permits. Third, however, whether the good sport *does* win is immaterial to her at the end of the day: her broader purposes are satisfied whether she wins or loses (provided the game is well designed in other ways). She is happy to high-five the other players and grab a drink once the game is over.

On the one hand, the good sport pursues the in-game "ends" via the in-game "means": she endeavors, for example, to be the first to send a colony into space in *Civilization* (i.e., to win) by developing the most scientifically advanced empire in the game (rather than by entering a cheat code, for example). On the other hand, the explanation for why she pursues that "end" is not that *achieving* it serves her broader purposes for playing in the first place. Her broader purposes are satisfied just by the engrossing *pursuit* that that "end", linked as it is with its "means", makes possible. An engrossing space race in *Civilization* is a great way for her to have fun or spend time with her friends, whether or not she wins it.

Call what the good sport does *striving play*. To condense the foregoing into something like a slogan, we can say that striving play is rule-based, purposeful pursuit without a concern for achievement.[7]

Many games are *designed for* striving play. Indeed, this seems to be one paradigmatic mode of design for modern games for adults, since such design typically involves creating fairly complex rule-sets that are interesting to navigate toward well-defined (cooperative or competitive) victory conditions.

Later, I'll say more of what such design comes to. In the meantime, notice that many *non-games* are also plausibly designed for striving play. Habitica, which we have seen, is one example. The platform enables users to perform tasks associated with self-defined "Good Habits", and avoid tasks associated with self-defined "Bad Habits", in order to acquire gold or experience, or to avoid losing health or magic points. Moreover, it is meant to be used by players to serve their broader purposes, such as to cultivate better habits and break bad ones. And those players are meant to follow the rules the platform provides and to genuinely try to acquire as much gold as possible and to level up their avatars as much as possible, because these attempts will serve their broader purposes. After all, players acquire more gold and level up their avatars *by* doing the activities they want to become habitual, and they avoid losing health or magic points *by* avoiding the habits they want to break. However, at the end of the day, how much money or experience points they acquire in the course of doing these things is immaterial as far as these broader purposes are concerned: it is the *pursuit* that serves those purposes.

Here are a few more examples of non-games that are plausibly designed for striving play.

Duolingo. Users complete language-learning exercises to gain "experience points" that accumulate to reach a new "Crown Level" of skill in a chosen language. Their Crowns can "break" from time to time, and then be "put back together" by reviewing the material for the corresponding level. They also have a "streak" that counts the number of consecutive days they have completed at least one lesson of instruction, which resets to zero upon missing a day.[8]

Weight-Loss Competition. A set of friends or coworkers agree to a competition to see who can lose the most weight within a relatively brief time frame, such as three months. They meet weekly to weigh in and encourage each other on. The person who loses the most by the end of the competition receives a nominal monetary prize, such as a small gift card.

Disney's Electronic Whip. A scoreboard is hung prominently in each of the laundry rooms of the Disneyland Resorts in California, which lists each worker present and highlights their name in relation to how close they are to hitting the current productivity target their boss sets from his office – red for seriously behind, yellow for falling behind, and green for meeting the target (see Gabrielle 2018).

The "ends" these artifacts provide are meant to be immaterial to their users outside of the artifact itself. Losing one's streak in Duolingo or seeing one's name in green on the laundry-room scoreboard, for example, are immaterial vis-à-vis the broader purposes for which people are meant to engage with these artifacts – developing competency in a secondary language or not getting fired.[9] However, *as* "ends" to the particular "means" the artifact connects them to, they make possible patterns of pursuit that *do* serve those broader purposes.

That non-games can also be designed for striving play helps us make theoretical progress. As I mentioned early on, the standard gloss on gamification is that it is the implementation of *game-like elements* to non-games. We are now in a position to postulate just what makes a feature of an artifact "game-like" in the relevant sense: it is an *inducement to striving play*.

In a moment, I will say more about what it is to *induce* striving play, and so what it is for an artifact to be *designed for* such a thing. In the meantime, a different question arises: when is implementing inducements to striving play *gamification*, and when is it *game design*?

The following path is perhaps most inviting but is ultimately a trap: identify when inducements to striving play are being implemented *in a game*, and when not; when they are, you're designing a game; when they are not, you're gamifying. This is a trap because it lures us into the brambles of saying just what a game is.[10]

I am not going to go that way. We have a good enough grip on the sorts of broader purposes for which artifacts might be used to keep gamification and game design separate as we go. Generally speaking, games proper are designed for entertainment and, at least in more ambitious cases, certain kinds of aesthetic experience in which the activity itself is the medium. In the latter sorts of cases, games are designed to produce aesthetically valuable forms of activity in collaboration with the players themselves, much as when a musician interprets a concerto, or an actor gives a scripted performance on screen (Nguyen 2020, ch. 6). By contrast, gamified artifacts are designed for other sorts of broader purposes, such as education, training, or losing weight. Our grasp on these differences will be enough for what follows because, as I mentioned at the outset, my main quarry is manipulative design in general, which in principle could happen in game design as well as in gamification. And while it will turn out that the broader purposes for which these artifacts are designed play a role in explaining why design is manipulative when it is, this will not hang on whether those purposes bear on the distinction between gamification and game design.

3 Inducements to Striving Play

What is it for an artifact to induce striving play? This question is like asking when, in general, a glove fits a hand. You can't answer that question without remarking on the general features of gloves *and* the hands for which they are made. Similarly, we cannot answer our question vis-à-vis artifacts without remarking on the general features of those artifacts and the users for which they are made.

I have already said something about the users: they are able to engage in striving play, in which the pursuit that the artifact makes possible serves their broader purposes. But it won't be enough for an artifact to simply *allow* for this. A sidewalk allows for striving play – remember "Don't Step on the Cracks" – but does not for that reason induce striving play. And it is too much for an artifact to *require* this. It is possible to engage with an app like Habitica without striving play. For example, one might be an alpha tester who is paid to run up one's experience, health bar, and so on as high as possible to see how the app behaves in extreme cases. And yet, Habitica still plausibly induces striving play.

To make progress, I need to enrich our picture of ourselves as thinking and acting agents. Across many areas of life, we use *heuristics* when reasoning about what to think and what to do. Heuristics are defeasible rules that recommend certain decisions on the basis of some circumscribed set of considerations. These rules, including their component decisions and sets of considerations, can vary from the concrete and particular to the abstract and general – though the rules themselves remain defeasible in all cases. For example, the rule to run twice each week is a highly concrete and particular heuristic: it recommends running, on any particular occasion, just on the basis of how many times one has already run that week. By contrast, the well-known Satisficing Rule is a highly abstract and general heuristic: it recommends choosing the first satisfactory option presented, abstracting from the particular details of the options presented or what sorts of considerations bear on satisfactoriness (cf. Simon 1955).[11]

An artifact can *enable* heuristic reasoning in a specific sense: by meeting the essential prerequisites of the relevant defeasible decision rule. To take a classic example, a cafeteria layout enables satisficing by arranging visitors' options in a sequence. This is because the Satisficing Rule requires such a setup essentially: one cannot choose the first satisfactory option presented unless the options are presented in sequence.[12]

So what it is for an artifact to *induce* a certain pattern of reasoning and downstream acting is in part for it to enable, in this specific sense, reasoning with a specific heuristic about what to do. In the case of artifacts that induce striving play, the heuristics in question have a means-end structure that need not be exhibited by all heuristics. The Satisficing Rule, for example, lacks such a structure: it recommends taking the first satisfactory option, end of story. It does not recommend taking some option *in order to* bring about something else. Gamified artifacts and games alike establish rules that link means and ends; this is essential to striving play. So the first feature of inducing striving play is as follows: that the artifact in question enables following a defeasible decision rule with a means-end structure.

Still, inducing striving play requires a bit more. This is easy to see when we consider the defeasible rules involved. Consider, for example, that Duolingo enables the rule to complete daily lessons in order to extend your streak, and a weight loss competition enables the rule to lose the most weight in order to win the prize money. Now, gamified artifacts like these often incorporate a lot of virtual elements, having to do with streaks, scoreboards, and so on, so the enabling conditions for these rules are rather more complex than those of the Satisficing Rule. But merely enabling these rules need not induce *striving play* rather than ordinary means-end motivations. For example, a weight-loss competitor can be pursuing the prize money when actually winning is immaterial to her purpose of losing weight, as the good sport would in striving play; or she can be trying outright to win the prize money as a means to buying birthday gifts. The rule itself is silent on how to be motivated in following it.

What additional features constitute inducing *striving play*, over and above enabling the relevant rule(s)? Two further features are needed. First, the artifact must offer or make salient putative justification for taking the *means* in the decision rule *independently of the ends* in it. Consider, by way of example, the means and ends embodied in decision rules provided by Duolingo and Habitica: Duolingo provides the rule to complete daily lessons (the means) in order to maintain one's streak (the end); and Habitica provides the rule to complete tasks on a daily basis (the means) in order to acquire more experience points for one's avatar (the end). In presenting itself as a language-learning tool, Duolingo makes salient putative justification for completing the daily lessons: to acquire and maintain competence in a secondary language. Habitica, similarly, makes salient putative justification for completing tasks on a daily basis: to develop habits one wants to have. These justifications have nothing to do with the ends inscribed in the rules these apps enable: they have nothing to do with extending one's Duolingo streak, or gaining experience within the Habitica virtual environment, for example.[13]

The second feature is that the artifact must offer or make salient putative justification for pursuing the *ends* in the decision rule that *depends on the putative justification for taking the means* in it. For example, it is easier to focus on, and be motivated by, maintaining one's Duolingo streak than to focus on, and be motivated by, developing a secondary language such as German in consistent increments. Now, facts about what is easier to attend to and be motivated to do do not, in themselves, justify attending to and pursuing what is easier. The easier option, after all, might still have nothing to be said in its favor. It is easier to drink three cans of paint than to drink four, for example. However, in a context where the easier option is linked to something for which there is such putative justification, such facts offer derivative justification to take the easier tack. With Duolingo, completing daily lessons (i.e., the means in the decision rule) *are* given such a putative justification (namely, that it supports secondary-language competence); in this context, the fact that it is easier to focus on maintaining one's streak (i.e., to focus on the end in the decision rule) offers derivative justification to do so.

Let us take stock. The decision rules in gamified artifacts have a familiar means-end structure: do *this* in order to do *that*. At the same time, however, these additional two features set the normative support running in the opposite direction by offering or making salient putative justification for taking the means that is independent of the end, and justification for pursuing the end that is dependent on the means. So an artifact induces striving play by enabling specific patterns of (heuristic) reasoning and acting, coupled with normative scaffolding that does *not* support *achieving* the ends involved in the heuristics themselves but rather what is served by the *pursuit* of those ends.

While tailored for striving play in games and gamification, this account provides a model for a more generic form of design, of which gamification is a species. Stepping back a bit, this genus of design can be seen as offering users *tools for reasoning*, supported by scaffolding that (putatively) justifies using those tools in specific ways or for the sake of specific purposes.

4 Deception Explains Manipulative Gamification

What makes designing things in these ways manipulative (when it is)? I will focus primarily on manipulative gamification to keep things tractable, but I have two further ambitions. The first is to offer an explanation that applies, with little emendation and as broadly as possible, to manipulative design in the form of scaffolded tools for reasoning, which I mentioned just a moment ago. The second is to offer an explanation that comports well with plausible explanations of manipulation in other kinds of cases, namely those in which the means of influence are not design at all.

My conception of gamification makes some candidate explanations more salient than others. Consider that I have placed heuristic reasoning front and center in my account: it is by enabling certain patterns of heuristic reasoning and providing some normative scaffolding around reasoning in that way that gamified artifacts induce striving play. So, to the extent that users use gamified artifacts as those artifacts were designed to be used, users are going to think and act in ways those artifacts recommend. These recommendations can be bad.

There are many ways such recommendations might be bad. One way is straightforwardly *procedural*: the recommended way of deciding might not be a good way to decide. It is true that gamified artifacts make use of tendencies in us to decide in fast-and-frugal ways – ways that, for example, prioritize avoiding loss over achieving gain, or substitute familiarity for normality, or favor the merely satisfactory over the optimal. To some, deciding in such ways is not deciding in a fully rational way, so, to the extent that gamification induces such decision-making, it might be viewed as subverting or bypassing rational decision-making and thus manipulative.[14]

The first thing to note is that this explanation does not comport well with the data itself. Many examples of gamification, such as Habitica or a weight-loss competition between friends, do not look prima facie to be manipulative. But it seems to me that this worry is motivated by distinctly theoretical considerations about the nature of rationality rather than intuitions over cases. The very fact that gamification works by enticing people to act without carefully and comprehensively considering, and weighing up, what speaks for or against their options, seems to drive the worry that they are being manipulated, perhaps even by themselves.

But here I must dig in my heels: gamification does not always bypass or subvert rational decision-making. Since I have argued for this at length elsewhere for heuristic reasoning in general, I will be brief (Parmer forthcoming).[15] I assume that reasoning is a matter of making inferences in a rule-governed way. And, while it is controversial what *good* reasoning is, I take it to have two hallmarks: first, the rules involved need to be generally *procedurally good rules* to follow; and, second, the agent herself needs to be able to be *flexible* vis-à-vis these rules when she takes the particular situation to call for it.

Using heuristics can exhibit both of these hallmarks. First, heuristics are often generally procedurally good rules to follow, when and because they recommend decisions on the basis of some normative considerations, and the decisions themselves are good enough (given the limitations of creatures like us). For example, the Satisficing Rule recommends a decision on the basis of the normative consideration of whether the next item encountered is sufficiently acceptable and choosing such an item is good enough considered against the limitations of working memory, time, and certainty creatures like us face. Second, we can be flexible vis-à-vis heuristics because of their defeasible character: we can disregard the rule when we take ourselves to have greater reason to do other than it recommends, even when the rule does not itself concern such reasons. To continue with the example, we can disregard the Satisficing Rule when if we believe our options are being presented to us in a sequence in which an exceedingly good item is deliberately delayed as long as possible.

This gives us some reason to think that heuristic reasoning can realize rational decision-making. People who harbor procedural worries, then, need to give us more; the *mere fact* that users of gamified artifacts are engaged in heuristic reasoning is not enough to show they are not deciding in a fully rational way, and so the fact that gamified artifacts induce such reasoning does not, on its own, look to be manipulative.

It is also worth noting that, if *games* induce striving play in the same way, this worry seems ill-formed because it is implausible that game designers manipulate us merely by inducing striving play in us. And, indeed, my account of inducing striving play is meant to be sufficiently general to cover striving play in games proper, as well as in gamification – such inducement is, after all, what makes gamification like game design.

A more promising view might be that gamification is manipulative when, and because, the designers are *indifferent to* the actual justificatory status of the striving play they induce and care only about the causal efficacy of their particular design choices (Coons and Weber 2014a; Gorin 2014a, 2014b).[16] Indeed, this view is commonly motivated by cases of manipulation that seem to proceed by offering *genuine* justification, such as Moti Gorin's *Trust Me* case, in which his imagined counterpart manipulates someone by offering them sound arguments and sensible advice so that, in two months' time, they will believe a lie of his that they otherwise would not have.[17] Such cases would be structurally analogous to cases of manipulative gamification that tap into heuristic reasoning that realizes rational decision-making.

The problem is that indifference does not provide a compelling explanation of such cases. Gorin's counterpart in Trust Me is *not* indifferent to the justificatory status of the considerations he is offering his victim: he's *relying on* their being genuine justifications. Extending this idea a bit, gamifiers can similarly rely on genuine justifications for using their artifacts as they are designed to be used and still manipulate their users through those artifacts. This is not indifference either.

Now, it might be replied that such manipulators are indifferent in the sense that, if they had thought that they could most effectively influence their victims *without* offering them genuine justifications, they would have taken that tack instead (Gorin 2014a, 59, 2014b, 93–94). The problem with appealing to this counterfactual, however, is that it comports just as well with an explanation of manipulation in terms of *deception about the influencer's purposes*. After all, if Gorin's counterpart manipulated you into trusting him without offering you good reasons to trust him, it is hard to see how we would do this, in the nearest possible situation, without deceiving you about the fact that he aims to get you to believe his lie in two months' time. Whatever efficacy his tactics had – despite their not being the offering of good reasons – would be undercut by his purpose being transparent; so he would have to deceive you at least in the minimal sense that he would have to keep his purpose opaque to you. Moreover, his tactics as presented in Trust Me clearly rely on such minimal deception about his purpose, since his sensible advice, sound arguments, and so forth would fall on deaf ears were his victim to see why he is offering them.[18]

So we need further argument to discriminate between these explanations. Now, Gorin has offered a putative counterexample to deception-based accounts of manipulation. In *Off the Wagon*, Adams wins a job promotion over his colleague Wilson by successfully leading her, a recovering alcoholic, into a relapse so that she will perform worse at work – all the while making no secret of what he is doing or why. According to Gorin, Adams has manipulated Wilson (Gorin 2014b, 80–81). However, *Off the Wagon* is not clearly an example of manipulation without deception. What is clear is that Adams is *exploiting Wilson's alcoholism*. However, because Wilson's control over her desire to drink is significantly attenuated, Adams's influence looks more *coercive* than manipulative. If that is right, this case is no counterexample at all.[19]

Moreover, the account of gamification I have developed is particularly amenable to an explanation in terms of deception. Gamification, recall, is designing artifacts to induce striving play. This involves, inter alia, scaffolding that putatively justifies the pursuit the artifact makes possible, for the sake of the user's broader purposes – be it losing weight, learning a secondary language, and so on. It is natural to say that this is manipulative, when it is, simply because this pursuit serves the designers' hidden purposes. And we can say much the same for the more generic form of design of which gamification is a species. Providing tools for reasoning, along with normative scaffolding that putatively justifies using these tools in particular ways, is manipulative when, and because, using those tools in those ways serves the designers' hidden purposes.[20]

5 Manipulative Design's Typical Impact on Meaningfulness

Let me wrap up by offering a preliminary account of one way in which manipulative gamification (and manipulative tools for reasoning with normative scaffolding, more generally) should concern us. I will argue that, typically, using such manipulatively designed artifacts hinders us from making our lives (more) meaningful.

First, some important stage-setting. Making one's life (more) meaningful involves dealing with the following basic challenge. We are agents who can *care about* all sorts of things in all sorts of ways, making them a focus of our emotional, attentional, and motivational inner lives. While we are fortunate that there is an abundance of practices, people, and objects in the world that befit our caring about them in various ways, we must work out how to properly care about any particular thing and how to balance that with our properly caring about every other thing we do. And yet, our mental resources are scarce regarding this task, not least because caring is a complex, multimodal, resource-intensive attitude. This constrains us to care about only some of the things we might and to care about them in only some of the ways we might. As a result, we are faced with the ongoing challenge of working out which things to care about, and how, constrained both by our own limitations and the other things we care about. As I prefer to put it, we face the challenge of *cultivating our cares* vis-à-vis their objects and ourselves. Cultivating caring is key to making our lives more meaningful than they antecedently are.[21]

Now, cultivating caring typically proceeds with a considerable amount of improvisation, experimentation, and reliance on others for guidance. This is for the simple reason that we are *working out* what to care about, and how, and so we will not have a clear picture, in advance, of the caring in question. Because we are still working out what to care about, and how, we lack the deep understanding that that caring would embody. So the caring we are after cannot guide our reasoning here and now. And there is little reason to expect that the manner in which we presently care about whatever we do can reliably develop that richer and deeper caring, for the caring we are developing could look very different from the caring we presently exhibit.

As I said at the end of Section 2, gamification is a species of design in which artifacts induce certain patterns of reasoning and acting by providing users tools for reasoning in the form of heuristics, supported by scaffolding that (purportedly) justifies using those tools in specific ways or for the sake of specific purposes. Providing one another with heuristics has a potentially helpful role to play in working out what to care about and how. For the use of such heuristics need not depend on any deep understanding embodied in caring; heuristic reasoning does not even depend on full-blown deliberation, though it often will require some normative judgment in the use of the heuristics themselves.[22] Of course, using such heuristics could just as well *hinder* us in working out what to care about and how.

To see how they might be a hindrance in this way, let us walk through an interesting case study: Duolingo's early business strategy.

It is perhaps not widely known how Duolingo once generated profits for its owners: it provided translation services to businesses that were done, for free, by its users in the form of exercises provided on the app.[23] As in the case of Trust Me, this strategy *relied* on offering its users genuine justification to use the platform to learn a new language: they could only get their users to provide high-quality, free translation labor if the app really did improve their secondary language(s). However, unlike in the case of Trust Me, it is implausible that, if the founders of Duolingo thought they could effectively procure free translation services *without* offering a genuinely effective learning platform, they would have done so. At least in this respect, their business model was built on providing free translation services *by* offering that platform.

It should be noted that, while this aspect of Duolingo's business model was not common knowledge, Duolingo made no secret about it: Duolingo's most prominent business partners in this setup were CNN and Buzzfeed, and this partnership was widely reported, including on Duolingo's own forums by the founder himself.[24] However, by the time Duolingo announced these partnerships and executed on this strategy (in 2013), it had already acquired approximately five million active monthly users and had amassed something of a reputation as *the* go-to, free-to-use, effective language-learning platform.[25] So even if Duolingo was not at that time deceptively profiting off its users' exercises, it is an open question to what extent users' foreknowledge of Duolingo's eventual profiteering strategy would have disincentivized them from engaging with the app to the extent that they did.

The general shape of this dynamic, however, highlights a more endemic feature of designing for profit, especially through gamified artifacts: a more enduring purpose, on the part of the designers, is ever-increasing monetizable engagement. Duolingo, for its part, aims to secure such engagement by offering an effective learning platform that has no off-ramp.[26] *How* the latter strategy serves the former is not always apparent, and the relationship itself can at times be obscured, especially during the early phases of rollout and adoption.

Most importantly, this more enduring purpose on the part of the designers stands in some tension with our need to cultivate our cares in order to live more meaningful lives. Recall what the challenge is: we have to work out which things to care about, and how, constrained both by our own limitations and the other things we care about. It seems to me that we make progress on this front with a considerable amount of idiosyncratic fine-tuning in response to our developing understanding of the proper ways to care about the things in our lives (Nguyen 2020, 207). And consider, in contrast, the normative support for acting that is embodied in gamified artifacts – the heuristics they enable, along with their surrounding normative scaffolding. These features are not amenable to such fine-tuning but stand together in the ways they do, to induce the sorts of activity they do, by their designers.

On the one hand, part of what makes these features so appealing is the simulacrum of clarity and simplicity they lend the pursuits we independently desire (cf. Nguyen 2020, 194–97). But, on the other, these features are also designed in part to advance the designers' purposes, such as ever-increasing monetizable engagement rather than to support any such idiosyncratic fine-tuning on the users' part. For both these reasons, we are poorly positioned to monitor and modify, as needed, our engagement with these artifacts. How they foster our ongoing engagement with them is often obscure, as is the manner and extent to which our ongoing engagement serves our purposes versus those of the designers.

Under such conditions, it is difficult to ascertain which aspects of these artifacts continue to serve us vis-à-vis the cares we need to cultivate, and which only serve to keep us hooked to using the artifacts. I have been making this point in the context of using a gamified artifact that is designed in part for ever-increasing monetizable engagement, but the point generalizes to many other designed artifacts that provide tools for reasoning with normative scaffolding. Just as Duolingo fosters ongoing, monetizable engagement *by* offering an effective language-learning platform, other artifacts advance their designers' purposes *by* providing users with scaffolded tools for reasoning that advance those users' purposes. The reader is invited, for example, to consider how grading rubrics or political platforms might function in this manner.

It is important to emphasize that this particular dynamic *as such* is not manipulative. As I said in Section 3, it is manipulative when and because these scaffolded tools for reasoning serve the designers' *hidden* purposes. The point I now want to make is rather simple. It is hard enough prising apart how, and to what extent, the various features of such designed artifacts serve the purposes for which they are designed to be used – indeed, their utility, as I've suggested, partly inheres in their ability to lend a simulacrum of clarity and simplicity to the pursuit of those purposes. When the designers' own purposes are themselves kept hidden, this task is even harder. We should thus expect it to be quite difficult to work out how to fine-tune our engagement with these artifacts, not merely vis-à-vis our purposes for using them, but vis-à-vis the caring we need to cultivate over time.

If these points are basically right, we are now in a position to see two principles of design ethics in force for providing scaffolded tools for reasoning, when we want these sorts of artifacts to support cultivating caring and living (more) meaningful lives. It should now be obvious what those are. The provided tools for reasoning need to come not only with normative scaffolding justifying the particular manner of their use but also with disclosure of the designers' purposes in providing those scaffolded tools in the first place, plus disclosure of how using those tools (as they are designed to be used) serves the designers' purposes.

6 Conclusion

A fuller treatment of these particular issues in design ethics will have to wait for a later occasion. Since my purpose in this chapter has been to make some headway on a variety of interlocking, but not well understood, issues, this sketch will have to do for now. Let me end with a brief recapitulation of the major themes.

The main focus of this chapter has been gamification, understood as a species of design. Indeed, I motivated giving gamification such an extensive treatment because it is a species of design through which we might better understand *manipulative* design. Manipulative design, in turn, merits careful consideration because it enables manipulation "at a distance" by manipulators who do not exhibit unified agency in any robust form. In Sections 1 and 2, I defended the view that gamification is inducing striving play for the sake of purposes beyond those typically found in games, such as to learn a skill, develop certain habits, and so forth. And I suggested we view gamification as one species of a more generic form of design that involves providing tools for reasoning along with scaffolding that purports to justify using those tools in certain ways.

This genus of design, moreover, offers us a useful starting point for understanding manipulative design more generally. In Section 3, I turned to the question of what makes gamification manipulative when it is, and sought a suitable explanation that also covers manipulative design more generally, and that stands as a plausible explanation for many cases of non-design-based manipulation. There, I defended the simple but compelling view that such manipulation should be explained in terms of deception, on the part of the manipulator, about her purposes. In the final section, I turned to an underexplored ethical dimension of manipulation to explain one way in which manipulative design can be dangerous. Here, drawing on my view that cultivating caring is a key part of how we make our lives more meaningful, I sketched an argument to the effect that manipulative design hinders users in making their lives more meaningful than they antecedently are.[27]

Notes

1. For a more thorough rundown, see https://habitica.fandom.com/wiki/Habitica_Wiki (retrieved 7 January 2020). Thanks to Niklaas Tepelmann for the example.
2. For this gloss, see, for example, Cherry (2012), and Deterding et al. (2011). For extensive discussion of the history and precursors of the notion, including "serious games", see Deterding et al. (2011) and Deterding (2014).
3. Strictly speaking, Zichermann does not appeal to the notion of "implementing game-like features", but to "using game thinking and game mechanics" (*ibid.*). But it is not all that clear if this marks a difference that should interest us, and anyway the point I'm about to make applies to his gloss just as well as it does to the one I provide in the main text.
4. Now, it is true that the evangelists for gamification point to particular features that they say are characteristic of *games* or of *game design* and that can be ported over into the design of non-game artifacts; but their proposals suffer from the same basic problem I've just given. By way of illustration, consider that McGonigal offers the following four traits as definitional of games: goals, rules, feedback systems, and voluntary participation (McGonigal 2011, 21). But

this "definition" falls to many counterexamples. Here is just one: cooking by a recipe, when one could just as easily order takeout.

5. In treating this as a distinct question, I assume that it is false that, as a matter of conceptual necessity, manipulation is pro tanto morally wrong. For defenses of this assumption, see, for example, Baron (2003), Coons and Weber (2014a), and Wood (2014).
6. See Nyholm (in this volume) for a similar evaluative approach. And see Gorin (in this volume) for a different account of gamification in terms of reasons, along with a different sort of explanation of when and why it is harmful, one given in terms of domination.
7. See Nguyen (2020, ch. 1), himself drawing on (Suits 1978). Nguyen goes further than I do and makes more substantive claims about the motivational psychology of striving players. In particular, he claims that they take up the in-game "ends" *for the sake of* the in-game "means". If true, this would mark a peculiar inversion as compared to ordinary life, in which we take up means for the sake of ends. It is controversial just how to characterize the psychology of striving play in further detail, but the further details will not matter for what follows. My thanks especially to David Heering, Annina Loets, and Richard Woodward for discussion on these points.
8. For more, see https://duolingo.fandom.com/wiki/Duolingo_Wiki (retrieved February 2, 2020).
9. In Disney's Electronic Whip, more is obviously going on: for one thing, the scoreboard generates its results by tracking the actual productivity of the workers. And *that* is not at all immaterial to whether or not they get fired. The point I'm making here is just that the colored *scoreboard* is a design feature that is, strictly speaking, immaterial to whether they get fired: the manager has their productivity data at his fingertips, and indeed sets the productivity target in light of that data, which in turn affects how the scoreboard displays the rankings. The *data* matters; the *interface* does not, except to the extent that it motivates the workers to work faster. And that is exactly the dynamic characteristic of striving play.
10. For an admirable attempt at this, see (Nguyen 2020, especially ch. 6). For present purposes, I will remain agnostic as to whether Nguyen succeeds, though I am highly sympathetic to his account. Contemporary skepticism about the possibility of giving a complete analysis of what it is to be a game is voiced most forcefully, of course, by Wittgenstein (1953 [2009]: esp. sections 66–71) – though his agenda in those passages is considerably more ambitious.
11. For other examples, see Gigerenzer and Gaissmaier (2011).
12. For the cafeteria example, see Thaler and Sunstein (2008, 1). As I have argued elsewhere, to enable rules like these is what it is for an artifact to be or contain a *nudge*; see Parmer (forthcoming).
13. Similarly, *games* that are designed for striving play make salient justifications about how enjoyable this pursuit would be or are perhaps aesthetic in character and concern agency: having to do, for example, with the elegance, difficulty, subtlety, etc. of this pursuit.
14. For this sort of worry as it relates to nudging in general, see Grüne-Yanoff (2012) and Wilkinson (2013). More generally, it is common to defend the view that manipulation *just is* the bypassing or subversion of our rational capacities; see, for example, Blumenthal-Barby (2012) and Wood (2014). Against this, consult Gorin (2014a).
15. A variety of recent work has responded to a structurally similar worry about nudges, such as Engelen (2019), Houk (2019), Levy (2019), Schmidt (2019), and Schmidt and Engelen (2020).

16. Especially Gorin's "non-paternalistic reasonable manipulation" (2014b). This explanation also bears similarities to Klenk's account, though his account states that manipulators are *careless* vis-à-vis genuine justification, and so it seems more demanding (2020, 2021). For other deployments of indifference to explain the *wrongness* of manipulative influences, see Jongepier and Wieland (in this volume) and Nyholm (in this volume), and Nys and Engelen (in this volume).
17. For this case and some discussion, consult Gorin (2014a, 58–59).
18. Many of the other examples Gorin gives are similarly amenable to explanations in terms of deception about purposes, such as his *Global Warming* example (Gorin 2014a, 58) and *Election* example (Gorin 2014b, 91–92). My thanks especially to Chris Bartel, Moti Gorin, and Kalle Grill for discussion on these matters.
19. My thanks to Moti Gorin, Fleur Jongepier, Lisa Vogt, and Richard Woodward for discussion on this point in my argument.
20. Recalling some of my introductory remarks, one might wonder whether this appeal to the designers' hidden purposes ascribes too much unified agency to them. My answer is: not really. The designers need to exhibit only as much unity as it takes to ascribe certain purposes for which they design the artifacts in the ways they do; they need not "wholeheartedly" endorse these purposes, nor need all aspects or divisions within the design process be supervised toward this purpose. Even in corporations with several different semi-autonomous design teams, for example, we can identify core directives for the products being designed: typically, some rather general strategy for maximizing profits – including some general sort of target clientele with a particular targeted need or desire, along with, perhaps, a few side constraints (such as those handed down from the ethics advisory board).
21. For more, see Parmer (2021). In that paper, I argue that the process of becoming more fulfilled, of which cultivating caring is a key part, makes our lives more meaningful for us independently of whether the things we care about have objective value. Here, my argument will not depend on that claim.
22. Indeed, to support the agent's capacity for self-guidance while reasoning with these heuristics, it *ought* to require her own normative judgments. For more on this, consult Parmer (forthcoming).
23. See https://producthabits.com/duolingo-built-700-million-company-without-charging-users/ (retrieved August18, 2020). Duolingo has since abandoned this strategy (see www.quora.com/Why-did-Duolingo-move-from-translation-to-certification-for-monetizing, retrieved February 2, 2020).
24. See https://forum.duolingo.com/comment/954969/Duolingo-now-translating-BuzzFeed-and-CNN (retrieved February 2, 2020).
25. See www.forbes.com/sites/parmyolson/2013/09/25/duolingo-takes-online-teaching-to-next-level-by-crowd-sourcing-new-languages/?sh=61e96d1e4dc2 (retrieved February 21, 2020).
26. Indeed, some of the features of Duolingo are plausibly designed to *prevent* you from setting the app aside, such as the mechanics in which your "Crowns" break from time to time and have to be put back together by completing exercises or the incorporation of "leagues" and leaderboards.
27. Earlier versions of this chapter were presented at Freie Universität Berlin and the *Manipulation Online* workshop series. My thanks to Barbara Vetter and Richard Woodward, and to Fleur Jongepier and Michael Klenk, for organizing those respective sessions and providing invaluable feedback. I would like to also thank the participants of those sessions for their feedback and conversation, especially Christopher Bartel, Moti Gorin, Kalle Grill, David Heering, Annina Loets, Sven Nyholm, Giacomo Figà Talamanca, and Lisa Vogt. Additional, special thanks are due to Michael Bratman, Berit Braun, and Niklaas Tepelmann for wide-ranging conversation that laid the groundwork for this chapter.

7 References

Baron, Marcia. 2003. "Manipulativeness." *Proceedings and Addresses of the American Philosophical Association* 77 (2): 37. doi:10.2307/3219740.

Blumenthal-Barby, J. S. 2012. "Between Reason and Coercion: Ethically Permissible Influence in Health Care and Health Policy Contexts." *Kennedy Institute of Ethics Journal* 22 (4): 345–66.

Bogost, Ian. 2014. "Why Gamification Is 'bullshit'." In Walz and Deterding 2014, 65–79.

Cherry, Miriam. 2012. "The Gamification of Work." *Hofstra Law Review* 40 (4): 851–58.

Coons, Christian, and Michael Weber. 2014a. "Manipulation: Investigating the Core Concept and its Moral Status." In Coons and Weber 2014, 1–16.

Coons, Christian, and Michael Weber, eds. 2014b. *Manipulation: Theory and Practice*. Oxford: Oxford University Press.

Deterding, Sebastian. 2014. "The Ambiguity of Games: Histories and Discourses of a Gameful World." In Walz and Deterding 2014, 23–64.

Deterding, Sebastian, Dan Dixon, Rilla Khaled, and Lennart Nacke. 2011. "From Game Design Elements to Gamefulness: Defining 'Gamification'." In *Proceedings of the 15th International Academic MindTrek Conference on Envisioning Future Media Environments – MindTrek'11*, edited by Artur Lugmayr, Heljä Franssila, Christian Safran, and Imed Hammouda, 9. New York, NY, USA: ACM Press.

Engelen, Bart. 2019. "Nudging and Rationality: What is There to Worry?" *Rationality and Society* 31 (2): 204–32. doi:10.1177/1043463119846743.

Gabrielle, Vincent. 2018. "How Employers Have Gamified Work for Maximum Profit." *Aeon Magazine*. Accessed August 23, 2021. https://aeon.co/essays/how-employers-have-gamified-work-for-maximum-profit.

Gigerenzer, Gerd, and Wolfgang Gaissmaier. 2011. "Heuristic Decision Making." *Annual Review of Psychology* 62 (2011): 451–83.

Gorin, Moti. 2022. "Gamification, Manipulation, and Domination." In *The Philosophy of Online Manipulation*, edited by Jongepier, F. and Klenk, M., 199–215. New York: Routledge.

Gorin, Moti. 2014a. "Do Manipulators Always Threaten Rationality?" *American Philosophical Quarterly* 51 (1): 51–61. Accessed June 04, 2019.

Gorin, Moti. 2014b. "Towards a Theory of Interpersonal Manipulation." In Coons and Weber 2014, 73–97.

Grüne-Yanoff, Till. 2012. "Old Wine in New Casks: Libertarian Paternalism Still Violates Liberal Principles." *Social Choice and Welfare* 38 (4): 635–45.

Houk, Timothy. 2019. "On Nudging's Supposed Threat to Rational Decision-Making." *Journal of Medicine and Philosophy* 44: 403–22.

Jongepier, Fleur, and Michael Klenk, eds. 2022. *The Philosophy of Online Manipulation*. New York, NY: Routledge.

Jongepier, Fleur, and J. W. Wieland. 2022. "Microtargeting people as a mere means." In *The Philosophy of Online Manipulation*, edited by Jongepier, F. and Klenk, M., pp. 156–179. New York: Routledge.

Klenk, Michael. 2020. "Digital Well-Being and Manipulation Online." In *Ethics of Digital Well-Being: A Multidisciplinary Perspective*, edited by Christopher Burr and Luciano Floridi. Cham: Springer. Accessed November 17, 2019. 81–100. add doi: 10.1007/978-3-030-50585-1_4.

Klenk, Michael. 2021. "Manipulation (Online): Sometimes Hidden, Always Careless." *Review of Social Economy*. 80: 1, 85–105. doi:10.1080/00346764.2021.1894350.

Levy, Neil. 2019. "Nudge, Nudge, Wink, Wink: Nudging Is Giving Reasons." *Ergo (Ann Arbor, Mich.)* 6. doi:10.3998/ergo.12405314.0006.010.

McGonigal, Jane. 2011. *Reality is Broken: Why Games Make Us Better and How They Can Change the World.* London: Random House.

Nguyen, C. T. 2020. *Games: Agency as Art.* Oxford: Oxford University Press.

Nyholm, S. 2022. "Technological Manipulation and Threats to Meaning in Life." In *The Philosophy of Online Manipulation*, edited by Jongepier, F. and Klenk, M., pp. 235–252. New York: Routledge.

Nys, Thomas, and Bart Engelen. 2022. "Commercial Online Choice Architecture: When Roads Are Paved With Bad Intentions." In *The Philosophy of Online Manipulation*, edited by Jongepier, F. and Klenk, M., 135–155. New York: Routledge.

Parmer, W. Jared. 2021. "Meaning in Life and Becoming More Fulfilled." *The Journal of Ethics and Social Philosophy* 20 (1): 1–29.

Parmer, W. Jared. Forthcoming. "Nudges, Nudging, and Self-Guidance Under the Influence." *Ergo*.

Schmidt, Andreas T. 2019. "Getting Real on Rationality – Behavioral Science, Nudging, and Public Policy." *Ethics* 129 (4): 511–43.

Schmidt, Andreas T., and B. Engelen. 2020. "The Ethics of Nudging: An Overview." *Philosophy Compass* 15 (4).

Simon, Herbert. 1955. "A Behavioral Model of Rational Choice." *The Quarterly Journal of Economics* 69 (1): 99–118.

Suits, Bernard. 1978. *The Grasshopper: Games, Life and Utopia.* Toronto: University of Toronto Press.

Thaler, Richard H., and Cass R. Sunstein. 2008. *Nudge: Improving Decisions About Health, Wealth, and Happiness.* New Haven, CT: Yale University Press.

Walz, Steffan P., and Sebastian Deterding, eds. 2014. *The Gameful World: Approaches, Issues, Applications.* Cambridge, MA: MIT Press.

Wilkinson, T. M. 2013. "Nudging and Manipulation." *Political Studies* 61 (2): 341–55. doi:10.1111/j.1467-9248.2012.00974.x.

Wood, Allen W. 2014. "Coercion, Manipulation, Exploitation." In Coons and Weber 2014, 17–50.

Zichermann, Gabe. 2013. *The Gamification Revolution: How Leaders Leverage Game Mechanics to Crush the Competition.* New York: McGraw-Hill.

12 Technological Manipulation and Threats to Meaning in Life

Sven Nyholm

1 Introduction

On January 6, 2021, what was supposed to have been a purely ceremonial confirmation of Joe Biden's election as the new US president was interrupted by a violent mob. Rioters stormed the US Capitol building, where the congress and Vice President Mike Pence had gathered to confirm Biden's election. They were egged on by a fiery speech by former President Donald Trump. He had called for his supporters to "fight like hell" and "march on the Capitol". There had also been a build-up of propaganda about a supposed need to "stop the steal" of the election. According to the *Associated Press*, many rioters were believers in the so-called QAnon conspiracy theory (Seitz 2021). This conspiracy theory had reportedly been amplified by fake social media accounts set up by Russian internet trolls (Menn 2020). It had also been boosted by polarizing algorithms of social media platforms that had created filter bubbles. Followers of the conspiracy theory allegedly believed that Trump was fighting dark forces while being counteracted by the establishment and something called the "deep state". The violent mob caused the death of at least five people. They disrupted a purely ceremonial part of a democratic process that by all accounts had been exceptionally well run. And they were seen by some as "desecrating the temple of American democracy".

This event is noteworthy and deeply regrettable for many reasons. But three observations are particularly relevant to the topic of this chapter. First, although details about what exactly happened during this fiasco remain unclear at the time of writing, many of the rioting mob members seem to have been manipulated into behaving like they did.[1] Second, if one assesses the storming of the Capitol building in terms of whether this was a positively meaningful thing, it seems that this sad event was utterly meaningless. More strongly, it can be seen as the polar opposite of a positively meaningful event. Third, one key consideration that helped to make this so meaningless or even the polar opposite of meaningful was precisely that many of these violent protesters had seemingly been manipulated to act as they did. This impression is made even stronger by a further consideration: much of

DOI: 10.4324/9781003205425-14

this manipulation seems to have been driven – or perhaps even partly perpetrated – by technologies that were supposedly created to "bring people together", like social media platforms and other algorithm- and AI-driven information and communication technologies.[2]

This example and these three observations help to illustrate the topic I discuss in this chapter. I discuss whether what I will call technological manipulation poses a serious threat to the values commonly associated with living a meaningful life. My thesis is that it does. Just as manipulation within interpersonal relationships threatens the values associated with meaningful human relationships, the ever-increasing manipulativeness of many technologies we use threatens the values associated with living a meaningful life.

Most discussions of apparent manipulation by technologies are about whether technological manipulation threatens human autonomy (e.g., Susser, Roessler, and Nissenbaum 2019a, 2019b; Klenk and Hancock 2019). There are not many published discussions particularly about whether technological manipulation threatens our opportunities to live meaningful lives.[3] However, I and others have recently written more broadly about whether AI, robots, and other emerging technologies and societal developments threaten our prospects for living meaningful lives, having meaningful relationships, or doing meaningful work (e.g., Campbell and Nyholm 2015; Danaher 2019; Smids, Nyholm, and Berkers 2020; Danaher and Nyholm 2020; Nyholm and Campbell 2022). Later, I draw on that other work. But I also draw on the recent work that has been done about how technological manipulation threatens human autonomy. After all, living an autonomous life is often thought to be a key aspect of living a meaningful human life (Smids, Nyholm, and Berkers 2020). Accordingly, my discussion in this chapter is not as far removed from some of the more common ways of approaching the topic of technological manipulation as it might seem to be.

What follows divides into the following sections. I start by briefly discussing manipulation in general and four different possible views about whether technologies can manipulate us (Sections 1 and 2). Next, I survey some widely shared views about what creates meaning in life (Section 3). I then formulate and defend my main argument to the effect that technological manipulation threatens to make our lives less meaningful (Section 4). Finally, I end with a brief concluding discussion (Section 5).

2 Manipulation by Humans and Technologies

There is no shortage of examples of what are claimed to be manipulative technologies: from deceptive social robots, to filter bubble-generating social media platforms, to recommender systems and other technologies steering and nudging us in different directions (Nyholm and Frank 2019; Sharkey and Sharkey 2020; Susser, Roessler, and Nissenbaum 2019a, 2019b; Frischmann and Selinger 2018; see also Jongepier and Klenk 2022). But can technologies themselves really be manipulative? How should we think of what we might call "technological manipulation"? In order to reflect on these questions, it is useful to have a more general account of manipulation to work with.

Rather than developing an account of manipulation of my own, I will rely on Marcia Baron's (2003) view. A key strength of Baron's account of manipulation is that it captures the broad spectrum of different forms that manipulation can take. All of them involve trying to steer other people in ways that are overly controlling. For this reason, Baron regards manipulativeness as a distinctive kind of vice. The types of manipulation she discusses include the following main kinds – and here I am strongly influenced by Allen Wood's (2014) useful summary of Baron's view:

1. *deception*,[4] including outright lying, false promises, encouraging false assumptions, or fostering self-deception advantageous to the manipulator's ends, and getting the manipulated person to see or interpret things in a way that favors the manipulator's aims;
2. *pressuring*, including intimidation, wearing down the manipulated person's resistance, creating potential embarrassment if the manipulated person does not do what the manipulator wants, and mild forms of threats; and
3. *playing upon emotions, needs, and character weaknesses*, including making the manipulated person feel guilty about something, making them feel an unwarranted sense of gratitude toward the manipulator, taking advantage of their fears and worries, and so on.

When people manipulate others in these ways, they fail to properly respect those others. This is morally objectionable, as Baron and Wood see things. I would add that manipulativeness also threatens one's capacity to have meaningful relationships with others of the most valuable sort. I follow Immanuel Kant (1998), Ronald Dworkin (2011), and others in viewing meaningful relationships of the most valuable sort as being based on mutual concern and respect. Relating to one another on the basis of manipulation (one-sided or mutual manipulation) is the opposite of a positively meaningful relationship.[5] More generally, a life in which one is constantly being manipulated – or a life in which one is constantly trying to manipulate others – strikes me as a life that is not a deeply meaningful one. This is part of why I am interested here in whether technological manipulation is another thing that might threaten the values commonly associated with living a meaningful life.

3 Can Technologies Manipulate People?

Let us now consider whether technologies can manipulate people. I will briefly discuss four different propositions that might be put forward about this issue. While all four have merits, I endorse some combination of the last two of the following propositions.

Consider first the proposition that *technologies can manipulate people*. One possible way of defending this idea is to follow Amanda Sharkey and Noel Sharkey (2020), by taking an outcome-based view of whether manipulation has occurred. They focus on deception in particular. As they see things, if the outcome of an interaction between a human and a technology is that the human ends up with false conceptions about something, or there is some general distortion of society's views about some issue as a result of the widespread use of a technology, then that technology has in effect deceived the humans involved. Whether the technology is able to have intentions to deceive is beside the point, on this view. If the just-described outcomes come about, that is enough for the human beings involved to have been deceived and thereby manipulated by the technologies in question.

Consider next the contrary proposition that *technologies themselves cannot manipulate people*. This is a little bit like the view that "guns don't kill people, people do". The idea behind this view – driven by the so-called instrumental theory of technology – is that there is no agency of the relevant kind in the technologies themselves (Gunkel 2018, 55–65). Technologies are, rather, tools with which people do things to others. On this way of seeing things, technologies cannot manipulate people, but people using certain technologies can manipulate other people. I am skeptical about a hard-lined version of the purely instrumental view of technology, which is wholly opposed to all attributions of agency to technologies (Nyholm 2020, ch. 2 & 3). Yet, it does make sense to say that the advanced kind of agency we typically associate with manipulation of the sorts that Baron describes is not something that any contemporary technologies are capable of. In other words, we might say, for example, that self-driving cars are a form of agents, since they can get from A to B in a functionally autonomous, goal-directed, and seemingly intelligent way. But at the same time, they are not – and nor are any other current technologies – sophisticated moral agents of the sort that can act or fail to act on the basis of moral reasons and be held responsible for their actions (Purves, Jenkins, and Strawser 2015; Nyholm 2020, 58–62). Similarly, since being manipulative is supposed to be a moral vice, for which moral agents can be held responsible, it might be thought that technologies cannot exercise the particular sort of agency associated with the vice of being manipulative.

Consider, however, this proposition: *technologies can relate to people in a manipulation-like way*. In other words, the idea might be that while technologies cannot be said to have the sophisticated form of agency associated with humans manipulating other humans, technologies can do some of the things we associate with human manipulation. In the same way, we might say that there are some aspects of human emotions (e.g., internal subjective experiences) that cannot be replicated in machines, even though there are other aspects of human emotions that can be replicated in machines (e.g., facial expressions, patterns of behavior) (Nyholm 2020, 143–47; Smids 2020). Just as a technology (e.g., a social robot) might have something that is emotion-like about its behavior, some technologies (perhaps the same social robot) could interact with a human being in a manipulation-like way. This might not be exactly like human manipulation of other humans. But it might overlap with key aspects of human manipulation and thereby be manipulation-like, so to speak.

Consider lastly this proposition: *technologies can be part of human–machine collaborations that manipulate people*. On this way of seeing things, technologies should never be seen as acting fully autonomously on their own, even if some technologies can be viewed as capable of a certain form of agency, which might involve some functional autonomy (Mindell 2015; Nyholm 2020, 62–65). ("Functional autonomy" refers to the capacity a technology might have to operate on its own for some period of time, without direct human steering.) For example, a military robot might operate on its own for some period of time and thereby exercise some functional autonomy. However, it will be part of a human–machine collaboration whereby certain humans have designed this technology, monitor its performance, sometimes update it, assess whether to continue using it, and so on – and whereby these humans are able to achieve their goals by "working together" with these technologies. This is a plausible way to think about most, if not all, autonomous technology systems: even when they are in their autonomous modes of functioning, they should always be seen as being part of human–technology collaborations aiming to achieve certain overarching goals had by certain humans (Mindell 2015). We could think of apparently manipulative technologies in this same way. That is, we could think of them as being part of human–machine collaborations that manipulate certain people. Sometimes, the technologies themselves might be doing most of the "work", so to speak. And they might be operating in a functionally autonomous mode. But we might still think that the best way to analyze what is going on is to say that this is a form of "team work" between humans and the apparently manipulative technologies (cf. Nyholm 2020, 64–65).

Which of these four propositions should we accept? Do we have to make a choice here? The two last propositions both have some plausibility to them. It is plausible to think that technologies can interact with humans in manipulation-like ways. It is also plausible to think that technologies can be part of human–machine collaborations that can manipulate people. So, it makes sense to speak of technological manipulation. By this expression, we can mean some combination of the last two propositions considered earlier. In what follows, when I speak of technological manipulation, I mean that technologies can relate to humans in manipulation-like ways and/or that technologies can sometimes be part of human–machine teams capable of manipulating people.

For example, a social robot that appears to have certain emotions and that apparently likes a certain human being (e.g., a sex robot designed to appear to love its user) might be said to relate to this person in a manipulation-like way (Nyholm and Frank 2019). The human being might end up being deceived, might even feel pressured to act in certain ways, and the technologies might play upon certain emotions, needs, and character traits of the human being – and this might be manipulation-like.

To give another example, the social media platforms operated by technology companies like Facebook or Twitter – or the recommender systems operated by companies like Amazon or Netflix – might be seen in an extended sense as being parts of these organizations. In this way, human users might be manipulated by these human–technology collaborations into exhibiting behaviors such as impulsive shopping or binge-consuming of endless streams of content. Similar things can be said about labor-nudging technologies such as those used by companies like Uber. This might sometimes be manipulative human–technology teamwork in the senses of deceiving, pressuring, or playing upon the emotions, needs, and character traits of human users in ways that can appear to be steering those human users in excessively controlling ways (cf. Baron 2003).

4 Meaning in Life and Technological Threats to It

Let us set technological examples aside for a moment and consider something completely different, namely the case of the Swedish teenager Greta Thunberg, who was named *Time* magazine's 2019 "person of the year". Back in 2018, Thunberg was upset about what she had been learning about human-caused climate change and most people's (including most governments') failure to respond decisively to this massive problem. To bring more attention to this issue in her native Sweden, Thunberg started skipping school on Fridays, to go and protest in front of the Swedish parliament. Before long, news spread first in Sweden, and then throughout the world, about this teenager who was protesting against the lack of decisive action on climate change. "Fridays for Future" was born. Within the course of a year, Thunberg "succeeded in creating a global attitudinal shift", influencing hundreds of thousands, if not millions, of young people to take part in climate activism to save the planet for future generations (Alter, Haynes, and Worland 2019). Thunberg has traveled the world (in an environmentally friendly way!) to spread this message, led peaceful protests, spoken to world leaders, and succeeded in communicating her message about this issue like none before her.

I mention this because it is a good example of a meaningful thing to have done with a year of one's life (Nyholm and Campbell 2022). In contrast with the storming of the US Capitol building in the introductory example, these just-mentioned peaceful marches aimed to bring attention to climate change for the sake of future generations also appear to be deeply meaningful. Moreover, doing these things could be part of an overall meaningful life. When I talk about meaningfulness and meaning in life in this chapter, I am using these expressions in a normative way to make normative judgments, and these just-considered examples are the sorts of things I am talking about (Wolf 2010). But as I see things, meaningful actions do not have to be grand, large-scale actions like those performed by Greta Thunberg. Nor does one need to start a worldwide movement for the common good in order to live a meaningful life. A life involving meaningful relationships, or in which one is able to do meaningful work (e.g., being a teacher, a nurse, a doctor, or whatever), can also be positively meaningful (Landau 2017).

In recent times, many philosophers working in the analytic tradition have gotten increasingly interested in meaningfulness. Authors like Susan Wolf (2010) and Thaddeus Metz (2013) have done highly influential work on this topic. Notably, much work by these and other authors has been quite abstract and meta-ethical in nature. Philosophers have discussed issues such as whether meaning is a wholly subjective notion; whether we should think of meaning as depending on objective features of one's life and actions that can be taken to have a not wholly subjective value; or whether we should perhaps accept some form of hybrid theory that understands meaning in life as having both subjective and objective components (Campbell and Nyholm 2015). Wolf (2010), for example, is well known for her thesis that meaning in life arises when one is passionate (= subjective component) about projects and activities that have value (= non-subjective or "objective" component).

Moreover, not only analytical philosophers are interested in this notion of living a meaningful life. Positive psychologists, to give another example, who empirically study human well-being and flourishing, are also interested in what is involved in living a meaningful life. Some leading voices in that field – such as Martin Seligman (2010) – also adopt partly non-subjective views about meaningfulness. Similarly, organizational psychologists study the idea of meaningful work, like some philosophers are also increasingly doing (Danaher 2019; Smids, Nyholm, and Berkers 2020). A leading idea in philosophical, psychological, and other discussions of meaningfulness is that in addition to seeking happiness, having various ambitions and so on, most people also desire to live meaningful lives, perform meaningful actions, have meaningful relationships, and do meaningful work (Seligman 2010; Metz 2013).

It should come as no surprise, accordingly, that one of the things that philosophers of technology have recently been interested in when thinking about emerging technologies is precisely the impact that these technologies might have on our opportunities to live meaningful lives, have meaningful relationships, and do meaningful work. Things like social media, robots and AI in the workplace, and social robots have appeared to some commentators to pose potential threats to the values we associate with living meaningful lives, having meaningful relationships, or doing meaningful work (e.g., Danaher 2019; Smids, Nyholm, and Berkers 2020).

I want to note here that, in my view, when we philosophize about this topic, we should not only concern ourselves with potential threats to meaningfulness in our lives. We should also investigate possible technologically mediated opportunities for new forms of meaningful relationships, meaningful work, or ways of living meaningfully (cf. Smids, Nyholm, and Berkers 2020). Like positive psychologists who argue that psychologists should not only investigate worries and problems related to mental health but also happiness and psychological flourishing (Seligman 2010), I think that philosophers should also investigate the potential for new technologies and new societal developments to make our lives more meaningful. In general, then, I adhere to a form of "cautious optimism" about what new technologies can do for the meaningfulness of lives and relationships (Danaher, Nyholm, and Earp 2018; Nyholm, Danaher, and Earp 2022). That being said, however, here my focus is on possible threats to meaningfulness posed by manipulation and manipulative technologies. Since we are increasingly surrounded by more and more technologies that appear to be manipulative, it is important to get clear on how such technologies can pose serious threats to the values we associate with living meaningful lives.

In investigating such potential threats, it is necessary to descend from the more abstract aforementioned meta-ethical level at which many analytic philosophers discuss meaning in life. We need to move down to a more practical level, where we work with substantive conceptions of what makes projects, relationships, work, lives, actions, activities, and so on meaningful. Notably, there is fairly wide agreement about what sorts of things are intimately associated with living a meaningful life, having meaningful relationships, doing meaningful work, and so on. The following types of considerations are often referred to in publications on this topic.

Autonomy: living a life that is self-directed, where one is afforded the space to make one's own choices and shape one's own life, is often thought to contribute significantly to making one's life more meaningful. Consider the contrast: being told what to do by others, not having any personal autonomy at work or at home, and so on. It is more meaningful, it is often thought, to enjoy a certain amount of autonomy in one's life. Some even go so far as to say that living an autonomous life is the most important aspect of living a meaningful life. For example, Jesper Ahlin Marceta (2021) defends what he dubs an "individualist" theory of meaning in life, according to which autonomy is the main characteristic of a meaningful life. This is surely exaggerated and not a complete theory of meaningfulness in life. But it is plausible that personal autonomy is a key component of a meaningful life.

Actively pursuing a purpose: whether we are talking about meaningful work, or meaning in life more generally, it is a commonly accepted idea that it is important that one does work, or leads a life, that allows one to actively pursue a purpose or set of purposes that one deems to be worthwhile. Again, the plausibility of this can be brought out by considering the contrast. Suppose you do not think that, say, the work you do for a living has any clear purpose that you find worthwhile or that you identify with.

You are then likely to find your work less meaningful than if you view your work as purposeful in a way that you find worthwhile and identify with. Part of this idea is also that one is being active – rather than passive – in how one relates to the purposes in question. The more passive we are in life, it is often thought, the less meaningful our lives become. For example, passively consuming light entertainment might be fun and relaxing. But it seems less meaningful than actively pursuing some purpose we see as having positive value (Nussbaum 2004).

Relating to others on the basis of mutual care, trust, and respect: being part of a mutually supportive community and having good personal relationships characterized by mutual care, trust, and respect are further aspects commonly associated with meaningfulness in life. Again, this applies both to life more generally and to more specific contexts, such as work (Danaher 2019; Smids, Nyholm, and Berkers 2020).

Being part of something "bigger than you", which is positively valuable: it is a commonly expressed idea that our lives become more meaningful when we participate in something bigger than ourselves that is a positive force for the good (Wolf 2010; Seligman 2010). Think again, for example, of the "Fridays for Future" movement. Many young people who are part of this movement might experience it as a meaningful thing to participate in precisely because it is something bigger than them that is of positive importance. Doing something together with others in order to try to help to save the world for future generations can almost seem like something that might be among the most meaningful things one could possibly do, especially if this should turn out to be a successful movement (Di Paola and Nyholm 2021; cf. Parfit 2011, 616). Even if something bigger than us that is a force for the good ends up ultimately not achieving its goal (e.g., because a gigantic asteroid hits the Earth and kills all life on Earth 100 years from now), being part of a movement like that, which is bigger than us as individuals and is a force for the good, can still seem like a very meaningful thing.[6]

Self-development and human achievement: another set of ideas commonly associated with meaning in life concerns the development of one's skills and talents, the fulfilment of human potential, and the realization of human achievement. This, too, is associated both with meaningful work and meaning in life more generally (Danaher 2019). In the context of work, for example, work is usually considered more meaningful if there is room to develop one's skills and talents in the workplace and if one's work involves room for achievement (Smids, Nyholm, and Berkers 2020; Danaher and Nyholm 2020). In life more generally, it is often thought that having and developing human capabilities is part of living a good and meaningful human life (Alkire 2002).

Insight and understanding: the last thing I will mention as a common idea about what it is to live a meaningful human life is that it will often involve having a certain amount of insight and understanding. This could be either self-knowledge or knowledge and understanding about the world around us (Hurka 2015). For example, Robert Nozick (1974) is appalled by the prospect of living in an "experience machine", in his famous thought experiment, partly because he thinks that if all of our experiences would be created via simulation – even a very pleasant simulation – we would lack proper knowledge and understanding of what is going on around us. Being able to make sense of ourselves and things around us is thought to be more meaningful than being deluded, misinformed, or otherwise misled about ourselves or about what is going on around us.

With the help of these ideas about what provides positive meaning in life – in life in general, in interpersonal relationships, at work, or in other parts of life – it is possible to systematically discuss whether technological developments pose serious threats to the possibility of living meaningfully. For each of the aforementioned aspects of meaning in life, we can ask whether technological developments pose threats to our opportunities for realizing these values.

5 Manipulative Technologies and Threats to Meaning in Life

Using the materials introduced in the previous sections, it is possible to formulate an argument to the effect that technological manipulation might threaten the values associated with a meaningful life. We can argue as follows:

1. If technological manipulation threatens one or more of (a) our autonomy (b) our capacities to actively pursue valuable purposes, (c) our capacities to relate to other people on the basis of mutual care, trust, and respect, (d) our opportunities to be part of things that are "bigger than us" that are good, (e) our opportunities for self-development and human achievement, or (f) our capacities for insight and understanding, then this technological manipulation thereby threatens our opportunities for living meaningful lives.
2. Technological manipulation poses significant threats to some, or perhaps all, of these different values associated with meaningful lives.
3. Therefore, technological manipulation poses significant threats to our opportunities for living meaningful lives.

How strong is this argument? I will now discuss the two main premises, first with three brief points about premise 1 and then a slightly longer discussion of premise 2.

The first thing I want to highlight about the first premise is that it speaks about *threats* to meaning in life. The premise does not assert anything about whether technological manipulation necessarily undermines meaning in life. It does not say that if we are subject to technological manipulation, we cannot possibly live meaningful lives, have meaningful relationships, do meaningful work, and so on. That would be too strong. Accordingly, this premise is about perceived or real threats to meaning in life. Yet, the threats I am discussing here are, as I see things, significant threats. In other words, the threats to the values associated with meaning in life coming from technological manipulation are not accidental or insignificant in magnitude. Rather, the nature of technological manipulation non-accidentally threatens the values associated with meaningfulness in life, according to premise one, and does so in a high-impact sort of way.

The second thing I want to acknowledge about the first premise is that although there is fairly wide agreement about what contributes to meaning in life, not everyone working on meaning in life regards all these aspects as being key ingredients in a meaningful life. This is why I have formulated this premise in a disjunctive way. I say that if technological manipulation threatens one or more of these things, this should be seen as posing a threat to our opportunities for living meaningful lives. Moreover, I do not think of the different criteria for meaning in life that I have put on the list of disjuncts as necessarily being wholly separate from each other. There might be partial overlap among some of them.

The third and final thing I will say about the first premise is that while it does list a number of different things commonly associated with meaning in life, it also leaves out some things sometimes associated with meaning in life. Earlier, for example, I mentioned the influential work of Thaddeus Metz. Those familiar with it will notice that while Metz (2013) relates meaning in life to "The True, The Good, and The Beautiful", the third of these – viz. The Beautiful – is mostly left out here. This is not because I disagree with Metz and others (e.g., Danaher 2019) that the beautiful can be a source of meaning in life. It is rather that I did not intend to cover absolutely everything that can sensibly be seen as sources of meaning in life. I instead simply leave some things out, such as The Beautiful. A more thorough discussion of whether technological manipulation threatens meaning in life would also deal with that consideration and any other ones that can also be seen as potential sources of meaning in life that might be under threat when we are subject to manipulation.

I turn now to the second premise. More can be said about it than I will be able to say here, but I hope that what I say will be enough to make this premise seem plausible. I will go through the aspects of a meaningful life mentioned in the first premise one by one. For each aspect, I will discuss whether technological manipulation poses significant threats to it.

Autonomy: as noted earlier, much discussion about technological manipulation has precisely been about whether it poses a threat to personal autonomy. It has been plausibly argued – in particular by Susser, Roessler, and Nissenbaum (2019a, 2019b) – that technological manipulation does indeed pose a threat to autonomy. When we are being manipulated, Susser, Roessler, and Nissenbaum argue, this threatens our ability to act on the basis of ends we adopt as our own, for reasons that we endorse as ones we want to act on the basis of. This is a threat to autonomy. Suppose that Susser et al. are right about this. Then since living autonomously is part of living a meaningful life according to premise one, we here have our first reason for accepting premise two. Notably, there are those – for example, Michael Klenk and Jeff Hancock (2019) – who argue that technological manipulation does not necessarily pose a threat to autonomy. But that is a much more controversial view than the view that it does pose such a threat. My inclination is to respond to Klenk and Hancock's view in a way that is similar to how I respond to Sarah Buss's view in Note 5; namely, if Klenk and Hancock can describe cases in which somebody is supposedly being manipulated but where this does not pose any threat to their autonomy, then most likely, "manipulation" is not quite the right word to describe what Klenk and Hancock are talking about. In other words, since I agree with Baron that being manipulative is to be too eager to steer or control others, I find it counterintuitive to say that somebody (or some technology) is being manipulative or does something manipulation-like without its being an instance of somebody's trying to steer another in an inappropriate way. I therefore take it that the view defended by Susser et al. is correct, though I acknowledge that there are those who disagree with it.

Actively pursuing a purpose: if Susser and co-authors are right that technological manipulation can lead us to act in the service of ends that are not our own, for reasons we may not endorse, this can also be seen as a threat to the second aspect of meaning discussed earlier, viz. the idea of actively pursuing a purpose we find valuable. Technological manipulation, moreover, can make us more passive,[7] with recommender systems and other technologies hooking us to our screens and making us passively consume content or trying to make us stay on some website as long as possible. We can think of this as partly being an "opportunity cost" argument. If it were not for the manipulative technologies designed to make us click on various links, remain as long as possible on some website, or passively binge-watch entertainment, and so on, we could be doing something else whereby we would in a much more active way be pursuing some valuable purpose we find important. I think it is a common feeling many share that if one has, say, passively spent too much time on manipulatively addictive social media platforms during a day, one has been "wasting one's time".

Relating to others on the basis of mutual care, trust, and respect: one of the things many online environments do – including ones designed to "bring people together" – is to create filter bubbles and echo chambers (Pariser 2011; Lynch 2017). People are manipulated into believing in conspiracy theories, their tribal instincts are triggered and run amok, and other perspectives are demonized. Go back to the initial example with the January 6, 2021, Capitol building storming, with people believing in the "QAnon" conspiracy theory and allegedly being manipulated by Russian trolls with fake social media profiles and polarizing online environments. Here, certain technologies – the social media platforms with their algorithms – can be interpreted as interacting with users in manipulation-like ways that undermine people's perhaps already fragile willingness to care about, trust, or respect those who are seen as members of out-groups. In this particular example, tensions also broke out within the US Republican party where the Trump faction started demonizing any members of the Republican party who were not staunch Trump loyalists (Murphy et al. 2021). According to some of the reporting of what led to all of this – such as the reporting by the *Associated Press* cited earlier – this was boosted by various forms of manipulation, including what I am calling technological manipulation.

Being part of something "bigger than us" that is good: the example of the January 6, 2021, mob violence can also be interestingly discussed in relation to the idea that meaning in life can involve being part of something bigger than us that is valuable. Certainly, the members of the mob who had been driven by conspiracy theories and manipulation into joining a mob and storming the US Capitol building can be seen as participating in something bigger than themselves. However, a crucial component of being part of something bigger than us that is *a force for the good* is missing here. These people were manipulated into joining something bigger than them that was bad, regrettable, and antithetical to the idea of joining something bigger than oneself that is good. So, if it is true that they were victims of technological manipulation, that technological manipulation posed a serious threat to their opportunities to act in a meaningful way on this occasion.

Self-development and human achievement: when it comes to whether technological manipulation can be viewed as posing threats to opportunities for self-development and human achievement, many of the remarks made earlier about threats to opportunities to actively pursue valuable purposes become relevant again. The more we are led to behave as we do because technologies relate to us in manipulation-like ways or because human–machine teams are manipulating us to behave as the humans in those teams want us to behave, the less room there may be for self-development and human achievement on our part. Elsewhere, John Danaher and I have written about whether automation, AI, and robots in the workplace might create an "achievement gap", whereby it becomes harder for humans to realize the value of achievement in the workplace (Danaher and Nyholm 2020). It can plausibly be argued that when work is partly driven by manipulative "labor nudges" of the sorts that Susser et al. Susser, Roessler, and Nissenbaum (2019a, 2019b) discuss, this poses serious threats to our opportunities for developing our skills and realizing human achievement in the workplace. So, with respect to this part of meaning in life as well, there is a plausible case to be made for the idea that technological manipulation may threaten meaning in life.

Insight and understanding: turning lastly to whether technological manipulation might pose threats to human insight and understanding, here again some of the previous discussion about technological manipulation and some people's being led to believe in things like absurd conspiracy theories becomes relevant once more. But we do not have to turn to this more extreme form of online polarization to have examples of how technological manipulation might threaten our opportunities for insight and understanding. The filter bubbles and echo chambers we are all manipulated by social media platforms into joining threaten to give us a very one-sided view of the world, as Michael Patrick Lynch (2017) describes in some detail in his striking book *The Internet of Us*. We can say, then, that insofar as positive meaning in life involves insight and understanding, and having a one-sided and polarized view of view the world is contrary to this goal, technological manipulation can be viewed as a threat to yet another aspect of a meaningful life.

Much more can be said about all of these issues. But based on this brief discussion, I submit that the second premise of the argument presented here enjoys strong support. Technological manipulation poses significant threats to all of the aspects of meaning in life considered earlier. Accordingly, the earlier-presented argument's general conclusion follows: technological manipulation poses significant threats to our opportunities to live meaningful lives.

6 Concluding Discussion

I have just argued that technological manipulation can pose serious threats to our opportunities to live meaningful lives, have meaningful relationships, or do meaningful work. It is appropriate to end with some remarks about limitations of my discussion and consideration of some possible objections that might be raised against it.

The first thing I should note is that I have not discussed possible differences in how grave the threats posed by different forms of technological manipulation to our opportunities to live meaningful lives are. It may very well be that threats to meaning in life posed by, say, technologies that help to manipulate people into believing wild conspiracy theories are much greater than the threats posed by, say, social robots that might be deceptive to some degree. It would be valuable to discuss particular examples in more detail and compare them with each other, to see which forms of technological manipulation pose the greatest threats to our opportunities to live meaningful lives.

A second limitation is that I have focused only on whether there might be threats to meaning in life posed by technological manipulation, without providing any corresponding discussion of what should be done to avert these threats. A fuller discussion would also consider this issue about possible defenses against these threats, again with a view to which of these threats are most severe. I have not done so here but hope to do so elsewhere. Having noted these two limitations of my discussion, I now turn to some possible objections that might be raised against it.

One potential objection might be a worry to the effect that discussing whether technological manipulation threatens meaning in life is a less pressing topic than that of when, and in what ways, technological manipulation might be most wrong, blameworthy, or otherwise morally problematic. My response to this is I agree that that might be a more pressing question – especially if we think of a case such as the pro-Trump mob's storming of the Capitol that was the opening example. But there is no need to discuss only the most pressing questions, leaving all other interesting questions aside. Moreover, in some, less dramatic cases, where it is not immediately clear that the manipulation involved rises to the level of seriously blameworthy wrongdoing, there might still be a lingering sense that there is something regrettable and problematic about the manipulation in question. In such cases, we need to turn to other ideas or concepts to assess what the issue is. And here a question such as whether our opportunities for living meaningful lives are being threatened is one of the crucial questions that we can turn to. Moreover, as I see things, whether technological manipulation poses a threat to meaning in life is an interesting and worthwhile question in its own right – even if some other questions, such as whether somebody has acted seriously wrongly or should be punished or blamed, might be more urgent under certain circumstances.

Another objection that might come up might be driven by an adherence, on behalf of the objector, to the instrumental theory of technology. Somebody who thinks that it makes no sense to view technologies as being manipulative – and who would insist that only human beings can manipulate – might question whether this whole discussion makes sense, given that I have been asking whether technological manipulation can be a threat to meaning in life. To such worries, my answer is to remind the reader that I have not been assuming that technologies themselves can be manipulative in exactly the way(s) in which human beings can be. Instead, I have been taking it that technologies can relate to human beings in manipulation-like ways – and that technologies can be part of human–machine teams that can be manipulative in the ways in which they relate to human beings. If either or both of those things are true, that is enough for it to be worthwhile to discuss whether either or both forms of manipulation might pose significant threats to the values commonly associated with meaningfulness in life.[8]

Notes

1. Notably, this assessment seems to be shared by some of the rioters themselves. For example, one member of the mob, who became known as the "QAnon Shaman" in the press because of his extravagant attire, felt that he had been "duped" by Donald Trump, according to the lawyer of this rioter (Kilander 2021).
2. According to the company Facebook's mission statement, for example, the aim of that social media platform is to "give people the power to build community and bring the world closer together". https://investor.fb.com/resources/default.aspx (accessed on August 3, 2021)
3. Michael Klenk (2020) suggests a causal connection between manipulation and a dent to well-being via a loss of autonomy and thus defends a view that is broadly

congenial with my main argument in this chapter. Klenk does not, however, explicitly discuss the impact of technological manipulation on the meaningfulness of people's life but instead formulates his argument in terms of claims about well-being.

4. According to Susser, Roessler, and Nissenbaum (2019a, 2019b), we should make a distinction between deception, on the one hand, and manipulation, on the other. I see no strong reason to distinguish between the two; I agree with Baron and Wood that deceiving people can be one of the ways in which we might manipulate them to behave in some way.
5. I am skeptical of Sarah Buss's (2005) intriguing claim that manipulation and deception are often key parts of at the least the initial stages of good romantic relationships. It seems to me that Buss is mischaracterizing the type of interaction she is talking about (e.g., trying to present oneself in the best possible light to the person one is trying to impress etc.) in calling it manipulative and deceptive. If we follow Baron's view, we can say that if some behavior (such as those Buss is discussing when she discusses the initial stages of romantic relationships) does not qualify as trying to steer another's behavior in an overly controlling way, then that behavior is not manipulative in the morally objectionable way or perhaps not manipulative at all.
6. That said, I do agree with Samuel Scheffler (2018) that if there would be no future generations and we would be the last generation of human beings, this would make our lives less meaningful than if, as most of us believe and hope, there will be others coming after us, who can carry on some of our projects and traditions, and who will also continue the development of humanity long into the future.
7. For an argument about how robots and AI threaten to make people less willing to be active moral agents (and more likely to take on the role of passive moral patients), see Danaher (2017).
8. Many thanks to Fleur Jongepier, Michael Klenk, and the participants of their online manipulation workshop series. My work on this chapter is part of the research program "Ethics of Socially Disruptive Technologies", which is funded through the Gravitation program of the Dutch Ministry of Education, Culture, and Science and the Netherlands Organization for Scientific Research (NWO grant number 024.004.031).

7 References

Ahlin Marceta, Jesper. 2021. "An Individualist Theory of Meaning." *Journal of Value Inquiry*. doi:10.1007/s10790-021-09803-3.

Alkire, Sabina. 2002. *Valuing Freedoms: Sen's Capability Approach and Poverty Reduction*. Oxford: Oxford University Press.

Alter, Charlotte, Suyin Haynes, and Justin Worland. 2019. "Greta Thunberg Is TIME's 2019 Person of the Year." Accessed August 24, 2021. https://time.com/person-of-the-year-2019-greta-thunberg/.

Baron, Marcia. 2003. "Manipulativeness." *Proceedings and Addresses of the American Philosophical Association* 77 (2): 37. doi:10.2307/3219740.

Buss, Sarah. 2005. "Valuing Autonomy and Respecting Persons: Manipulation, Seduction, and the Basis of Moral Constraints." *Ethics* 115 (2): 195–235. doi:10.1086/426304.

Campbell, Stephen M., and Sven Nyholm. 2015. "Anti-Meaning and Why It Matters." *Journal of the American Philosophical Association* 1 (4): 694–711. doi:10.1017/apa.2015.9.

Danaher, John. 2017. "The Rise of the Robots and the Crisis of Moral Patiency." *AI & Society*, 1–8. doi:10.1007/s00146-017-0773-9.
Danaher, John. 2019. *Automation and Utopia: Human Flourishing in a World Without Work*. Cambridge, MA: Harvard University Press.
Danaher, John, and Sven Nyholm. 2020. "Automation, Work and the Achievement Gap." *AI and Ethics*, 1–11. doi:10.1007/s43681-020-00028-x.
Danaher, John, Sven Nyholm, and Brian D. Earp. 2018. "The Quantified Relationship." *American Journal of Bioethics* 18 (2): 3–19. doi:10.1080/15265161.2017.1409823.
Di Paola, Marcello, and Sven Nyholm. 2021. *Climate Change and Anti-Meaning*. Under review.
Dworkin, Ronald. 2011. *Justice for Hedgehogs*. Cambridge, MA: Harvard University Press.
Frischmann, Brett M., and Evan Selinger. 2018. *Re-Engineering Humanity*. Cambridge: Cambridge University Press.
Gunkel, David J. 2018. *Robot Rights*. Cambridge, MA: MIT Press.
Hurka, Thomas. 2015. *The Best Things in Life: A Guide to What Really Matters*. Oxford: Oxford University Press.
Jongepier, Fleur, and Michael Klenk. 2022. "Online Manipulation: Charting the Field." In *The Philosophy of Online Manipulation*, edited by Jongepier, F. and Klenk, M., 15–48. New York, NY: Routledge.
Kant, Immanuel. 1998. *Groundwork of the Metaphysics of Morals*. Cambridge: Cambridge University Press.
Kilander, Gustaf. 2021. "QAnon Shaman feels 'Duped' after Trump doesn't Pardon Him." *The Independent*. www.independent.co.uk/news/world/americas/qanon-jacob-chansley-arrest-trump-pardon-b1791365.htm.
Klenk, Michael. 2020. "Digital Well-Being and Manipulation Online." In *Ethics of Digital Well-Being: A Multidisciplinary Perspective*, edited by Christopher Burr and Luciano Floridi. Cham: Springer. 81–100. doi: 10.1007/978-3-030-50585-1_4.
Klenk, Michael, and Jeff Hancock. 2019. "Autonomy and Online Manipulation." *Internet Policy Review*. https://policyreview.info/articles/news/autonomy-and-online-manipulation/1431. Accessed February 28, 2020.
Landau, Iddo. 2017. *Finding Meaning in an Imperfect World*. New York, NY: Oxford University Press.
Lynch, Michael P. 2017. *The Internet of Us: Knowing More and Understanding Less in the Age of Big Data*. New York, NY: Liverlight.
Menn, Joseph. 2020. "Small but Growing Russian Support for QAnon Conspiracies Seen Online." *Reuters Media*, August 24. Accessed August 24, 2021. www.reuters.com/article/usa-election-qanon-russia/small-but-growing-russian-support-for-qanon-conspiracies-seen-online-idUSL1N2FP0DM?edition-redirect=u.
Metz, Thaddeus. 2013. *Meaning in Life*. Oxford: Oxford University Press.
Mindell, David A. 2015. *Our Robots, Ourselves: Robotics and the Myths of Autonomy*. New York, NY: Penguin Books.
Murphy, Paul P., Gregory Wallace, Ali Zaslav, and Clare Foran. 2021. "Trump Supporters Confront and Scream at Sen. Lindsey Graham." https://edition.cnn.com/2021/01/08/politics/lindsey-graham-donald-trump-supporters-airport/index.html.
Nozick, Robert. 1974. *Anarchy, State and Utopia*. Oxford: Blackwell.

Nussbaum, Martha. 2004. "Mill between Aristotle & Bentham." *Daedalus* 133 (2): 60–68.

Nyholm, Sven. 2020. *Humans and Robots: Ethics, Agency, and Anthropomorphism*. London: Rowman & Littlefield.

Nyholm, Sven, and Stephen M. Campbell. 2022. "Meaning and Anti-Meaning in Life." In *Oxford Handbook on Meaning in Life*, edited by Iddo Landau, 277–91. Oxford: Oxford University Press.

Nyholm, Sven, John Danaher, and Brian D. Earp. 2022. "The Technological Future of Love." In *Love: Past, Present, and Future*, edited by Natasha McKeever, Joe Sanders, and Andre Grahle, 224–39. London: Routledge.

Nyholm, Sven, and Lily E. Frank. 2019. "It Loves Me, It Loves Me Not." *Techné: Research in Philosophy and Technology* 23 (3): 402–24. doi:10.5840/techne2019122110.

Parfit, Derek. 2011. *On What Matters*. Oxford: Oxford University Press. Volume II.

Pariser, Eli. 2011. *The Filter Bubble: What the Internet Is Hiding from You*. New York, NY: Penguin Books.

Purves, Duncan, Ryan Jenkins, and Bradley J. Strawser. 2015. "Autonomous Machines, Moral Judgment, and Acting for the Right Reasons." *Ethical Theory and Moral Practice* 18 (4): 851–72.

Scheffler, Samuel. 2018. *Why Worry About Future Generations?* Oxford: Oxford University Press.

Seitz, Amanda. 2021. "Mob at U.S. Capitol Encouraged by Online Conspiracy Theories." *Associated Press*, April 29. Accessed August 24, 2021. https://apnews.com/article/donald-trump-conspiracy-theories-michael-pence-media-social-media-daba3f5dd16a431abc627a5cfc922b87.

Seligman, Martin. 2010. "Flourish: Positive Psychology and Positive Interventions." *The Tanner Lectures of Human Values*. https://tannerlectures.utah.edu/_resources/documents/a-to-z/s/Seligman_10.pdf.

Sharkey, Amanda, and Noel Sharkey. 2020. "We Need to Talk about Deception in Social Robotics!" *Ethics and Information Technology*. doi:10.1007/s10676-020-09573-9.

Smids, Jilles. 2020. "Danaher's Ethical Behaviourism: An Adequate Guide to Assessing the Moral Status of a Robot?" *Science and Engineering Ethics* 26 (5): 2849–66. doi:10.1007/s11948-020-00230-4.

Smids, Jilles, Sven Nyholm, and Hannah Berkers. 2020. "Robots in the Workplace: A Threat to-or Opportunity for-Meaningful Work?" *Philosophy & Technology* 33 (3): 503–22. doi:10.1007/s13347-019-00377-4.

Susser, Daniel, Beate Roessler, and Helen Nissenbaum. 2019a. "Online Manipulation: Hidden Influences in A Digital World." *Georgetown Law Technology Review* 4 (1): 1–45. Accessed February 27, 2020.

Susser, Daniel, Beate Roessler, and Helen Nissenbaum. 2019b. "Technology, Autonomy, and Manipulation." *Internet Policy Review* 8 (2): 1–22. doi:10.14763/2019.2.1410.

Wolf, Susan R. 2010. *Meaning in Life and Why it Matters*. Princeton, NJ: Princeton University Press.

Wood, Allen W. 2014. "Coercion, Manipulation, Exploitation." In *Manipulation: Theory and Practice*, edited by Christian Coons and Michael Weber, 17–50. Oxford: Oxford University Press.

13 Digital Manipulation and Mental Integrity

Geoff Keeling and Christopher Burr

1 Introduction

This chapter is about software agents that influence the behaviour of internet users by deploying personalised content. Here, software agents can be understood as computer programmes that exhibit goal-directed behaviour in the sense of adjudicating between candidate options relative to some utility function.[1] Companies like Amazon, Google, and Facebook use software agents to adjust features of users' online environments such as ad content, mobile notifications, suggested videos, and prices for goods and services, with the aim of maximising time-on-site, click-through-rate, user spending, or neighbouring parameters (Burr, Cristianini, and Ladyman 2018; Burr and Cristianini 2019; Milano, Taddeo, and Floridi 2020). Some software agents learn via experimentation what content to deploy and at what times so as to tailor content to the behaviours of individual users or groups of relevantly similar users.

Software agents can influence behaviour in ways that are morally permissible. For example, YouTube's recommender engine might influence a user to explore a novel musical genre after predicting that the user is likely to be receptive to that genre given their tastes and the tastes of relevantly similar users.[2] Other cases are more pernicious. For example, an online casino might use predictors of gambling addiction such as a user's betting frequency or betting variance to selectively deploy pop-up 'free bets' to gambling addicts each time their cursor movements suggest they are about to exit the game (Finkenwirth et al. 2020; LaBrie and Shaffer 2011). Yet more cases are such that the behavioural influence in these cases is neither obviously permissible nor obviously impermissible. For example, a video sharing app may employ a rapid auto-cue feature such that new and targeted content appears momentarily after old content is consumed. The repeated use of this strategy will for a broad class of users result in those users spending more time on the platform than they originally anticipated.

What distinguishes morally permissible from impermissible behavioural influence strategies by software agents? We argue that morally impermissible instances of behavioural influence by software agents typically involve

DOI: 10.4324/9781003205425-15

manipulation, coercion, and deception, and that the wrongness of these strategies admits analysis in terms of *mental integrity* and *authentic choice*. Roughly, an individual's mental integrity is compromised if the conditions required for them to make authentic choices are compromised. These conditions include, inter alia, having options to choose between and having the capacity to enact them; being in an environment that permits rational assessment and evaluation of the available options; having a stable set of beliefs and values that facilitate the pursuit of objectively worthwhile ends; and having a suitably stable sense of who they are,[3] which is appropriate given the social relations in which they stand. We argue that impermissible instances of behavioural influence by software agents undermine the mental integrity of users, and in so doing, diminish their capacity for authentic choice. In contrast, morally permissible instances of behavioural influence by software agents respect the mental integrity of users. The concepts of mental integrity and authentic choice at issue here will be clarified and qualified in due course.

In Section 2, we introduce the technologies with which we are concerned and argue that existing accounts of manipulation, deception, and coercion are unsuitable for behavioural influence by software agents. In Section 3, we develop a novel account of impermissible behavioural influence on the part of software agents. In Section 4, we consider practical implications. In Section 5, we conclude.

2 The Problem

In this section, we characterise the technologies with which we are concerned. We then articulate and clarify the problem of demarcating morally permissible from morally impermissible strategies available to software agents for influencing user behaviour. Finally, we argue that morally impermissible instances of behavioural influence by software agents typically involve manipulation, coercion, or deception but argue that standard accounts of these moral concepts are unsuitable for behavioural influence involving software agents. We take this to motivate the need for a novel account of impermissible behavioural influence involving software agents influencing the behaviour of human users.

2.1 *The Technologies*

We are interested in software agents that influence features of a human user's online environment.[4] These features include ads, recommended media content such as videos and news stories, notifications, and the prices for goods and services (Baird 2017; Ezrachi and Stucke 2016; Larson and Surya Mattu 2015). The agent's goal is to maximise some parameter (or set of parameters) such as user click-through-rate, time-on-site, or total amount spent on purchases. The agent receives feedback for its decisions, such as whether the user clicks on a particular ad or purchases a recommended product, and, over time, refines its content choices to better realise its goal. The agent in effect learns the user's content preferences via repeated experimentation, such that in the long run the agent can deploy content that is maximally or close to maximally conducive to its goal. Agents of this stripe are integral to the business models of Google, Facebook, and Amazon. In what follows, we provide a rigorous characterisation of these agents, and then make precise some of the strategies they employ to influence user behaviour.

First, our concern is with software *agents* that perform actions to influence an *environment* (Russell and Norvig 2020, 36–39; Sutton and Barto 1998, 47–53). At each time-step, the environment is in a particular *state*, and the agent must choose some act from a non-empty non-singleton set of alternatives.[5] Once the agent has performed its action, the environmental state will change. State changes are stochastic. What this means is that the agent's act and present state do not uniquely determine the next state. Rather, state changes are governed by a probabilistic transition function that returns the probability of transitioning to a given state conditional on the agent performing a particular act in its present state. The agent has a *reward function* that returns a real-valued reward for performing an act in a given state and then transitioning to a new state. The agent's goal is to maximise *expected* reward. Here the expected reward for performing a given act in a given state is the weighted sum of rewards received conditional on transitioning to different states where the weights are given by the probability of each transition.

How does this bare-bones characterisation of an agent map onto our *intuitive* picture of an agent that deploys personalised content to influence user behaviour?

First, the act space represents the agent's content options (e.g., which price to set for a product, or which video to recommend). This set includes whatever options are available to the agent. Second, the reward function represents "what matters" to the agent. If the agent's role is to select ads to display to the user, then an appropriate goal for the agent is to maximise the user's click-through-rate (i.e., the number of ads that the user clicks on divided by the total number of ads displayed to the user). The more ads that the user clicks on, the better things are going from the perspective of the agent. In contrast, if the agent determines the price for some product, then the agent's reward might be proportionate to the total amount of money that the user spends on the product. Such an agent would then need to adjudicate between the probability of the user purchasing a product at different prices (i.e., the user is less likely to buy the product the more expensive it is) and also the amount of money made if the user buys the product (i.e., it is better from the algorithm's perspective if the user pays more for the product than less). Finally, the environmental states encode information about how the human user responds to features of their online environment. For example, a state might encode the choice of a user to click on a particular link or their failure to do so. It might also include other data such as how long the user spends on a page or facts about their cursor movements (Huang, White, and Dumais 2011).

The model as we have presented is simplified for brevity. However, we will make precise some of the ways in which the model is simplified in order to give a non-technical overview of the more sophisticated technologies of which we are concerned. First, the model takes it for granted that the agent knows the probability of transitioning to each state conditional on its performance of each act in its present state. This cannot be assumed. The agent's environment is a formal representation of the user's online environment and their interactions. The agent does not know in advance what the user will do in response to particular content. Rather, it has to make a model of how the user will respond to content with different features, so that it can predict, say, how likely the user is to click on an ad given relevant feature of the ad and relevant features of the user's behaviour. This model is based on what the agent observes about the user's behaviour, and that of relevantly similar users, in response to historical content choices, known as 'relevance feedback'.

Relatedly, the agent does not interact with the user only once. There is a succession of interactions. Here, the agent must adjudicate between two competing aims. On the one hand, it matters to the agent that it maximises its reward (e.g., by giving the user ads that they are likely to click on). But, on the other hand, to do this effectively in the long run the agent needs to gather more information about the kinds of ads that the user is disposed to click on. Hence, experimentation is required. The agent's aim is not to maximise expected reward *in each decision* but rather to maximise expected reward *over the long run*. What this means in practice is that, say, an agent that determines what ads to show on a social media platform may from time to time display ads that the user is perceived to be unlikely to click on, as if the user does click the ad, information is gained which may enable the software agent to enhance its click-through rate in the long run.

The morally salient feature of the technologies with which we are concerned is their capacity to influence human behaviour. There are three ways in which agents can do this (Burr, Cristianini, and Ladyman 2018, 743–45; see also Faden and Beauchamp 1986, 261–62). First, there is deception. To influence human behaviour in a way that is deceptive is to provide false or misleading information to that person such that the resultant false beliefs, or miscalibrated expectations, at least partly explain their decision to choose one option over another. A straightforward example of deception is a case in which an agent deploys a 'click bait' ad that conveys false information in order to induce the user to click on the ad. Instances of deceptive ads are particularly malign when the false or misleading information is targeted at users who are most likely to be receptive to such ads given their psychological vulnerabilities (e.g., individuals who suffer from bipolar disorder and may be more prone to make impulsive decisions during a manic phase).

Second, there is coercion. To coerce someone to perform an act is either to deprive them of the choice not to perform the act or to adjust their choice architecture in such a way that the non-performance of the act carries a significant cost. How serious the cost needs to be in order for the influencing act to count as coercive varies in proportion to the overall pay-off structure of the choice. For example, a user who tries to watch a video on a social media platform may have to view a personalised advertisement as a condition on their being able to watch the video. This need not qualify as coercion if the benefit of watching the video is small and the cost to the user of watching the ad is also relatively small. However, imposing more costly barriers to watching the video, such as the requirement that the user discloses their email address and consents to regular emails, may qualify as coercion – especially if there is a significant social cost to not being on the platform.

Third, agents can use persuasive strategies. We take the class of persuasive influencing acts to be those which are non-deceptive and non-coercive. For instance, so-called nudges or persuasive techniques that target an individual's cognitive biases are frequently used online in ways that are not necessarily deceptive or coercive. In addition to these three respects in which software agents can influence behaviour, software agents can indirectly influence behaviour through second-order effects. Second-order effects include changes to the user's utility function through prolonged exposure to certain kinds of content (e.g., behavioural addiction) and changes to their doxastic attitudes (e.g., political polarisation) (see Burr, Cristianini, and Ladyman 2018). Understanding which behavioural influence strategies are permissible, and in what circumstances, is obviously of great significance for the design and regulation of these technologies.

2.2 The Demarcation Problem

The demarcation problem is the problem of separating the morally permissible from the morally impermissible instances of behavioural influence by software agents. What makes the demarcation problem challenging is that the class of behaviour influencing acts is morally complex. What this means is that these acts lack a common moral status. That an act is an *influencing act* is neither a right-making nor a wrong-making feature of that act.

There are paradigmatically *morally wrong* instances of software agents influencing human behaviour (e.g., influencing user voting preferences through targeted misinformation campaigns). There are also paradigmatically *morally permissible* cases (e.g., YouTube influencing users to explore new musical genres). Thus, acts of behavioural influence differ from, say, acts that involve the breaking of promises. All promise-breaking acts are wrong pro tanto. That is, necessarily, the fact that an act involves breaking a promise grounds a moral reason not to perform that act, even though that reason may in principle be outweighed by countervailing reasons (Broome

2013, 51–62; Kagan 1989, 17 fn 13). The regulation and governance of software agents thus requires a criterion for distinguishing the permissible cases of behavioural influence from the impermissible cases.

The demarcation problem is additionally challenging because the moral permissibility of influencing acts does not straightforwardly track the *kinds* of software agents (or algorithms) at issue nor the *context* in which the agents feature. Consider, for example, a machine learning algorithm that predicts whether or not particular users of an online casino website suffer from a gambling disorder. In predicting whether a given user has a gambling disorder, the algorithm may take into account a range of factors such as betting intensity and frequency (Braverman et al. 2013; Finkenwirth et al. 2020; LaBrie and Shaffer 2011; Nelson et al. 2008). Imagine two online casinos. The first uses a software agent to display ads for free mental health services to individuals predicted to suffer from behavioural addiction (e.g., problem gambling). The second uses a software agent to supply users with free bets if those users are predicted to suffer from gambling disorders. The technologies in these two cases are the same – both attempt to influence the behaviour of users. But the behavioural influence is in one case morally permissible, perhaps even morally required, and in the other case the behavioural influence is impermissible.

Hence, what we take to be the object of moral evaluation is neither merely the kind of technologies at issue, nor merely the context in which they are used, although both may be relevant as explanatory factors. Instead, the objects of evaluation here are particular kinds of software agents (A), deploying particular strategies (S) that influence particular users (U) to behave in particular ways (B). For example, 'Facebook's targeted advertising agent (A) deploying persuasive public health information (S) to vaccine hesitant users (U) to take the COVID-19 vaccine (B)' is the kind of thing we have in mind when we say that a particular interaction is either morally permissible or impermissible. Whilst we leave open the possibility that certain kinds of agents in particular use cases may almost always be used in ways that are morally permissible or impermissible, our objects of evaluation are maximally specific.

2.3 Manipulation, Deception, and Coercion

What is plausibly the most straightforward approach for addressing the demarcation problem is to find a wrong-making feature that is shared by all morally impermissible instances of behavioural influence by software agents. However, what we find on inspection is a plurality of features that explain, or partially explain, the wrongness of particular behavioural influence strategies. Some strategies, such as click-bait ads or targeted emails that generate false and user-specific expectations of reward if some personal data is shared, are wrong because they are *deceptive*. Other strategies, such as the use of targeted mobile notifications that exploit users' cognitive vulnerabilities to ensure sustained use of an app, are wrong because they are *manipulative*. For example, a hook-up app such as Grindr might selectively deploy notifications to lapsed users in the late evening to inform them how many available people are in the local area. A second example is the selective deployment of paid upgrade offers in a game to users whom the software agent predicts are addicted to the game. Yet more strategies are wrong because they are *coercive*. For example, imposing mandatory email subscription for continued use of a service once a user is predicted to become reliant on the service.

There are, however, two barriers to analysis of impermissible behavioural influence strategies in terms of deception, manipulation, and coercion. On the one hand, analyses of these moral concepts are typically formulated for *interpersonal* contexts. For example, cases in which one *person* deceives another. Accordingly, the conditions under which an agent is said to engage in acts of, for example, manipulation, typically involve reference to mental state terms such as intent. To illustrate: Marcia Baron (2014, 103) argues that X manipulates Y only if X intends Y to do what X wants; although, on Baron's view, an agent's intending that *p* does not imply that the agent knows or is aware of their intention that *p* (Manne 2014, 228–29; see also Klenk 2020). Similarly, Robert Noggle (1996, 48) holds that manipulation requires 'a certain kind of insincere, conniving intention.' Attribution of mental states such as these to software agents is at best dubious – despite the agential language we employ. Hence, it is at best unclear that deception, manipulation, and coercion, as they are standardly formulated, are suitable for the context of software agents.

On the other hand, and relatedly, concepts such as deception, manipulation, and coercion provide a *solution* to the demarcation problem only if the concepts at issue are moralised. Here, for example, moralised accounts of manipulation hold that, necessarily, manipulative acts are morally wrong or at least *pro tanto* morally wrong (Baron 2014; George 2010; Macklin 1982).[6] In contrast, non-moralised accounts of manipulation hold that an act's being manipulative is consistent with that act being permissible or impermissible (Faden and Beauchamp 1986, 354–55; Wood 2014, 19–20). Accordingly, we can use concepts such as manipulation to tease apart permissible and impermissible instances of behavioural influence by software agents only if those concepts are moralised. To exacerbate the problem further, what typically distinguishes moralised accounts of manipulation, coercion, and deception is malign intent or at least indifference to the interests of another on the part of the actor. These considerations motivate the need for a novel analysis of what explains the wrongness of behavioural influence strategies exercised by software agents that are wrong because they are manipulative, coercive, or deceptive. We turn to this task in the next section.

3 Mental Integrity and Authentic Choice

We have argued that impermissible behavioural influence strategies by software agents typically involve manipulation, deception, and coercion. However, these moral concepts are standardly formulated in *interpersonal* terms and thus make ineliminable reference to mental states such as belief and intention which it is not obvious software agents can possess. In this section, we provide a unified account of what the wrongness of impermissible behavioural influence strategies consists in, which is better suited to the context of software agents. The salient feature of our account is that it seeks to clarify how and in what way people are *wronged* when they are manipulated, deceived, or coerced and thus focuses the moral evaluation on the receiver of the act rather than on the agent performing the act. This allows us to circumvent problematic mental state attribution for software agents.

The account we defend holds that impermissible behavioural influence strategies by software agents are wrong because they undermine the mental integrity of users and thus undermine their capacity for authentic choice. People who are manipulated, deceived, or coerced into performing certain acts are in an important sense not the authors of those acts – someone (or something) else is. Thus, the morally problematic feature of manipulation, coercion, and deception is that such acts deprive people of authorship over their actions. We propose that software agents can deprive people of authorship over their actions, and that in these cases, people are wronged in much the same way that victims of coercion, deception, and manipulation are wronged in interpersonal cases. Although software agents cannot influence user behaviour with malign intent, they can influence user behaviour in a way that diminishes their capacity for authentic choice. Hence, impermissible behavioural influence strategies by software agents are wrong *in the same way* as interpersonal cases of manipulation, deception, and coercion.

3.1 Authentic Choice

We begin with authentic choice. On our view, manipulative, coercive, and deceptive acts undermine the capacity of people to make *authentic choices*. This raises two questions: what is authentic choice? And, why is it good for people's choices to be authentic?

The ethical ideal of authentic choice, as we understand it, takes as its starting point the idea of a *socially situated* individual. By this we mean an individual that is situated within a social environment, comprising distinct norms, values, or practices, which partially determines how the individual relates both to themselves and to others (e.g., friends, family, colleagues, strangers).

Standardly, what characterises a failure of authenticity is a disparity between the *public self* (i.e., the convictions and values that the individual presents themselves as having to the outside world) and the *private self* (i.e., the convictions and values that the person in fact has). Charles Taylor (1992, 29) suggests that authenticity, so understood, 'accords crucial moral importance to a kind of contact with . . . [one's] own inner nature, which it sees as in danger of being lost.' A choice may fail to originate from the private self either due to 'pressures towards outward conformity' (e.g., alignment with externalised norms) or through individuals conceiving of themselves in such a way that they have 'lost the ability to listen to [their] inner voice' (e.g., internalisation of social values that supplant individual values). Thus, on this view, authentic choosing consists in an individual's making choices of which they, as opposed to someone else or some imagined projected self, are the author.

This account of authentic choice, as Taylor highlights, is not entirely satisfactory. First, there is an assumed tension between the individual and the society that they reside in, such that the individual's being socially situated distorts the choices that they would otherwise make outside the social environment (cf. Trilling 1971). The Rousseauian image of a person living outside social circumstances, untainted by the distorting influence of the social environment, is quite implausible as a moral ideal. Humans are social creatures, and any plausible account of authentic choosing needs to register how our commitments, values, and ultimately 'true selves' arise naturally in a socially situated context. Second, there is no obvious reason to suppose that an individual's merely making choices in accordance with their convictions and values is necessarily a good thing (cf. O'Neill 2003, 6). As Taylor (1992, 37) suggests, '[in] stressing the legitimacy of choice between certain options, we very often find ourselves depriving the options of their significance.'

We propose to understand authentic choice along the following lines. First, we follow Taylor in supposing that individual *choosing* does not matter *as such*. Rather, the moral significance of individual choosing is parasitic on the value of the options themselves. As Taylor (1992, 39) puts it, 'unless some options are more significant than others, the very idea of self-choice falls into triviality.' Second, authentic choice matters morally because each person has the potential to be human in *their particular way*, and the realisation of this potential is part of what is involved in human flourishing.

Recall that the socially situated individual is within an environment comprising distinct norms, cultures, values, or practices – perhaps conditional upon myriad social roles (e.g., professional employment setting versus familial roles). Over the course of the individual's life, they will iteratively and reflectively engage in a process of both internalising some values or practices (e.g., allowing some decision-making to become habitual) and a process of externalising norms (e.g., displaying disapproval for certain norms, and perhaps working to alter them). Both aspects of these social and moral dynamics are at play when enacting authentic choices. As such, a life lived in accordance with the ideal of authenticity involves choosing and pursuing goals that are fulfilling to the individual given what they care about, and also objectively worthwhile given who the agent is and the social relations in which they stand.

3.2 Mental Integrity

We now turn to mental integrity. There are certain conditions for the selection and pursuit of the aforementioned goals. An individual's *mental integrity* consists in the obtaining of these conditions. In turn, the capacity for authentic choice is diminished to the extent that mental integrity is diminished. We shall break mental integrity down into three conditions.

The first condition is *optionality*. It matters for the pursuit of meaningful and objectively worthwhile goals that individuals have options to choose between. Greater optionality does not consist in the individual merely having more options to choose between at any given time. That is, the quantity of options does not matter. Instead, what is important is that there are *some* options available, that those options facilitate the pursuit of meaningful and objectively worthwhile projects, and that the options available to the individual can be traced back to their decisions rather than circumstances beyond their control.

Consider an example. An individual may decide to go to university to study computer science. This constitutes a restriction in certain options that are available to them over time. For instance, it can restrict the kinds of careers that will be available to them (e.g., not becoming a veterinary surgeon), how they will spend their time, and the kinds of projects that will be available to them to pursue in the future given their skillset. None of this is intrinsically bad. On the one hand, the individual's options are constrained in such a way that the available options facilitate the pursuit of meaningful and objectively worthwhile goals. However, if rather than studying computer science, the individual had instead experimented with heroin and formed a serious drug addiction, then the restriction in options would be such that the individual could not easily pursue meaningful and objectively worthwhile goals given the options available to them. On the other hand, the restriction in the individual's options is best explained by a choice made by *them*. It is not merely a matter of circumstance that the individual's options are restricted in the relevant way, as it is with the heroin addict. Rather, the individual who chooses to study computer science is responsible for, and in an important sense the *author of*, the sorts of options that will be available to them given the choices they have made.

The second condition is having an accurate *representational model* of one's environment, including the interpersonal relations in which one stands, which facilitates the individual's predicting the consequences of their actions with reasonable precision.[7] This condition can fail to be satisfied in one of two ways. First, the individual's environment may be unpredictable. This might be true if, for example, the individual lives in an unstable political environment or in an environment with scarce resources. In such environments it is difficult to select and pursue meaningful and objectively worthwhile goals because the success of any particular plan is highly contingent on unpredictable circumstances beyond the individual's control. In short, part of the process of pursuing goals involves making one's plans robust against contingencies. But if the contingencies are unpredictable and often severe, the formation and execution of any kind of long-term plan is impossible. Second, certain psychological features of the individual may impede their capacity to model the consequences of their actions moving forward into the future. For example, if the individual suffers from severe anxiety, they may be disposed to distort the probable consequences of their actions, in particular, by assigning undue significance to possible outcomes that are unlikely but, in some way, catastrophic if they obtain.

The final condition is *value alignment*.[8] Individuals have values. There are things which matter to agents. Individuals use values to adjudicate between options. Deciding what to do in accordance with one's values is not in itself a good thing. Rather, it matters that the values used to adjudicate between the options align with what is objectively good for the individual given who they are, the social relations in which they stand, and their responsibilities given the social roles which they occupy. Value alignment has both *cognitive* and *conative* elements. On the one hand, cognitive value alignment obtains when the individual's beliefs about what is important, for example, the climate, align with things that matter stance-independently. On the other hand, conative value alignment obtains when the individual's affective responses, for example, approval and disgust responses, track that which is stance-independently good or bad for the individual. In both cases, value alignment is realised through moral education in childhood, and ultimately, through dialogue about what is important with those in one's social sphere.

This notion of value alignment is also reflected in contemporary views in the cognitive sciences, most notably in recent translational research that attempts to develop a bridge between evolutionary and neuroscientific perspectives and process-based views of mental health in psychology and psychiatry (e.g., Sterling 2014). This matters, as our account of mental integrity is designed to complement ongoing research in the empirical sciences, offering generative potential to interdisciplinary research into the effects of novel data-driven technologies on the mental health and well-being of human users (Burr and Floridi 2020). As such, a brief digression into how the concept connects with related concepts in the cognitive sciences is worthwhile.

According to Sterling (2014, 2019), mental disorder is typically understood in *homeostatic* terms, as a process of physiological regulation. From this perspective, mental disorder consists in the *deviation* of certain synaptic parameters from 'normal' values (Sterling 2014). To illustrate, consider selective serotonin reuptake inhibitors (SSRIs), which are prescribed in the treatment of common mental health disorders, such as depression. The rationale behind the use of SSRIs, according to the homeostatic view, is that there is a 'normal' level of serotonin reuptake and that certain mood disorders are characterised by a deviation from this normal level. As such, SSRIs are intended to correct for this deviation holding the parameters to a set point.

Sterling rejects this homeostatic conception of mental disorder in favour of what is known as an *allostatic* model. Whereas homeostasis is a model of regulation on which 'normal' parameter values are *maintained* according to some set point, allostasis is a model of predictive regulation on which parameter values change flexibly in response to anticipated demand from the environment. According to Sterling, '[p]arameter values vary widely . . . not because they are "inappropriate" . . . but because the brain predicts changes in need and continuously retunes the parameters to keep them exactly appropriate' (Sterling 2014, 1192). On this view, mental disorder consists in a diminished capacity to anticipate demand on one's affective responses and adjust one's mental parameters accordingly. Conversely, mental health is the 'capacity to choose among thoughts and shift flexibly between them; it is the capacity to match mood and affective expression to the immediate situation' (Sterling 2014, 1193).

This understanding of mental health stands in relation to mental integrity, as accounts of bodily health stand in relation to bodily integrity. Whereas the latter concepts are inherently normative in their scope, they are grounded in and complementary to their respective empirical accounts. An allostatic view of mental health provides robust empirical and theoretical support for our view of mental integrity, and further elucidates the conditions we discussed earlier (e.g., optionality as capacity to choose; accurate representations and value alignment as a process of adaptive regulation to the immediate situation). Moreover, this connection is suggestive of a potential for pursuing the concept within the empirical sciences (i.e., operationalising the concept of mental integrity), in an effort to determine *to what extent* software agents may affect mental integrity, in line with the various objects of evaluation outlined in Section 2.2.[9]

3.3 Returning to Manipulation, Coercion, and Deception

With the concepts of authentic choice and mental integrity in place, we can now explain how they support a unified account of impermissible behavioural influencing strategies by software agents on human users. Recall that morally impermissible instances of behavioural influence by software agents typically involve manipulation, coercion, and deception. However, reductive analyses of these normative concepts are standardly formulated in interpersonal terms and thus make ineliminable reference to mental state terms such as intention. This is problematic because software agents cannot obviously possess such mental states. What authentic choice and mental integrity allow us to do is reverse the direction of analysis. Rather than focusing on the properties that are necessary and jointly sufficient for an *agent* to count as deceiving, coercing, or manipulating another, the locus of analysis is on the *patient* (or, user). In particular, how and in what respect an individual is *wronged* when they are deceived, coerced, or manipulated. We suggest that such acts undermine the mental integrity of individuals and thus diminish their capacity for authentic choice.

Digital Coercion: Digital coercion consists in restricting a user's optionality. Often, this involves setting up a choice so that the user must select, say, one of two options neither of which is in their best interests given who they are, the relations in which they stand to others, and the social role which they occupy. Restricting an individual's optionality need not consist in reducing the *quantity* of options available to an agent. Rather, diminished optionality is consistent with the agent being afforded a range of options that fail to facilitate the pursuit of worthwhile and meaningful goals (e.g., persistent recommendation of clickbait videos designed to solicit compulsive viewing). Because optionality is a precondition on authentic choice, influencing strategies that target a user's optionality undermine their mental integrity.

Digital Manipulation: Digital manipulation consists in adjusting the user's online environment so as to render the user's subjective values divergent from what in fact matters given their identity and the social relations in which they stand. Digital manipulation is typically achieved through the creation of incentive structures, which lead agents to act in ways that are contrary to their own interests or those to whom they have special obligations. These strategies include, *inter alia*, incentivising a user to invite other users to a platform by rewarding the user for doing so, and exploiting temporal discounting on the part of users (e.g., offering to remove the mild inconvenience of subscription notifications in exchange for signing up to receive emails from a given service, which in turn advertise subscription). Importantly, given that the concept of manipulation at issue here is a moralised concept, behavioural influence by software agents qualifies as manipulative only if the incentive structures lead users to act in ways that are contrary to their interests or are inappropriate given their social roles. Only these incentive-based strategies undermine users' mental integrity.

Digital Deception: Digital deception consists in diminishing the accuracy of a user's representational model of the online environment. Deceptive strategies target the user's ability to predict the consequences of their actions in the digital sphere. Such strategies include, *inter alia*, click-bait ads and targeted emails containing false or misleading information. The introduction of deceptive content into the user's online environment renders the user unable to deliberate in a way that is consistent with authentic choice-making insofar as the consequences of their actions in the digital sphere are objectionably unpredictable. This constitutes an attack on the user's mental integrity.

Influencing strategies that are coercive, manipulative, or deceptive, each target different aspects of a user's mental integrity, viz. optionality, value alignment, and representation, which are preconditions on authentic choice. In doing so, such strategies erode the user's authorship over their choices and are for this reason morally impermissible. These considerations facilitate a unified condition on permissible behavioural influence by software agents in the digital sphere. A behavioural influencing strategy is permissible only if and because it affords due respect to the user's mental integrity. That is to say that the strategy does not restrict the user's optionality, nor misalign their subjective values from what is best for the agent given their identity and the social relations in which they stand, nor undermine the agent's representational model of their environment. What is important for the permissibility of behavioural influencing strategies, then, is that preconditions for authentic choice on the part of the user are not undermined.

4 Practical Implications

We have argued that certain behavioural influence strategies by software agents on human users are morally impermissible, and that what explains the wrongness of these strategies is that they undermine the mental integrity of users. The effect is to diminish the user's capacity for authentic choice.

In this section, we spell out some of the practical implications of our account for the design and development of software agents that influence user behaviour. We begin with a case.

> Charlie is a 25-year-old single mother. She suffers from bipolar disorder and has struggled with behavioural addiction in the past. Charlie's friend Dean is a regular user of online casinos. Dean is offered free £50 bets for himself and a friend if he invites a friend to an online roulette website. He invites Charlie to play roulette in order to get the free bet. Charlie has never played roulette before but uses the £50 bet. She has an initial winning streak but continues to gamble with her own money for several hours, ultimately making a £100 net loss. She vows not to use the online casino again. The next morning, she receives a targeted email, automatically selected by a software agent, incentivising her to win back the £100 with a £25 free bet. Charlie clicks the link, and has a winning streak, ultimately winning back the £100. Again, Charlie decides to call it quits. She moves her cursor to close the window, and another software agent generates a targeted ad that pops up: 'You're on a roll! Have a free £25 bet on the house.' Charlie takes the free bet, wins a few more times, and then continues to play for another few hours. Eventually, Charlie stops playing after an unduly optimistic bet results in a £250 loss.

The twin concepts of mental integrity and authentic choice allow us to better diagnose and explain the wrongness of the behavioural influence strategies in this case. We shall make three points. First, Dean is incentivised to invite friends to play online roulette through the persuasive framing of a targeted message. It is, presumably, neither in Dean's interest nor the interests of his friend Charlie for Charlie to play online roulette. However, the roulette website sets up Dean's decision problem in such a way that it *appears* to be in his interest that he invites a friend to the website. In particular, it uses the instant gratification of a free bet to incentivise Dean to make a choice that, if given due reflection, he is unlikely to believe is in his or his friend's interests. The sense in which this act of behavioural influence undermines Dean's mental integrity, and thus his capacity for authentic choice, has to do with *value alignment*. There are certain things that – whether or not they matter to Dean – *ought* to matter to Dean given who he is and the social relationship in which he stands to Charlie. The roulette website, in forcing the choice between a free bet and inviting a friend to play roulette, causes Dean to mis-evaluate the moral and prudential significance act of inviting Charlie. This undermines Dean's mental integrity, in the sense of depriving him of authorship over a choice.

Second, the content of the email sent to Charlie is personalised, based on her gambling behaviour, and it is intended to generate a false expectation of winning back the money she lost. This personalised message undermines Charlie's mental integrity insofar as it distorts her representational model, making an improbable outcome of her choice (i.e., winning back the lost money), seem more probable than it is. Indeed, the expected value of playing a game of online roulette is negative. The website's software agent rationally expects to make money out of Charlie. In directing Charlie's attention towards an improbable outcome that is good for her, the email subverts Charlie's ability to accurately model the decision. Charlie's mental integrity, and thus the authenticity of her choices, is diminished.

Third, perhaps the most subversive behavioural influencing strategy used in the case is the deployment of free bets when Charlie is expected to exit the game. Here, the effect is to diminish Charlie's *optionality*: When she makes a *rational* decision in attempting to exit the game, the software agent alters the options available to her. Here, Charlie's mental health and history of behavioural addiction become even more salient. It is well known that mental health disorders, such as bipolar disorder, are risk factors for problem gambling and subsequent financial difficulties (Holkar and Lees 2020). A coping strategy that is widely adopted for myriad forms of behavioural (and substance) addiction is the implementation of environmental constraints, which work by limiting the optionality of the individual. For instance, Charlie may choose to set time limits on her web browser or spending blocks through her bank that stop her from gambling at certain times (e.g., during manic episodes) or from overspending (Nelson et al. 2008). While the agent may not directly infer psychopathological features of the site's users,[10] the use of a 'timely' messaging strategy[11] can nevertheless exploit psychological vulnerabilities, which in the present case further undermines Charlie's authentic choice by weakening the desired efficacy of the external environmental constraints that she has rationally put in place to support her mental integrity.[12]

We have illustrated how the concepts of mental integrity and authentic choice can elucidate and explain how and in what way users are wronged by software agents that deploy subversive strategies to influence their behaviour. What is important to note, however, is that in seeking to understand digital manipulation, coercion, and deception, in terms of mental integrity and authentic choice, we have not provided anything like an algorithm for determining whether token instances of behavioural influence by software agents are permissible or impermissible. Rather, what we have provided is an explanatory account of the features of behaviour influencing strategies that make them wrong *qua* acts of manipulation, deception, and coercion. Nevertheless, while the moral status of particular influencing strategies cannot straightforwardly be read off our account, the account can inform the kinds of ethical deliberation that is appropriate when designing and deploying these technologies. In particular, mental integrity and its components (i.e., optionality, representation, and value alignment) provide a plausible framework in which to evaluate the likely or possible impact of software agents on users. This framework, minimally, offers the resources to guide discussion about how and in what way these technologies might impact users in respects that are morally wrong. For example, by providing false expectations about the consequences of their decisions or by misaligning what the agent values with what is in fact good for the agent in their social context.

5 Conclusion

This chapter considered a demarcation problem: what distinguishes morally permissible from morally impermissible behavioural influencing strategies by software agents on human users? We argued, first, that impermissible influencing strategies typically involve deception, coercion, and manipulation. Second, we developed an analysis of the wrongness of these kinds of influencing acts. On the account defended, morally impermissible instances of behavioural influence by software agents undermine the mental integrity of human users and in doing so diminish their capacity for authentic choice. Accordingly, we argued that strategies for behavioural influence by software agents are permissible only if and because those strategies afford due respect to the mental integrity of the user.

Notes

1. Our use of the term 'software agent' throughout this chapter is based on the various definitions of agents in Russell and Norvig (2020), and is also more permissive than the concept, 'intelligent software agent' that we discuss in greater

detail in Burr, Cristianini, and Ladyman (2018). We reserve usage of the prefix 'intelligent' for instances of software agents that employ some form of artificial intelligence (AI) in their operation, such as machine learning (ML). While this will lead to a fuzzy boundary, there is nevertheless a conceptual need to differentiate between these two classes and recognise that all members of the class, 'intelligent software agents' are, by definition, also members of the class, 'software agents'. Our focus in this chapter will be the larger class, though we acknowledge that many of the normatively significant concerns arise due to the implementation and use of novel intelligent software agents. We discuss the terms more fully in Section 2.1.

2. The set of morally permissible acts includes all acts the non-performance of which is not wrong. Hence permissible acts include acts that are justified, that is, there is a positive moral reason to perform the act, and acts that are unjustified but not morally wrong.
3. The use of 'suitable' here is to acknowledge that some life events may temporarily disrupt or perturb this stability (e.g., transformative experiences such as pregnancy (Paul 2014)) or that an individual's natural progression over the course of a life results in notable changes to beliefs and values. However, in the latter case, these changes tend to be gradual, and so can still be characterised as 'stable.'
4. For the sake of brevity, we will refer to 'software agents' as 'agents' and 'human users' as users.
5. Here we assume for simplicity that the software agent is situated within a discrete-time choice framework. The ethical discussion is intended to apply both to discrete-time software agents and to continuous-time agents (see Sutton and Barto 1998, 48).
6. To say that acts of manipulation are *pro tanto* wrong means that an act's being manipulative is a reason that counts against performing it, but that reason can in principle be outweighed by countervailing considerations (Broome 2013, 51–62; Kagan 1989, 17, fn 13).
7. Much will depend of course on fleshing out what 'reasonable precision' amounts to here, as an agent can always pursue increasing precision and representational veracity in their model. However, establishing a criterion for determining the appropriate level of precision in a given case cannot be specified a priori, as it will depend on specific contextual factors, including pragmatic considerations of the agent.
8. Note that in presenting this condition we will speak in realist terms about value, but we remain neutral on the dispute over the correct meta-ethical theory. The view we defend is compatible with a range of realist and constructivist positions.
9. We offer this suggestion as a possibility for further research only, but do not, for instance, suggest ways in which the concept could be operationalised or how it could generate possible hypotheses that could be tested.
10. Though see (Burr and Cristianini 2019) for a survey of the methods employed by intelligent systems to infer mental states or traits, including personality, emotions, psychopathology, and more. We make no claim about the psychological validity of such techniques.
11. The use of context-based timing is a well-known persuasion technique, with roots in the Ancient Greek notion of Kairos – the right or opportune moment (Ham et al. 2017).
12. This claim reflects a position adopted by advocates of *situated cognition*, which views the external environment as a scaffold for our cognitive processes (e.g., decision-making). As such, we further reinforce a view of mental integrity as a (socially) situated property of individuals.

6 References

Baird, N. 2017. "Dynamic Pricing: When Should Retailers Bother?" *Forbes* www.forbes.com/sites/nikkibaird/2017/04/18/dynamic-pricing-when-should-retailers-bother.

Baron, Marcia. 2014. "The Mens Rea and Moral Status of Manipulation." In Coons and Weber 2014, 98–109.

Braverman, Julia, Debi A. LaPlante, Sarah E. Nelson, and Howard J. Shaffer. 2013. "Using Cross-Game Behavioral Markers for Early Identification of High-Risk Internet Gamblers." *Psychology of Addictive Behaviors* 27 (3): 868–77. doi:10.1037/a0032818.

Broome, John. 2013. *Rationality Through Reasoning*. Malden, MA: Wiley-Blackwell.

Burr, Christopher, and Nello Cristianini. 2019. "Can Machines Read our Minds?" *Minds and Machines* 29 (3): 461–94. doi:10.1007/s11023-019-09497-4.

Burr, Christopher, Nello Cristianini, and James Ladyman. 2018. "An Analysis of the Interaction Between Intelligent Software Agents and Human Users." *Minds and Machines* 28 (4): 735–74. doi:10.1007/s11023-018-9479-0.

Burr, Christopher, and Luciano Floridi, eds. 2020. *Ethics of Digital Well-Being: A Multidisciplinary Perspective*. Cham: Springer.

Coons, Christian, and Michael Weber, eds. 2014. *Manipulation: Theory and Practice*. Oxford: Oxford University Press.

Ezrachi, Ariel, and Maurice E. Stucke. 2016. *Virtual Competition: The Promise and Perils of the Algorithm-Driven Economy*. Cambridge, MA: Harvard University Press.

Faden, Ruth R., and Tom L. Beauchamp. 1986. *A History and Theory of Informed Consent*. New York, NY: Oxford University Press.

Finkenwirth, S., K. MacDonald, X. Deng, T. Lesch, and L. Clark. 2020. "Using Machine Learning to Predict Self-exclusion Status in Online Gamblers on the PlayNow. Com Platform in British Columbia." *International Gambling Studies*, 1–18.

George, R. T. de. 2010. *Business Ethics*. Hoboken, NJ: Prentice Hall.

Ham, J., J. van Schendel, S. Koldijk, and E. Demerouti. 2017. "Finding Kairos: The Influence of Context-Based Timing on Compliance with Well-Being Triggers." In *Symbiotic Interaction*, edited by L. Gamberini, A. Spagnolli, G. Jacucci, B. Blankertz, and Jonathan Freeman, 89–101. Cham: Springer.

Holkar, M., and C. Lees. 2020. "A Safer Bet? Online Gambling and Mental Health." www.moneyandmentalhealth.org/wp-content/uploads/2020/07/A_Safer_Bet.pdf.pd.

Huang, Jeff, Ryen W. White, and Susan Dumais. 2011. "No Clicks, No Problem." In *Proceedings of the SIGCHI Conference on Human Factors in Computing Systems*, edited by Desney Tan, Geraldine Fitzpatrick, Carl Gutwin, Bo Begole, and Wendy A. Kellogg, 1225–34. New York, NY: ACM Press.

Kagan, Shelly. 1989. *The Limits of Morality*. Oxford: Oxford University Press.

Klenk, Michael. 2020. "Digital Well-Being and Manipulation Online." In Burr and Floridi 2020. 81–100. doi: 10.1007/978-3-030-50585-1_4.

LaBrie, Richard, and Howard J. Shaffer. 2011. "Identifying Behavioral Markers of Disordered Internet Sports Gambling." *Addiction Research & Theory* 19 (1): 56–66.

Larson, J. A., and Jeff Surya Mattu. 2015. "The Tiger Mom Tax: Asians Are Nearly Twice as Likely to Get a Higher Price from Princeton Review." *ProPublica*.

www.propublica.org/article/asians-nearly-twice-as-likely-to-get-higher-price-from-princeton-review?token=mEhYb9WcJtcod4qoPNYwJm31y7HHaLA.
Macklin, Ruth. 1982. *Man, Mind, and Morality: The Ethics of Behavior Control.* London: Prentice Hall.
Manne, Kate. 2014. "Non-Machiavellian Manipulation and the Opacity of Motive." In Coons and Weber 2014, 221–46.
Milano, S., M. Taddeo, and Luciano Floridi. 2020. "Recommender Systems and their Ethical Challenges." *AI & Society*, 1–20.
Nelson, Sarah E., Debi A. LaPlante, Allyson J. Peller, Anja Schumann, Richard A. LaBrie, and Howard J. Shaffer. 2008. "Real Limits in the Virtual World: Self-Limiting Behavior of Internet Gamblers." *Journal of Gambling Studies* 24 (4): 463–77. doi:10.1007/s10899-008-9106-8.
Noggle, Robert. 1996. "Manipulative Actions: A Conceptual and Moral Analysis." *American Philosophical Quarterly* 33 (1): 43–55.
O'Neill, Onora. 2003. "Autonomy: The Emperor's New Clothes." *Aristotelian Society Supplementary Volume* 77 (1): 1–21. doi:10.1111/1467-8349.00100.
Paul, L. A. 2014. *Transformative Experience.* Oxford: Oxford University Press.
Russell, Stuart J., and P. Norvig. 2020. *Artificial Intelligence: A Modern Approach.* Harlow: Pearson.
Sterling, Peter. 2014. "Homeostasis vs Allostasis." *JAMA Psychiatry* 71 (10): 1192. doi:10.1001/jamapsychiatry.2014.1043.
Sterling, Peter. 2019. *What is Health? Allostasis and the Evolution of Human Design.* Cambridge, MA: MIT Press.
Sutton, R. S., and A. G. Barto. 1998. *Reinforcement Learning: An Introduction.* Cambridge, MA: MIT Press.
Taylor, Charles. 1992. *The Ethics of Authenticity.* Cambridge, MA: Harvard University Press.
Trilling, L. 1971. *Sincerity and Authenticity.* Cambridge, MA: Harvard University Press.
Wood, Allen W. 2014. "Coercion, Manipulation, Exploitation." In Coons and Weber 2014, 17–50.

Part III

Epistemic, affective, and political harms and risks

14 Is There a Duty to Disclose Epistemic Risk?

Hanna Kiri Gunn

1 Introduction

Filter bubbles and echo chambers metaphorically elicit the images of epistemic isolation, but that is not primarily their manipulative quality. Rather, filter bubbles and echo chambers are often taken to be manipulative for the ways that they seize our attention and, following this, direct us towards content that, while it might be deeply appealing to us, does not in fact enable us to exercise epistemic agency *responsibly*. Put differently, these phenomena are an easy means for failing to live up to one's own epistemic expectations and norms: what directs one's involvement in epistemic life is not truth but clickbait.

In this chapter, I consider whether there is a duty to disclose particular kinds of epistemic risk that seemingly come hand in hand with a personalised internet experience. What is threatened by the kind of epistemic risks to be discussed are not only the cognitive attitudes of individuals but the health of the epistemic community as a collective. Thus, the kinds of epistemic risk discussed here are not only the familiar topics about chances of acquiring false beliefs or missing out on true ones; in addition, they are those risks that threaten to undermine our ability to develop and maintain the kind of epistemic community we desire.

As William James rightly pointed out, and epistemologists of varying stripes have developed and formalised since, if we aim to avoid falsehoods we will get different results than if we aim to maximise our true beliefs. We will also plausibly act differently: I can avoid all falsehoods by never believing anything, and you can maximise true beliefs by doing the opposite. In a similar way, our collective epistemic activities are also guided by the norms that we take to maximise epistemic goods or values.

If we aim for truth alone as a collective then we will get different results, plausibly, than if we optimised for the epistemic health of that same collective – individually and as a group. To do this latter task involves attending not only to auditing the cognitive attitudes and dispositions of the members of the collective, for example, by understanding the influence of bias on processes of belief acquisition. Caring for the health of the epistemic

DOI: 10.4324/9781003205425-17

community also entails caring for the processes and activities of knowledge creation and dissemination. Such work involves attending to the interpersonal aspects of epistemic life and the epistemic and communicative norms that shape participation in such activities.

This raises a natural question: in what ways would our conduct change as individual knowers and as a collective epistemic community if we aimed for the health of our epistemic community and not only for truth? The duty to disclose the kind of epistemic risks I am concerned with in this chapter is, I propose, the kind of duty that becomes intuitive when we take the health of our epistemic community to be a basic epistemic value. Filter bubbles and echo chambers are epistemically risky for individuals beliefs, but they are also epistemically risky to us as an epistemic community.

The discussion proceeds as follows. In Section 2, I motivate a broader sense of epistemic risk. In Section 3, I explain internet personalisation by introducing research into selective exposure, homophily in networks, and polarisation. In Section 4, I explain how such personalising technologies may be manipulative. In Section 5, an existing argument about the duty to disclose adverse effects to participants in clinical trials is presented, and I explain how epistemic risks would seem to pose similar threats to the autonomy of persons that justify this duty in a more straightforwardly moral domain. Finally, in Section 6 I close with a discussion of responsibility and the many hands problem to highlight some immediate issues in identifying a duty bearer.

2 Developing a Sense of "Risks to Healthy Epistemic Community"

In his essay, "The Will to Believe", James explicitly discusses two distinct epistemic duties that have corresponding risks:

> There are two ways of looking at our duty in the matter of opinion, – ways entirely different, and yet ways about whose difference the theory of knowledge seems hitherto to have shown very little concern. We must know the truth; and we must avoid error, – these are our first and great commandments as would-be knowers; but they are not two ways of stating an identical commandment, they are two separable laws.
>
> (James 2009 [1896])

These "first and great commandments" can conflict, and thus many contemporary discussions about epistemic risk are about how we ought to balance them and the consequences of privileging one over the other.

The goal of this chapter, though, is to ask if there is a more expansive view of epistemic risk available given a range of concerns that we find in many discussions about the effects of the internet on epistemic and communicative practice. An underlying assumption made here is that things are risky because they threaten some value or ways of promoting values, as in the case of norms. While we can motivate caring about such risks from the value of truth, the way that our epistemic character affects our wider epistemic community is something to be considered in its own right. For instance, one's dispositions to arrogance or credulity plausibly impact this wider community by influencing one's (dis-)information sharing behaviours online.

To motivate an expansion of our potential range of epistemic risks, let us begin with what might be the most applied of epistemic subjects: education. And, we will also attend to a regrettably neglected concept in analytic epistemology: listening.[1] John Dewey wrote extensively about education and its role in developing a healthy democracy. For our purposes, what matters are Dewey's comments about how the norms, practices, and physical spaces of classrooms affect the quality and kind of educating that takes place.

In *The School and Society*, Dewey offers the following anecdote:

> Some few years ago I was looking about the school supply stores in the city, trying to find desks and chairs which seemed thoroughly suitable from all points of view – artistic, hygienic, and educational – to the needs of the children. We had a great deal of difficulty in finding what we needed, and finally one dealer, more intelligent than the rest, made this remark: "I am afraid we have not what you want. You want something at which the children may work; these are all for listening".
>
> (Dewey 1900)

The dealer intuits what Dewey comes to call "one-way" or "straight-line" listening. This is the kind of listening that can be imagistically described as a kind of mere osmosis: fill a room with children at desks that are difficult to get out of and ask a teacher to talk at them from the front of the room while hoping something trickles in one ear.

What Dewey is after, and what he proposes we ought to collectively seek, is active listening in conversation (Waks 2011). The distinction is that active listening in conversation is a collaborative process between unique interlocutors. Learning to actively listen in conversation to the diversity of viewpoints and opinions in classrooms is presented as training in a civic culture that democracies are meant to value.[2] Such a culture involves particular kinds of communicative practice like sincere and open-minded listening to one another. These practices in turn rely on shared commitments to testimonial practices and norms. These norms do not merely exist to make civic deliberation pleasant, they are also epistemic: these norms support testimonial knowledge exchange. When we systematically design schools to hold students quietly en masse as they perhaps learn some facts, we miss out on the opportunity to instill in them the practical values and practices that help to support and maintain the kind of society – in both moral and epistemic dimensions – that we desire.

It is certainly not the case that our experience of the online world is that it is organised to promote and encourage us to be passive listeners. If anything, the risks of the online environment are very much in the opposite direction. We are invited to advertise our convictions across almost all platforms that we can sign-up for, and further, to self-sort into collections of like-minded individuals. This is not to say that they promote practices of active listening in conversation, because, of course, one can also fail to listen well by simply never doing it. Insofar as we are encouraged to speak loudly and often to as many people as we can reach online, we are not also encouraged to attend carefully, sincerely, and individually to their replies. Such an environment is conducive to processes like group polarisation, whereby a like-minded collective are driven to more "extreme" beliefs by interacting with one another and not because of better information.

Unlike in the context of video games, where we take our actions to be irrelevant for our moral character, it is implausible that the ways we engage in online conversations or conduct research (or, "research" as in cases of mere Google-knowing) have no bearing on our epistemic and communicative character more broadly. If antagonistic trolling is one's default mode in online conversations, we should ask how long it takes – or under what conditions – for this to become one's default mode in conversations generally. Similarly, we might wonder how our ability to search for and locate relevant information is impacted by Google's search algorithm – when we ask the technology to do more for us, we do less for ourselves.

What is important to take away from this brief look at Dewey's writing on the connection between (value-)theory and practice in the classroom is that the structure and practice of communicative and epistemic life are importantly formative for the kinds of speakers and listeners that we go on to become. Even if we are interested in only maximising true beliefs (or minimising false ones) we ought to consider how we are learning to achieve this – and we ought to attend to the ways that our epistemic and communicative environments may have a "hidden curriculum".

The notion of the "hidden curriculum" refers to processes of learning that take place inside classrooms, but not because they are in the lesson plan. The hidden curriculum is instead constituted by the processes of socialisation that take place alongside formal educating. Such lessons might impart norms of intellectual autonomy and epistemic humility, convey social expectations about how one should look and act, or emphasise the value of particular skills, like abstract thinking, over others, like listening. The sorts of epistemic risks that are involved in personalising our internet experiences, I propose, may be sites for "hidden" learning in this sense. The concern being that, if we are not working out what "lessons" are being passed on during our time online in personalised spaces, we may fail to recognise how we are being re-trained as epistemic (and communicative) agents.

Much like the idea of the hidden curriculum in formal education, the hidden curriculum of our online environments is not unknowable. In fact, we have good reason to think that many companies behind, for example, social media websites do actually possess knowledge about the epistemic and communicative effects that their platforms have on users. When teachers or professors learn that the hidden curriculum of their syllabi negatively impacts their students, we expect them to revise it. Of course, teachers and professors have different obligations to their students than a company like Facebook does to its users. Nonetheless, we might think that if it is known that one's social media platform, say, makes users more credulous and susceptible to false information, then there is at least *a duty to warn* prospective users that they may be changed in these ways.

In the rest of the chapter, I will use "social-epistemic" to refer to our interpersonal epistemic character, actions, and activities, for example, providing testimony, engaging in debates, engaging in collaborative research. I am interested in understanding the norms that ought to govern the social-epistemic actions of persons who are invested in the welfare of their epistemic community as a whole – and this is the emergent value that I will entertain for the remainder of the discussion and that I take to be evinced by our concern over various aspects of the personalised internet experience.

In this section, we have considered some ways in which, for example, practices of listening might be more conducive to civic communities that engage in productive public debate. Such a culture of listening and productive debate are facts about the epistemic community, borne out in the actions and cognitive attitudes of its members. In the next section, I explore some of the research on filter bubbles and echo chambers in order to identify some epistemic risks for social-epistemic agents online.

3 Discourse, Online Discourse, and Epistemic Risks From Online Environments

If the preceding discussion has been compelling, then it is not too much of a stretch to assert that the qualities of our epistemic community affect our ability to engage in conversations. These qualities belong both to the agents in such a community and to the wider structure and background environment that serve as the medium for those agents to participate in epistemic and communicative activities. In turn, this affects our ability as individuals to participate in social-epistemic activities like listening to and sharing the advice of, for example, public health experts. Most of what we know, we know because we learned it from others. This is an inherently risky, but unavoidable, practice. It's very easy to go astray in placing trust in the word of someone else. Hence, we have safeguards in place across many of the social-epistemic domains of life.

In academia, the safeguards of peer review, statistical standards for journals, replication standards, and qualifications serve as institutional safeguards that enable individuals to place epistemic trust in the testimony of others without having to vet the testifier themselves. In traditional media, broadcasting standards and advertising standards are two kinds of social-epistemic safeguards that protect consumers (for content and of products) so that they are able to place trust in testimony without having to do the legwork themselves. In something as familiar as day-to-day conversations we have norms also – for instance, the norms of quality and quantity – that are widely taken to guide communicative practice and secure epistemic goods.[3] The idea of a duty to disclose epistemic risks is motivated by similar protectionist principles.

The steadily and at times rapidly growing research into the structure of online discourse and networks have raised numerous concerns about how the virtual world is leading to increasingly ideologically distinct, homophilous, and estranged communities. The persistent concern over phenomena like these are more familiarly referred to with the labels "filter bubbles" and "echo-chambers".

Of course, just mentioning these terms automatically invites their association with social media. And social media use is now pervasive, with recent Pew Research polling of the United States showing that sizeable majorities in the age ranges of 18–29 (88%) and 30–49 (78%) use social media regularly and 64% of 50–64-year-olds and 37% of those aged 65 use some form of social media (Smith and Anderson 2018). While the majority of people are apparently *comfortable* with using social media, that does not entail that there are not a range of concerns about how this is changing our relationships with one another at a local and personal level and at the society-wide level.

Recent research into online discourse in the United States has presented results including a general decline in trust for traditional media sources (Pew Research Center 2020), partisan disparities between online communities in their relative exposure to fake news (McBrayer 2021), and the perhaps surprising conclusion that trolls could be any one of us (Cheng et al. 2017). There is, then, a widespread sentiment that we are collectively losing the ability to have quality public discourse and that this is importantly related to our online lives.

This is partly due to the actions of individuals: trolling, online shaming and abuse, sharing of false or fake information, and uncharitable debates.[4] It is also plausibly partly due to the automated algorithms that structure our online experience that nudge us into like-minded groups and supply a constant feed of tailored content. These results, though, are complicated by studies on online networks that show that personalising internet tools in fact support some people in accessing a wider range of quality content than they otherwise would. The simplicity of the filter bubbles and echo chambers debates in popular media are misleading, in reality the matter is still somewhat opaque.

One observation to make at this point is that we are generally happy to "offload" some of our epistemic responsibility to such automated algorithms. It is simply not possible to vet and research all testifiers and their testimony online. A power of personalisation technologies is that they have the potential to do a lot of the epistemic heavy lifting for us. However, indiscriminately relying on digital tools is epistemically poor practice. As a minimum constraint, it is probably permissible to offload epistemic responsibility in this way if it in fact meets one's epistemic standards. If we do not know the personalisation is happening, we cannot check whether it meets our epistemic standards.

To appeal just to the value of truth for a moment, the internet that we want is one where we can justifiably rely on such personalisation technologies to deliver us results because they uphold values like truth; and it is one where we can rely on the testimony of others because they are relevant experts, not merely popular influencers. It is unclear that current personalisation algorithms meet this minimum standard. It is clear that most users lack an adequate understanding of the technology such that we could say that they are offloading epistemic work in an informed and responsible way (OECD 2016).

The term "filter bubble" came into colloquial usage with the publication of Eli Pariser's book of the same name (Pariser 2011). In Pariser's usage, filter bubbles are the product of poor digital media curation. Such curation is poor because it aims to capture our attention rather than aiming to present us with true or at least well-justified content. "Filter bubble" is somewhat recent as a term referring to poor access to content online, two earlier terms include both "splinternet" and "cyberbalkinazation" (Bozdag and van den Hoven 2015). Similar concerns are presented in Cass Sunstein's books about the internet and democracy in his description of the "Daily Me" (Sunstein 2017). If one's preferences were all truth-indicating, such personalisation might not be problematic, but we are not so fortunate.

All of these names gloss over two important processes of selective exposure that it is worth distinguishing. Selective exposure refers generally to the ways that we consume only some of the media on offer. In the online mediasphere, it is common that users prefer to access their media using indirect methods like getting links via search engines or social media feeds (Cardenal et al. 2019). These platforms are, of course, the public face of not only personalised content delivery platforms but also platforms that are optimised to capture our attention. This need not, however, take place because the platform itself is selecting what we see.

Voluntary selective exposure is one of those phenomena that is surely not "new" with online technologies like Really-Simple-Syndication (RSS). This does not entail that it might not play out in importantly different ways online. At a minimum, and given that we generally take psychological biases like confirmation bias to be worthy of widespread disclosure because of their negative epistemic consequences, it is intuitively plausible that users of online technologies should be so informed as well. Again, while not new, it is certainly far easier to curate much more niche, that is, exclusionary, media diets on the internet.

As Cardenal et al. (2019) discuss, though, it is the *involuntary* forms of selective exposure that are typically appealed to as filter bubbles. In these cases, what is driving a curated media experience is an automated process not one driven by the consumer. Cardenal et al. (2019) present findings that suggest that selective exposure in fact varies across media platforms, for example, personalised search engines like Google Search may in some cases reduce selective exposure, and that some social media websites like Facebook may have no effect on selective exposure.[5] So, this suggests that we have motivation for investigating the more fine-grained details about how selective exposure driven by automated processes affects our consumption of information online. One of the advantages of selective exposure of both kinds is that the risks of informational isolation are fairly easy to correct once one is aware of what is taking place.

Similar things may not be able to be said about echo chambers. While some use filter bubbles and echo chambers co-referentially, it is advantageous to use them in more specific ways.[6] Here I will explain echo chambers in terms of homophily in networks. Homophilous networks describe networks of like-minded or similar individuals. The causal arrows between online personalisation and homophily are unclear, especially in discussions of echo chambers in popular media sources. It's possible that homophily is reproduced online from offline networks, that it is caused by online processes, and, naturally, that it might be some combination of both.

Some research indicates, for example, that the levels of political homophily observed on Twitter are fairly close to those observed offline (Halberstam and Knight 2016). While the appearance of like-minded political groups online and offline may be similar, this entails just another epistemic environment in which our access to political information is limited. Such results count against the techno-optimism of the recent past that the internet would obliterate such partisan differences and unite us at some more basic level of shared values.[7]

Such statements may not generalise, however, in light of research suggesting that, although political conversations do typically take place in homophilous groups online, conversations about non-political subjects do not (Barberá et al. 2015). In addition, it seems that one's risk of ideological segregation is importantly related to one's level of news consumption, with those who are on the higher end of the scale consuming a broader and more diverse, thus less risky, range of content (Flaxman, Goel, and Rao 2016; Cardenal et al. 2019).

The social-epistemic risks of being too selective or segregated in one's online networks are fairly clear: a reduction in sources increases the risks that one will miss out on relevant information. Of course, there are important cultural and social consequences that flow more directly from how we are organised online and these consequences affect the interpersonal qualities of epistemic communities.

As McBrayer (2021) reviews, there is an abundance of results, again from the United States, that indicate a connection between partisanship and so-called fake news. These include notable decreases in the amount of overlap between Republicans and Democrats with respect to party policy and rapidly rising rates of cross-partisan antipathy. There is, for example, an eightfold increase amongst Republicans and a sixfold increase amongst Democrats who report they would be upset if their child married across party lines, and this is alongside roughly half of each political affiliation believing supporters of the opposite party are simply evil. The causes of antipathy McBrayer discusses are deeply connected to the consumption of partisan media and a decline in the availability of quality local (and non-partisan) news (importantly displaced by the rise of online media).

We cannot point to personalising algorithms as the exclusive cause here, and such a claim has not been made. However, it is not news to any of us that a decline in quality – and, relative to opinion columns and sponsored content, expensive – investigative journalism, investment in neutral public broadcasting, and other similar bastions of traditional media have struggled to maintain meaningful existence with our rapid shift to online media alternatives. The uneasy popularity of the concept of "post-truth" in 2016 is one indicator that as an epistemic community we are not exactly celebrating the arrival of these changes. Such upset is for good reason given that these networking patterns would seem to work against (or, at least not for) our ideals for public discourse. High rates of antipathy are not fertile grounds for open-minded and sincere debates about matters of public policy, or, as the effects become more pronounced at more local levels, even for discussions about how to tackle a public health crisis in the midst of a pandemic.

So far, we covered some ways in which the mediation of our online experience may undermine explicitly epistemic values like truth via the ways it shapes our access to content. The social and cultural results discussed also attest to some consequences for valuable dispositions like intellectual humility and open-mindedness. Such dispositions intuitively help to promote the health of the epistemic community in that they help to safeguard testimony (by, e.g., securing sincere listeners and promoting particular methods of expressing one's commitments).

But the internet is not only a place where we go to give and ask for reasons. It is also a place where we go simply to discover and express ourselves, our emotions, and our achievements – via emojis, internet memes, and 280-character-limited messages. These are not necessarily incompatible ends. Moreover, if we return briefly to the classroom, it is intuitive that the best formal education spaces are not those that aim to leave their students with the most true beliefs while gaining as few false ones along the way. The vision of formal education provided by some of the most well-known philosophers of education, such as bell hooks and Paulo Freire, in fact bring to the fore the role of atmosphere, of culture, and of meaningful personal relationships (Hooks 1994; Freire et al. 2018).

There is a tension, though, between our social-epistemic ideals and the goals of many of the companies and individuals who are providing our online experiences. And an apparently live question is whether or not our social-epistemic expectations ought to outweigh these goals of companies.

4 Untangling Epistemic Risk, Autonomy, and Manipulation

For the sake of the discussion here, I will assume that social-epistemic agency is a subset of one's autonomy generally. To be autonomous in one's life is to be in some sense self-authoring of that life. One's social-epistemic agency can be understood as how well one is achieving autonomous direction in two domains of action: the epistemic and the communicative.

Social-epistemic agency should be understood as a degreed property and one that reflects the extent to which one is self-authoring over one's participation "locally" in particular epistemic activities and "globally" via the social and professional roles that one can occupy in one's life. One assumes that such agency is relationally determined at least by causal processes of development and socialisation – including, as may be obvious, one's formal and informal education. A duty to disclose epistemic risks of the sort discussed in this chapter is a duty grounded in the value of autonomy both as it is found locally through specific epistemic and communicative acts, and globally through the roles one can occupy as a social-epistemic agent.

Intuitively, manipulation undermines the value of autonomy. It does so because it challenges the view that the best – as in, appropriate – way to change someone's mind or compel them to do something is through rational persuasion. To manipulate someone's cognitive attitudes or their choices fails to both respect them as a rational agent and to allow them to act from their rational capacities. Thus, one may both be harmed by the limited ability to exercise their autonomy and be simultaneously harmed by an act of disrespect.

The personalisation of online platforms is thus risky from the perspective of autonomy. When an online retail company personalises one's experience of their products, say, one's attention is directed to some products and not other products based on the *assumptions* that the model driving the personalisation makes.

A secondary concern about online personalisation and manipulation, and the one more centrally of concern in this chapter, is that online personalisation is manipulative because it actively shapes one's social-epistemic agency. The hidden curriculum idea from the second section is tied to this second concern. Put differently, online personalisation is itself a process that contributes to the development and maintenance of one's social-epistemic agency. So, online personalisation may be manipulative and thus harmful for the ways that it limits choice-making, but these instances of manipulative action may also be risky for the ways that they *cause* changes in us – modifying our desires, beliefs, or the norms we act in accordance with.

Instead of understanding online personalisation as impacting autonomy in this direct way, we might try to understand the personalisation of online experience as manipulation qua trickery and/or pressure. As noted at the close of the previous section, a well-recognised fact about, say, social media companies, is that they aim to maximise engagement in their user-base rather than align with whatever a user's motivations for being there might be. The manipulative quality in trickery and pressure accounts crops up in the way one might expect from the name: a person is manipulated when they are tricked or conned into doing one thing under false pretences.

At a more insidious level, and as explored in documentaries like *The Social Dilemma* through interviews with software engineers, there are attempts to leverage various psychological dispositions shared by all of us in order to get more of our attention. These include, for example, gamification and gambling-like qualities that make the user experience one that always promises that after just the next doom-scroll or Farmville plot placement something will be complete.

Due to the fact that none of these techniques in fact closes down options or choices, it is implausible that they are coercive. However, they do seem to "get us to act" in particular ways that are most conducive to the ends of the company and not to ourselves (as in manipulation-as-pressure views, e.g., Feinberg 1989). Alternatively, we might take aim at the lack of good faith attempts by these companies to get what they want – namely, our attention – by actually providing us highly valuable and enriching experiences and instead appealing to cognitive and psychological tricks to get more of our time from us.

As explained by Noggle (2018), manipulative action aims or intends to get one to act, believe, or want on grounds that go against one's norms or ideals. Such accounts are also connected to autonomy, insofar as we are interested in preserving our ability to live and act genuinely or authentically. To the extent that the kinds of epistemic risks canvassed here are risky because they threaten the values of truth and healthy epistemic community, then online personalisation that uses trickery undermines these values. Thus, they would appear to be manipulative for the ways that, in practice, cause us to behave in ways that we would not endorse because they are not in line with our underlying epistemic values.

If manipulative action is just that action that aims or intends to get one to act against one's norms or ideals, then we will run into some issues in evaluating involuntary (i.e., automated) processes of online personalisation. Such processes lack an agent with aims or intentions. However, we might still maintain that many processes of online personalisation do in fact cause us to act against our epistemic norms and ideals. This may happen directly because of the kind of action we may be drawn into by bespoke advertising, but it may also happen over time as our dispositions as speakers, listeners, and learners are modified by online personalised spaces. We cannot point the blame only at the technology, as users of the internet and its many platforms, we are also not holding ourselves and one another accountable in ways that would help overcome nudges to share content as fast as possible, with hyperbolic tweaks for chances at internet fame.

Social-epistemic agency, I have proposed, is a subset of one's autonomy. Autonomy and manipulation are concepts with deep ties given that the intuitive harm of manipulation is that it subverts one's ability to be autonomous. Online personalisation occurs through many processes and in many guises, and at least some of these seem plausibly manipulative just because of this connection to our autonomy. When they undermine our autonomy within the epistemic and communicative domain they undermine social-epistemic agency. To the extent that online personalisation is affecting individuals at the level of cognitive attitudes or dispositions like their humility or open-mindedness, then autonomy is still implicated – and manipulation still relevant – because it is shaping one's social-epistemic agency. This is to interfere not with someone's ability to *exercise* their agency but to interfere with their agency itself. At the collective level, such changes to constituents' agency are risky for the epistemic community.

5 Disclosing Risks of Adverse Clinical Trials and Autonomy

As noted in the previous section, autonomy and rational persuasion are conceptually linked. Informed consent in biomedical ethics is a standard aimed at supporting the principle of autonomy, which itself has at least four components in this area: that participants in treatment or research act intentionally, in a well-informed manner, are sufficiently free from internal constraints that would undermine rational decision-making, and are sufficiently free from external constraints that would similarly undermine them.

Liao, Sheehan, and Clarke (2009) make the following argument about the requirement to disclose adverse risks in clinical trial results to potential participants in later trials. First, individuals have a human right not to be put at risk of harm without first giving informed consent. Therefore, second, there is a moral duty not to put others at risk of harm without their informed consent. Third, if adverse results from clinical trials are not disclosed to prospective participants in future trials, then they are placed at risk of harm without their informed consent. Finally, there is therefore a moral duty to disclose adverse clinical trial results to prospective participants in clinical trials. If this argument is persuasive, and if the earlier discussion motivating several kinds of epistemic risk is convincing, then it would seem we need an explanation for why the principles underlying this argument are not typically taken to apply in these cases of epistemic risk.

The ideal of informed consent is a controversial standard for digital applications. Nonetheless, a duty to disclose epistemic risks sounds very much in the spirit of pre-existing norms for informed consent. The idea is narrower and more particular: the duty to disclose the epistemic risks of some personalising digital platform requires that the particular risk(s) be made known to potential users. We can use the following case to perhaps get a sense of what it might look like to discharge such an epistemic duty online: articles on Wikipedia include disclaimer snippets to notify consumers that the entry is under-cited or that it includes too much content from a single source. These snippets are warnings about epistemic risks: the risk that one may be led to form an unjustified or unwarranted belief.

By relying on social media for news as so many of us do, we risk exposure to a range of social-epistemic agency undermining sources. These include exposure to information pollution, to biased sources, being funnelled into homophilous groups, and also the testimony of propagandistic bots. The issue here is not merely that one is exposed to such risks – after all it is not an issue that one is exposed to risks in clinical trials. The issue is that, while we can choose to put ourselves at risk, our standard for making risky choices is that it be an *informed* one. A duty to disclose the social-epistemic risks of online personalisation serves this informational need. The demandingness of this duty is fairly modest as the aim is merely to enable informed participation and use of online tools and platforms, and this is importantly different from any proposal to remove or ban various epistemically risky online technologies.

So, we risk undermining our ability to be in reasonable control of our epistemic capacities in some online spaces. Online life is rife with non-rational persuasive efforts: advertising, personalisation of newsfeeds, and search engine results, being exploited by bots to spread fake news, emotional persuasion (see Kramer, Guillory, and Hancock 2014), and design features that appeal to or exploit cognitive biases and heuristics. There are also reasonable concerns about the ways that prolonged engagement with online platforms encourage dispositions to dogmatism, close-mindedness, and arrogance.

While companies like Facebook stipulate how they expect you to engage with their platform and the ways that they will mediate content, they do not warn prospective users about the social-epistemic harms they might experience as a result of using the platform. Thus, we may be tempted to make an analogous argument to the previous one: 1. Individuals have a human right not to be put at risk of harm without first giving informed consent; 2. Therefore, there is a moral duty not to put others at risk of harm without their informed consent; 3. If the social-epistemic risks of online services are not disclosed to prospective users, then they are placed at risk of harm without their informed consent; 4. Therefore, there is a moral duty to disclose the social-epistemic risks of using online services to prospective users.

In the preceding discussion I have been proposing that there are specifically epistemic risks both to the value of truth and that of building and maintaining a healthy epistemic community from personalised internet technologies. The earlier argument is presented for consideration, for it might be that one is willing to accept some variation of the idea of social-epistemic risk but prefers to conceive of these as moral risks. My aim here is only to get the idea of such a duty into the conversation for evaluation. If one does want to conceive of these as moral risks to epistemic aspects of life, then the active literature on moral encroachment is a key place to begin.

In the final section, I consider some of the applied issues in attributing responsibility and thus finding candidates to disclose such epistemic risks.

6 How Many Hands Minimise Responsibility Claims

Intuitively, one is responsible for the morally significant impact that one has on the world but only when those impacts are the outcome of voluntary action. This includes both the intended consequences that result from voluntary actions and at least some cases where a choice leads to a moral outcome that is significant but perhaps is not what one was intending to bring about.

The familiar Trolley Problem thought experiment provides a useful tool for demonstrating different cases of responsibility. As the speeding train threatens to end the lives of multiple people, there are three apparent options open to a bystander: to divert the train onto a second track where only one life will be taken, to actively choose to do nothing, or to passively choose to do nothing in an attempt to absolve oneself of responsibility for the situation. The active and passive distinction is the subject of much debate – in the Trolley Problem and beyond – and it is an attempt to more finely individuate one's actions by appealing to the intentions behind those actions so that in some cases the morally significant outcomes will not be such that one is responsible for them. Tied tightly to this intuitive picture conceptually is blame, such that one can be held blameworthy for these morally significant outcomes of one's voluntary actions.

Understanding the moral responsibility of online technologies is complicated by multiple distinct issues. First, the closest causal agent for many of these online risks may be non-human and instead an algorithm, and without an agent to point to, it is hard to make use of normative notions like responsibility and blame. This may lead us to point to the person(s) who created the algorithm.

However, the programmer(s) who creates the algorithm driving some manipulative process or state of affairs may have been operating under the instructions of someone else, thus pushing us to understand responsibility in this domain as a species of collective or distributed responsibility. Deborah G. Johnson (1992) provides an early and helpful discussion of the social responsibilities of engineers generally. As she explains we might look in multiple places to ground such responsibilities for engineers including social contract accounts or by simply appealing to our ordinary morality. In a subsequent paper, Johnson (2017) provides a new approach to understanding the responsibilities of engineers in terms of accountability. All three of these proposals offer ways for future progress in understanding the nature of responsibility for internet technologies.

A well-known problem in this area of computer ethics is known as the problem of many hands – and this counts as a barrier to accountability. As explicated by Nissenbaum (1994, 75),

> Most computer systems in use today are the products not of single programmers working in isolation, but of groups, collectives, or corporations . . . while our conceptual understanding of accountability directs us to "the one" who must step forward . . . collective action presents a challenge.

Thus, in any attempt to hold someone responsible for causing risks we will plausibly run into some species of this problem. If we tie the duty to disclose risk to being responsible for causing it, then we will hit a fairly immediate barrier here.

On that note, it is also possible that there is no party directly responsible for such risks to our epistemic values – but this does not undermine a duty to disclose risks in all cases. If one lives in an earthquake prone region, for example, we generally take ourselves to be owed warnings about things like the possibility of landslides. Due to the fact that this is a threat to us all, we generally take an entity like the state to be the one who ought to take on the burden of ensuring that this risk is adequately communicated. When we consider risks to the epistemic community as a whole, we might take a similar stance and suggest that the duty to disclose the epistemic risks of new media and the internet more broadly ought to be discharged by the state.

In closing, to the extent that *we* may be implicated in causing some of these epistemic risks we will have to take seriously that we may end up partially accountable, in some way, for some of these potentially manipulative states of affairs.

Notes

1. For an overview of philosophical theories of listening including Plato, Aristotle, Jean-Jaques Rousseau, Johann Herbart, John Dewey, and Martin Buber, see Haroutunian-Gordon and Laverty (2011).
2. As Waks (2011) explains, for Dewey

> [the] consequences of the one-way pattern of listening are severe. As one-way, straight-line communications do not invite – and, indeed, leave no room for – response, listeners habituated to them remain passive and lax, irresponsible, thoughtless, fickle, emotionally susceptible, shortsighted, amusement-seeking, and shiftless, imbued neither with the courage or energy to speak nor the intellectual power to say anything worth listening to.
>
> (Waks 2011, 193)

3. See Grice (1991, 21–40).
4. See Ronson (2015); Norlock (2017).
5. See also Dubois and Blank (2018).
6. The identification of echo chambers with political identity occurs in Jamieson and Capella's (2008) influential account of the US Conservative media network.

7. As an example of such statements, I refer to the Declaration of the Independence of Cyberspace: "We believe that from ethics, enlightened self-interest, and the commonweal, our governance will emerge. Our identities may be distributed across many of your jurisdictions. The only law that all our constituent cultures would generally recognize is the Golden Rule".

7 References

Barberá, P., J. T. Jost, J. Nagler, J. A. Tucker, and R. Bonneau. 2015. "Tweeting From Left to Right." *Psychological Science* 26: 1531–42.

Bozdag, Engin, and Jeroen van den Hoven. 2015. "Breaking the Filter Bubble: Democracy and Design." *Ethics and Information Technology* 17 (4): 249–65.

Cardenal, A. S., C. Aguilar-Paredes, C. Galais, and M. Pérez-Montoro. 2019. "Digital Technologies and Selective Exposure: How Choice and Filter Bubbles Shape News Media Exposure." *The International Journal of Press/Politics* 24: 465–86.

Cheng, Justin, Michael Bernstein, Cristian Danescu-Niculescu-Mizil, and Jure Leskovec. 2017. "Anyone Can Become a Troll: Causes of Trolling Behavior in Online Discussions." *CSCW: Proceedings of the Conference on Computer-Supported Cooperative Work. Conference on Computer-Supported Cooperative Work* 2017: 1217–30. doi:10.1145/2998181.2998213.

Dewey, John. 1900. *The School and Society*. Chicago: University of Chicago Press.

Dubois, E., and G. Blank. 2018. "The Echo Chamber is Overstated: The Moderating Effect of Political Interest and Diverse Media." *Information, Communication & Society* 21: 729745.

Feinberg, J. 1989. *Harm to Self*. New York, NY: Oxford University Press.

Flaxman, S., S. Goel, and J. M. Rao. 2016. "Filter Bubbles, Echo Chambers, and Online News Consumption." *Public Opinion Quarterly* 80: 298–320.

Freire, Paulo, Myra B. Ramos, Donaldo P. Macedo, and Ira Shor. 2018. *Pedagogy of the Oppressed*. New York, NY: Bloomsbury Academic.

Grice, H. P. 1991. *Studies in the Way of Words*. Cambridge, MA: Harvard University Press.

Halberstam, Yosh, and Brian Knight. 2016. "Homophily, Group Size, and the Diffusion of Political Information in Social Networks: Evidence from Twitter." *Journal of Public Economics* 143 (2016): 73–88. doi:10.1016/j.jpubeco.2016.08.011.

Haroutunian-Gordon, S., and M. J. Laverty. 2011. "Symposium: Philosophical Perspectives on Listening." *Educational Theory* 61 (2): 117–237.

Hooks, Bell. 1994. "Teaching to Transgress: Education as the Practice of Freedom." *New York Times Book Review*, 27.

James, William. 2009 [1896]. *The Will to Believe and Other Essays in Popular Philosophy*. The Gutenberg Project. www.gutenberg.org/files/26659/26659-h/26659-h.htm.

Jamieson, Kathleen H., and Joseph N. Cappella. 2008. *Echo Chamber: Rush Limbaugh and the Conservative Media Establishment*. Oxford: Oxford University Press. http://lib.myilibrary.com/detail.asp?id=234632.

Johnson, D. 1992. "Do Engineers Have Social Responsibilities?" *Journal of Applied Philosophy* 9: 21–34.

Johnson, D. 2017. "Rethinking the Social Responsibilities of Engineers as a Form of Accountability." In *Philosophy and Engineering: Exploring Boundaries,*

Expanding Connections, edited by Diane P. Michelfelder, Byron Newberry, and Qin Zhu. Cham: Springer.

Kramer, A. D. I., J. E. Guillory, and Jeff Hancock. 2014. "Experimental Evidence of Massive-scale Emotional Contagion Through Social Networks." *Proceedings of the National Academy of Sciences* 111: 8788–90.

Liao, S. M., Mark Sheehan, and Steve Clarke. 2009. "The Duty to Disclose Adverse Clinical Trial Results." *American Journal of Bioethics* 9 (8): 24–32.

McBrayer, Justin P. 2021. *Beyond Fake News: Finding the Truth in a World of Misinformation*. New York, NY: Routledge.

Nissenbaum, Helen. 1994. "Computing and Accountability." *Communications of the ACM* 37: 73–80.

Noggle, Robert. 2018. "The Ethics of Manipulation." In *Stanford Encyclopedia of Philosophy: Summer 2018*, edited by Edward N. Zalta. Summer 2018.

Norlock, Kathryn J. 2017. "Online Shaming." *Social Philosophy Today* 33: 187–97.

"OECD Skills Studies: Further Results from the Survey of Adult Skills." 2016. www.oecd-ilibrary.org/education/skills-matter_9789264258051-en.

Pariser, Eli. 2011. *The Filter Bubble: What the Internet Is Hiding from You*. New York, NY: Penguin Books.

Pew Research Center. 2020. "U.S. Media Polarization and the 2020 Election: A Nation Divided." www.journalism.org/2020/01/24/u-s-media-polarization-and-the-2020-election-a-nation-divided/.

Ronson, Jon. 2015. *So You've Been Publicly Shamed*. London: Picador.

Smith, A., and M. Anderson. 2018. "Social Media Use in 2018." www.pewinternet.org/2018/03/01/social-media-use-in-2018/.

Sunstein, Cass R. 2017. *#Republic: Divided Democracy in the Age of Social Media*. Princeton, NJ: Princeton University Press.

Waks, Leonard J. 2011. "John Dewey on Listening and Friendship in School and Society." *Educational Theory* 61 (2): 191–206.

15 Promoting Vices

Designing the Web for Manipulation

Lukas Schwengerer

1 Introduction

It is Friday evening and you are exhausted from another day of overtime.[1] You want to relax by watching a TV series. Say you want to watch *The Wire*. You wonder where you can access episodes of *The Wire*. You do not have the box set, nor is it running on any channel right now. So you decide to find out which streaming service offers episodes. You open Google Search, and the field for your search input jumps right to your attention. You start to type "The Wire Stream" into the box. As soon as you reach the "W" Google Search suggests to autocomplete to "The Wire". As soon as you reach the "S" Google suggests your intended query "The Wire Stream", so you hit enter. Immediately, Google not only presents you websites that likely tell you which streaming service provides the opportunity to watch *The Wire*, Google Search itself presents all available choices directly at the top. Clearly visible. Impossible to miss. Even when you are exhausted you can find the right streaming service in a matter of seconds. It is that easy.

The Google Search website is a paradigmatic example of so called *user-friendly design* – design that makes it particularly quick, easy and efficient to use a website for the task the user aims to complete. User-friendly design is intended to increase processing fluency in users and successful user-friendly design makes cognitive processing of a website faster and easier.[2] I will focus on tasks aimed at gathering information, but the same ideas can also apply to other forms. The website is designed in a way that makes all required information as obvious as possible and avoids features that are distracting or difficult to access. You likely have also experienced websites that did not include speed and ease of use as a goal during their design process: flickering colours and moving backgrounds fighting for your attention, low contrast in colour that renders text illegible, dead links that prompt frustration and the dreaded Papyrus as the font of choice. And most likely you avoid any such website at all costs. It seems natural to claim that the user-friendly site has epistemic and practical advantages over the user-unfriendly one. Just take the earlier example: it is a lot easier to find out where I can find a streaming service including "The Wire" with a website featuring user-friendly design.

DOI: 10.4324/9781003205425-18

That's the good news, and I will not deny these benefits. But the bad news is that the same design principles also come with a particular epistemic problem: user-friendly design tends to promote an intellectually vicious attitude towards a website. It tends to promote an *overly trusting attitude*.

The guiding idea in the background is that trust is an unquestioning attitude I can have towards a person or object (Nguyen forthcoming). When I trust a friend I will usually not question their information, nor their intentions. Often, my trust is related to a person – such as the friend I trust. But unquestioning attitudes are possible towards objects also. A climber trusts a rope, and I trust the bridge I am walking on. In these cases objects are unquestioned in performing their intended function. Trusting a website can be read in both ways. I can trust an author of a website, or I can trust the website itself. I will work with the latter reading, but an argument with the same structure is available with the former reading. The object reading is easier to combine with a framework of cognitive integration, and it has the benefit of being applicable even when the website would be largely created by algorithms without much deliberate human input.

I can compare the degree of trust we have towards an object or person with the degree of trust the object or person deserves: their trustworthiness. Sometimes, I do not question a person as a source of information, even though I should. They do not know what they claim to know. Similarly, sometimes I trust a rope that I should not trust. I do not question its stability, but it is already damaged and cannot hold my weight reliably. If the trust towards something exceeds its trustworthiness I will speak of an overly trusting attitude. My aim is to show that such a mismatch between trust and trustworthiness can arise for websites because of user-friendly design. The ease and speed with which I parse information from a website changes the trust I have towards a website in a way that is unrelated to the trust the website deserves. For instance, I take the information that Google Search provides me at the top of the results to be obviously correct even though it does not deserve such an unquestioning attitude. I develop this attitude in part because Google Search is especially easy and quick to use, and my psychology functions such that ease and speed of processing induces an increase in trust. Hence, I end up trusting the Google Search result more than I ought to. I have an overly trusting attitude. An attitude that is not justified by the trustworthiness of the website itself that makes me vulnerable to manipulation.

The road to manipulation is straightforward. If a manipulator can induce an unquestioning attitude of trust, then they will be able to manipulate the beliefs of the trusting person without much effort. Hence, if someone wants to manipulate via a website, they can use psychological effects of user-friendly design for the website to generate a gap between the trust users assign to the website and the trust the website deserves. And in doing so a manipulator makes users more intellectually careless and their beliefs easier to manipulate. It is this danger of users being exploited via psychological features affecting trust judgements and the mechanism involved that will be the target of my discussion. User-friendly design is also manipulation-friendly design in this specific way – or so I will argue.

My plan is the following: I will start with a sketch of the argument against user-friendly design based on cognitive integration. Then, in Sections 3 and 4, I develop the groundwork for a refined version of the argument. In Section 3, I give an overview of virtue and vice epistemology, showing how an overly trusting attitude is detrimental for intellectual virtues. In Section 4, I present my preferred version of the extended mind thesis and cognitive integration. Section 5 features the expanded argument in detail with a focus on the empirical support that brings us from user-friendly websites and cognitive integration to the overly trusting attitude. I conclude in Section 6 with directions of how to limit the epistemic badness of user-friendly design without giving up on the benefits.

2 The Argument From Cognitive Integration Against User-Friendly Design

My aim is to show how a website[3] promotes an overly trusting attitude based on its user-friendly design. I do this by treating the website as an artefact that can be cognitively integrated – that can be part of an extended mind. I thereby treat a website akin to a tool that can be used to enhance the abilities of an agent. In particular, I am interested in the epistemic abilities and the epistemic actions of agents. Here an epistemic action is understood as an action with a function to improve an agent's cognition such that some cognitive tasks become easier or in some cases become possible in the first place (Kirsh and Maglio 1994). For instance, we can use pen and paper to enhance our ability to perform arithmetic and then perform epistemic actions in writing and reading numbers and symbols on the paper. Pen and paper become part of the information-processing system and therefore deserve part of the credit for successful performance of a task (Clark and Chalmers 1998). This sort of cognitive integration is independently plausible with significant explanatory power (Hutchins 1995; Clark and Chalmers 1998; Sutton 2006; Clark 2008, 2010; Menary 2010; Heersmink 2015) and when applied to a website constitutes the basis for my argument against user-friendly design. Using the framework of cognitive integration also provides us with a good way to explore the fine-grained mechanisms that lead to an overly trusting attitude.

The core of the argument was first pointed to by Smart (2018, 297) and starts with the assumption that a website is a particular kind of artefact which can be integrated in a cognitive system to varying degree. The explicit reference to degrees of integration already hints at my preferred theory of cognitive integration: a second wave theory of the extended mind (Sutton 2010; Heersmink 2015). In Section 4, I provide further details on this account of cognitive integration. For now, all I need is the idea that cognitive integration of artefacts comes in degrees. Artefacts that are relied on frequently and without much conscious effort (e.g., a white cane or a smartphone) are integrated to a higher degree than artefacts with one-off uses that require a significant conscious effort in interaction (e.g., a ticket terminal at an airport). This idea also provides the basis for the second premise in the argument. User-friendly design of a website leads to higher cognitive integration because it lowers the effort necessary to engage with the website. Of course, user-friendliness is not the only factor. However, given that as a design principle user-friendliness aims at making the user experience as effortless and quick as possible, and that effort and speed of the engagement with an artefact partially constitute how cognitively integrated an artefact is, it seems straightforward to conclude that user-friendly design also promotes cognitive integration. To make my argument work I now need to bring a dimension of trust into the picture. Hence, the third premise is an empirical claim that higher cognitive integration generally comes with higher trust towards the artefact for non-epistemic reasons. Crucially, that trust is not fully warranted as it is not built on a proper epistemic basis.[4] Hence, I conclude that user-friendly design promotes an overly trusting attitude.

In Section 5, I will modify the argument from cognitive integration slightly based on the particular account of cognitive integration I work with. But for now this general structure is sufficient:

1. Websites are artefacts that can be cognitively integrated.
2. User-friendly design promotes cognitive integration.
3. Generally, cognitive integration promotes trust in an artefact to a degree that is not fully epistemically warranted.
4. C: Generally, user-friendly design promotes an overly trusting attitude towards a website and its content.

The work in this argument is done primarily by premise 3, which can be established by paying attention to empirical research on judgements of trust and confidence in relation to the speed and ease of processing information. As I will show later in detail, there is considerable evidence that points towards an increased feeling of trust and the assignment of higher credence purely because information is processed more easily (Alter and Oppenheimer 2009). As Smart (2018, 297) suggests, this empirical research on judgements of trust in relation to the fluency of processing information shows us that properties that constitute higher cognitive integration also come with higher trust in an artefact and its outputs. User-friendly design aims at speed and ease of user interactions, increases cognitive integration and leads to an overly trusting attitude towards a website. This is exactly the conclusion I aim at.

I have now provided a general argument from cognitive integration showing that user-friendly design leads to an overly trusting attitude towards a website.[5] For the rest of the chapter I want to spell out and support the argument in detail. Moreover, I provide an analysis of the mechanisms involved. To do so, I need to build on the not-yet-fully explained notions of trust and cognitive integration before looking at the empirical evidence supporting premise 3.

3 Trust, Intellectual Virtues, and Intellectual Vices

Before arguing for the plausibility of all premises in my argument I need to provide some background on the problems with an overly trusting attitude. After all, I want to show that user-friendly design should worry us epistemically because it leads to an overly trusting attitude. But what is so bad about trusting a website too much?

First of all, it is not all that clear how trust applies to websites. Usually, trust is taken to be an interpersonal affair (Baier 1986). How can I trust a website if I do not treat it as a form of testimony? Trust is often distinguished from mere reliance (e.g., Baier 1986; Hawley 2019). For instance, trusting a chair not to break seems to be mere reliance. Trust proper seems to be normatively laden in ways that trusting the chair is not. I will not blame a chair for breaking – or at least only in jest. And neither are chairs praised for being trustworthy when they do what they are supposed to do. On the other hand, if I trust your word I do not merely rely on your testimony. I blame you as a person if you betray my trust with a lie. When discussing cognitive integration trust has to be understood as a more general term that also applies to artefacts. One option would be to simply stipulate that the term trust here refers to both reliance and trust proper. However, I think a more motivated solution is to analyse trust with Nguyen (Forthcoming) as an unquestioning attitude: trust is a suspension of deliberation. When we trust someone or something we leave aside all questions of whether the person or artefact will be reliable. Trust in this sense is not necessarily targeted towards agents. Nguyen's notion fits well with the notion of trust that is in play in the debates on cognitive integration (Clark and Chalmers 1998; Heersmink 2015). Importantly, this does not commit me to a binary notion of trust. As Nguyen explains, "[o]ne can trust with varying degrees of unreservedness, since one can hold the dispositions with varying degrees of force" (Nguyen forthcoming). This is important for my argument, because theories that allow for degrees of cognitive integration also demand a gradual notion of trust.

I now have an adequate notion of trust in place and can look at a theoretical foundation of the epistemic badness of trusting too much. My suggestion here is the following: an agent's overly trusting attitude leads to behaviour that is epistemically improper. It promotes vicious epistemic behaviour over virtuous epistemic behaviour. To spell out this idea I rely on some general ideas of virtue epistemology as a theory of knowledge that puts the agent and its role in acquiring knowledge at the centre of attention. To understand knowledge – so the virtue epistemologist – I ought to look at what makes potential knowers good or bad thinkers (Battaly 2008). I am limiting myself to epistemic responsibilism,[6] which pays attention to character traits that constitute intellectual virtues and vices (e.g., Zagzebski 1996; Baehr 2015). Intellectual virtues help in acquiring knowledge whereas vices are obstacles to knowledge.

Intellectual virtues in the responsibilist sense are directly impacted by an overly trusting attitude. Take intellectual carefulness. An agent is intellectually careful when they avoid intellectual errors, including the formation of false beliefs (Baehr 2015). To be intellectually careful one has to be aware of the risks of any particular belief-forming process. I need to know in which ways I might go wrong in forming beliefs in order to avoid mistakes. I need to know how easily I could fail in acquiring knowledge through a particular source. Only then I can properly judge how careful I have to be and only the appropriate amount of care is virtuous. Suppose I am reading a newspaper. Being overly careful in forming beliefs based on the newspaper's content is not virtuous because it leads to missing out on knowledge. The newspaper might be a good source of information for the results of the latest football matches, but I am reluctant to base my beliefs about football results on the newspaper. I miss out on knowledge. But being overly careless is not virtuous either, because it leads to false beliefs. Suppose the newspaper has an insufficiently funded science section and frequently misrepresents scientific studies. If I am careless and base my beliefs on the newspaper's science content I end up with false beliefs. I need to be careful to the proper degree – the degree that this particular source of belief deserves. But my judgement on how careful I ought to be can be influenced by the amount of trust I put into a source of beliefs. When I trust the newspaper I will be rather careless because I will not question it as a source of knowledge. This is fine if the newspaper is worthy of my trust, if it is indeed a good source of information. Then my unquestioning attitude usually leads to knowledge. However, if I overly trust a source – if I trust it more than the source deserves – I will not be careful enough. An overly trusting attitude destroys the virtue of intellectual carefulness.

There are similar worries for trust in relation to other virtues. A high amount of trust will lead us to give up on intellectual autonomy to an extent that we ought not to. It will lead to us being less thorough than we ought to and less open-minded. If we highly trust a source, we stop enquiries early and are not willing to take other sources into consideration. All these problematic influences of trust on intellectual virtues stem from the same source. Intellectual virtues all aim at manifesting a character trait to a particular degree in a particular situation. The ideal intellectually virtuous agent is as careful as the situation requires, as autonomous as the situation requires, as open-minded as the situation requires. Rarely anyone fits the ideal, but that is at least what agents should aim for, and what they can get reasonably close to. The ideal is set by the situation the agent is in, and the further we diverge from the ideal the worse epistemic agents we are.

If we consider the effects of trust on intellectual virtues, we can capture the relevant properties of the situation in terms of the trustworthiness of artefacts[7] involved: an intellectually virtuous agent will act in ways that are partially determined by the trustworthiness of relevant artefacts. The amount of intellectual carefulness required is set by the trustworthiness of the artefact. An agent will act intellectually careful if they put trust in the artefact roughly equal to the trust the artefact deserves. Trust and trustworthiness have to match. Whenever they are too far apart, the agent will end up acting in an intellectually vicious way. Even if they might be generally intellectually virtuous, the virtues will be unable to manifest in the concrete situation because of the mismatch between trust and trustworthiness. This in turn leads to epistemically bad consequences: the formation of false beliefs or missing out on knowledge. An overly trusting attitude therefore qualifies as an epistemic vice – it gets in the way of knowledge (Cassam 2019).

I have now shown why an overly trusting attitude should worry us and therefore why user-friendly design should worry us. Putting more trust into an artefact than it deserves leads to intellectually vicious behaviour. It stops us from being appropriately intellectually careful by misguiding us in our judgements. And being intellectually careless makes us a target for manipulation. A website's author can influence beliefs and resulting actions more easily if they can prompt the user to be careless in their belief formation. Careless users form their beliefs in ways that they would not deliberately endorse. This sort of careless belief formation fits with a general idea of classifying "an effort to influence people's choices . . . as manipulative to the extent that it does not sufficiently engage or appeal to their capacity for reflection and deliberation" (Sunstein 2016). By pushing users towards carelessness these users cannot sufficiently manifest their capacities for reflection and deliberation anymore. Hence, they stop forming beliefs virtuously. With these general results in place, I can now come back to developing the argument from cognitive integration in detail. To start, I will expand on my preferred theory of cognitive integration.

4 Cognitive Integration

Humans are proficient in using and shaping the environment to make their lives easier. We do our calculations on paper. We use post-it notes, notebooks or smartphones to remember important tasks. Humans excel in outsourcing cognitive work to the environment. Clark and Chalmers (1998) were the first to use this observation to argue that in all these cases the environment is part of the cognitive process, labelling their view *the extended mind thesis*. Cognition and mental states are not limited to the brain and skull. They leak into the environment.

This thesis is not uncontroversial. Opponents of cognitive integration models suggest that cases used to motivate the extended mind thesis are better explained otherwise because they are too different from our internal cognitive processes (cf. Rupert 2004; Sterelny 2004) or lack features that our internal mental states have (Gertler 2007). Perhaps there are even differences in the nature of content (Adams and Aizawa 2010). I will not discuss these objections here. If you find objections to cognitive integration compelling I can still retreat to the argument from testimony hinted at in Note 5. In this case you can skip directly to Section 5 and the empirical evidence that supports both the argument from testimony and the argument from cognitive integration.

Clark and Chalmers focus on a parity between the functional role the environment plays and the role that something could play inside our brain as the deciding factor for extended minds. For instance, an extended belief has to be functionally on par with a biological belief. In contrast, a second wave of theories of the extended mind (e.g., Sutton 2006, 2010; Menary 2010; Heersmink 2015) focuses on artefacts that expand the cognitive realm and allows humans to succeed in cognitive tasks that often were not possible at all without these artefacts.[8] Besides focusing on the complementary nature of extended cognitive processes, the second wave theorists also leave the largely binary nature of Clark and Chalmers's (1998) model behind. They argue that we can describe our relations to artefacts more appropriately if we think of cognitive integration as covering different dimensions, not on all of which an artefact has to be equally integrated. Take for instance, Heersmink's (2015) suggested framework of dimensions of cognitive integration. In this framework we can evaluate how integrated an artefact is among eight different (although related) dimensions. I take the shorter descriptions of these dimensions from Schwengerer (2021); for the extended presentation, see Heersmink (2015, 582–92):

Information Flow – the directions that information is passed on between an agent and an artefact.
Reliability – the frequency an artefact is used to impact the agent's cognitive processes.
Durability – the permanence of one's relation to an artefact.
Trust – the degree to which one takes the information provided by an artefact to be correct.
Procedural Transparency – the degree of fluency and effortlessness in interacting with an artefact.
Informational Transparency – the degree of fluency in receiving, interpreting, and understanding information from the artefact.
Individualisation – the degree to which an artefact is personalized or can be used by anyone.
Transformation – the degree to which the cognitive capacities of an agent change in virtue of using an artefact.

These dimensions allow a more fine-grade analysis of the human–artifact relationship. For instance, think of a notebook I take with me whenever

I leave my home. My notebook might have a two-way information flow. I write in the notebook and read information from it. I use my notebook only every once in a while, so the notebook is not highly integrated on the reliability dimension. It ranks higher on durability, because I keep the same notebook with me for a long time. It also ranks highly on trust. I rarely doubt what is written in my notebook. If I read that I have to finish this chapter on Friday, I believe that to be the case. Both transparency dimensions are also satisfied to a high degree. There is barely any effort required to open and read my notebook. Given that I usually have only a few recent entries that matter, I can also find the relevant entries quickly and easily. Moreover, because it is written in my own language and in my own style of talking and thinking it does not take much effort to interpret and understand the content either. How the notebook ranks on individualisation is unclear. On the one hand, it is not especially individualised, because anyone can read the contents of, or write into, my notebook. But on the other hand, everything in the notebook is written by me and for me. Finally, the notebook ranks relatively low on the transformation category. All – or at least most – of what I use the notebook for could be achieved by me without the notebook as well. Just with a little less convenience.

For the rest of the chapter I will work with this picture of cognitive integration suggested by Heersmink. The additional flexibility allows this theory of cognitive integration to deal with objections more easily. For instance, a general worry for theories of the extended mind is that too much becomes part of one's mind. The dimensions of integration framework can make this problem more palatable by suggesting that most things around us are integrated to only a very small degree on particular dimensions. They are not fully part of one's mind. More importantly, for my purpose, Heersmink's theory fares a lot better if I want to combine it with virtue epistemology. Whereas for Clark and Chalmers, only highly trusted artefacts can be integrated, Heersmink allows for integration of artefacts even while I do not trust the artefact fully. In his framework, I can distinguish between epistemic dimensions – which consist of only the trust dimension – and the other, non-epistemic dimensions.[9] For instance, a website can be highly integrated on reliability, durability, procedural transparency and informational transparency but still shows only a low integration on the trust dimension. Hence, I can cognitively integrate a website that is not very trustworthy and still remain intellectually careful – an option not available in the Clark and Chalmers account. The integration just has to be limited to the non-epistemic dimensions. And this is exactly what I aim for: cognitive integration that allows one to frequently, quickly and easily perform an epistemic action, without sacrificing on epistemic virtues and standards.

Unfortunately, this is possible only in theory. In practice, humans are a lot worse in isolating the trust dimension from other dimensions of integration. Cognitive integration spills over from the non-epistemic dimensions to the sole epistemic dimension of trust. This empirical claim is at the core of the argument from cognitive integration. And I am now in a position to show why this is the case, before looking for ways that help us isolate different dimensions of cognitive integration.

5 How User-Friendly Design Promotes Vices – The Expanded Argument From Cognitive Integration

Let me start by restating the initial argument from cognitive integration:

1. Websites are artefacts that can be cognitively integrated.
2. User-friendly design promotes cognitive integration.
3. Generally, cognitive integration promotes trust in an artefact to a degree that is not fully epistemically warranted.
4. C: Generally, user-friendly design promotes an overly trusting attitude towards a website and its content.

I am now equipped to modify the initial premises in light of Heersmink's theory of cognitive integration. The first premise can stay as is, but premises 2 and 3 have to be modified. Both premises 2 and 3 are too general with regard to cognitive integration. The argument needs to allow for the conceptual possibility of cognitive integration without an overly trusting attitude. And the dimensions of integration framework make this possible by distinguishing between epistemic dimensions and non-epistemic dimensions. Only in virtue of formulating premise 2 solely with non-epistemic dimensions in mind the full force of the argument will be present. If premise 2 already included high integration on the trust dimension without showing specifically that they result from non-epistemic factors the argument would be question-begging at best. Similarly, what premise 3 aims at is that non-epistemic factors in cognitive integration usually impact the extent to which one trusts an artefact. Only if this connection is established I can conclude that whatever trust is generated by cognitive integration is not fully epistemically warranted. Hence, premises 2 and 3 have to be reformulated as follows:

2. User-friendly design promotes cognitive integration on non-epistemic dimensions (dimensions other than trust).
3. Generally, cognitive integration on non-epistemic dimensions promotes an increase in cognitive integration on the trust dimension in a way that is not fully epistemically warranted.

Perhaps, the additional clause "in a way that is not fully epistemically warranted" is not required, given that the non-epistemic dimensions are responsible for the difference in the trust dimension. However, the clause is still a safeguard against the idea that some of the non-epistemic dimensions could be potentially used as an indicator of the care put into a website – and hence also as an indicator for truth conduciveness.[10] We can now state the extended argument from cognitive integration:

1. Websites are artefacts that can be cognitively integrated.
2. User-friendly design promotes cognitive integration on non-epistemic dimensions (dimensions other than trust).
3. Generally, cognitive integration on non-epistemic dimensions promotes an increase in cognitive integration on the trust dimension in a way that is not fully epistemically warranted.
4. C: Generally, user-friendly design promotes an overly trusting attitude towards a website and its content.

And given the discussion on epistemic virtues and vices in which I showed that intellectual virtues are incompatible with an overly trusting attitude I can now reach a further conclusion:

C2: Generally, user-friendly design promotes an intellectually vicious engagement with a website and its content.

This is the worry that I have been following throughout the chapter. If the argument is sound we should be apprehensive about websites with user-friendly design because they foster a form of user interaction that makes users intellectually vicious – intellectually careless. What is still missing is the evidence for premise 3. Why should one believe that the trust dimension cannot be isolated from other dimensions of cognitive integration? Why should high integration on non-epistemic dimensions spill over to the epistemic dimension of trust?

My answer here is an empirical claim. Human beings have a psychological make-up that makes it difficult to prevent non-epistemic dimensions from contaminating the epistemic one. Our psychology cannot, or at least not easily, keep the ease and speed of cognitive processes apart from a judgement of trust. When some process comes quickly and easily to a person, they tend to trust the result of that process more, purely for the epistemically irrelevant aspects of speed and ease of processing. Aspects that by and large[11] have no relation to the truth of the output given by that process.

The main source of evidence for this claim are studies about the influence of processing fluency on judgements of trust and credence. The effects of processing fluency are well researched and support a general conclusion that the easier it is to process information, the more likely we are to believe that information (Alter and Oppenheimer 2009). Let me look at a small selection of these studies to illustrate the point, before applying the observations to user-friendly design.

Reber and Schwarz (1999) provide evidence that statements that are easier to read are taken to be more likely to be true. They presented subjects with statements in colours that made them easier or more difficult to read.

For instance, they showed statements in the form of "Town A is in country B" (e.g., "Lima is in Peru") and varied the visibilities of the colours used. Blue and red were highly visible on a white background but yellow or light blue less so. The experimenters ensured that statements for all visibility ranges were balanced – statements in red were not more obviously true than statements in yellow. After presentation of a statement the subjects had to decide whether the statement was true or false. Subjects were told the colours were meant to measure reaction times with different colours to disguise the actual goal of the study and prevent manipulation. The results show that statements written in colours that could be read more easily were endorsed significantly more frequently than statements written in colours that were less visible. In other words: subjects judged statements to be more likely to be true, merely because they had an easily readable colour. The most plausible explanation is that the information processing was more fluent – it was easier and faster to read the visible colours.

McGlone and Tofighbakhsh (2000) observe a similar effect of processing fluency in the effects of rhyming. Subjects were confronted with aphorism that they were not familiar with that they had to judge on their accuracy on a scale of 1 (not at all accurate) to 9 (very accurate). The complete list of aphorisms featured pairs of rhyming and non-rhyming versions such that for each pair the experimenters could compare the accuracy judgement for the rhyming and the non-rhyming versions. For instance, the list included "Woes unite foes" and "Woes unite enemies". As a control measure they also included pairs in which neither version was rhyming. For instance, "Good intentions excuse ill deeds" and "Good intentions excuse ill acts". It turned out that if the subjects were not warned of potential effects of rhyming, they assigned higher accuracy to aphorisms that did in fact rhyme. They propose that "this effect is a product of the enhanced processing fluency that rhyme affords an aphorism such as 'What sobriety conceals, alcohol reveals' relative to a semantically equivalent nonrhyming version" (McGlone and Tofighbakhsh 2000, 427). Again, speed and ease of processing comes with an increase in perceived accuracy.

Finally, Oppenheimer (2006) provides evidence that easier to process texts are deemed to be written by more intelligent authors. In particular, he shows that using overly complex words comes with being judged of lower intelligence. This relationship held regardless of the quality of the text in question. This result might not be completely surprising, given that every writing guide suggests simple prose, but it is again further evidence that processing speed and ease impacts judgements about the epistemic merits of some perceived informational content. Oppenheimer explicitly states the results of these judgements are best explained by considering processing fluency (Oppenheimer 2006, 151).

These examples are a mere glimpse at the evidence available. Alter and Oppenheimer's (2009) meta-analysis includes an abundance of similar studies that all point in the same direction: human psychology infers from processing fluency broadly epistemic features, even when such an inference is not justified. Most importantly, processing fluency leads to more trust and giving higher credence to information processed fluently. I can import these results directly into Heersmink's cognitive integration framework. Processing fluency – the speed and ease of processing – is captured by procedural and informational transparency. I thereby have identified evidence for at least two non-epistemic dimensions of cognitive integration that spill over to the trust dimension. An increase in cognitive integration on transparency dimensions also leads to an increase on the trust dimension, as supported by the empirical evidence. And it seems clear that these non-epistemic dimensions have no relation at all to truth. Take the mentioned colour effect in Reber and Schwarz (1999). It seems obvious that the colour a statement is written in has no connection to the truth of that statement. These effects are exactly what I am looking for to establish premise 3: cognitive integration on non-epistemic dimensions leads to an increase in cognitive integration on the trust dimension in a way that is not fully epistemically warranted. The ease of reading a text increases the integration on the non-epistemic dimension of procedural transparency in a way that also increases the integration on the trust dimension.

Taking a step back the same idea can be applied more generally to user-friendly design – design that makes it particularly quick, easy and efficient to use a website for the task the user aims to complete. Making a cognitive process particularly quick, easy and efficient is nothing else than increasing processing fluency. And given that processing fluency increases perceived trust, premise 3 is established, and I can conclude that user-friendly design leads to an overly trusting attitude towards a website and its content.

One might wonder whether there is an alternative reading available. Perhaps, fluency does not lead to an overly trusting attitude, but lower fluency leads to a lack of trust. This does not seem to be the right interpretation, because studies of repeated presentation of the same content point towards processing fluency influencing judgements of trust beyond the trustworthiness of a source (Hasher, L., Goldstein, D. and Toppino 1977; Begg, I. M., Anas, A. and Farinacci 1992). Hence, there is clear evidence of trust due to processing fluency exceeding trustworthiness of a source.[12]

Of course, even though this is bad news, it need not be terrible news yet. All that I have established is that user-friendly design promotes an overly trusting attitude and therefore also promotes an intellectually vicious engagement with websites. But nothing has been said to the extent of excess in trust and intellectual viciousness. I have not established that user-friendly design always leads to high agential gullibility, the kind of gullibility in which we too eagerly accept an artefact and its processes as trustworthy (Nguyen forthcoming). Trusting a website a little more than the site deserves is perhaps not that big a problem. But the worry looms that developments to make the user experience even faster, even easier and more comfortable brings us to a larger and larger gap between our trusting attitudes and the trust a website deserves. A gap that could be exploited by people aiming to manipulate us for their gain by eliciting false beliefs that prompt actions that we would not have otherwise performed. How can we stop that? This is what I will address in the final part.

6 Fixing the Web

I have established that user-friendly design leads to an overly trusting attitude towards a website. This should worry us, even if I have not shown that the excess of trust is already at a particularly dangerous level. There are at least three different responses available that I will sketch in turn.

First, we could abandon user-friendly design principles. Stop making websites accessible, use illegible fonts and colours. Remove all forms of personalisation that increase the ease of using a website. But obviously this cannot be the way to go. It is a clear case of throwing the baby out with the bathwater. We should not sacrifice all epistemic benefits we get from websites just because of a worry of an overly trusting attitude. Moreover, economic pressures make this option practically impossible. The market forces will always promote user-friendly websites over completely unusable ones.

Second, we could limit user-friendliness. The aim here would not be to stop us from being overly trusting completely but to limit the extent to which our trust exceeds the trust the website deserves. As long as the gap between an agent's trust in a website and the website's trustworthiness is not too big, the potential damage is also limited. An agent might end up with some false beliefs and miss out on some knowledge, but by and large the agent's belief formation will be truth conducive because the agent's behaviour is not too far off from that of an ideal, intellectually virtuous agent. The agent can still be close enough to the required intellectual carefulness. Maybe that is good enough for all our purposes.

How these limits on user-friendliness look in practice is a difficult question. To give you one example of such a limit, consider a law that restricts a website's use of personalisation via tracking cookies. If the website cannot personalise efficiently, then the website loses a tool in increasing user-friendliness. It can no longer predict efficiently what a user wants to do. Hence, the user will likely be required to take an extra step and reduce their processing fluency.

Finally, third, we could look for strategies that stop or compensate the spill from non-epistemic dimensions of integration to the epistemic dimension of trust. This is the ideal solution. It allows to increase user-friendliness with all its benefits while it prevents the design to influence trust in a website. Strategies here might be available on a structural level and on the level of the individual user. On a structural level one approach is to provide means that artificially lower the integration on the trust dimension.[13] The aim here is to counteract the spill from non-epistemic to epistemic dimensions. This can be achieved by providing some sort of psychological defeater to the agent when they visit a website: a consciously available reason that decreases justification with regard to the contents of the website. In fact, in the European Union there is already a version of this approach established – although likely not with this goal in mind. The General Data Protection Regulation forces website providers to make their personalisation via tracking cookies obvious and explicit. Websites have to inform users in the European Union of their tracking mechanisms and users can choose to continue to the website while declining those tracking cookies that are not necessary for the core functioning of the website. By being presented with a pop-up pointing to the tracking cookies, the mechanisms behind the website become more salient to users with enough background information. The necessity of accepting tracking cookies functions as a warning that can decrease trust in a website and thereby compensates some of the effects of user-friendly design. As is, there are still some hurdles for the effectiveness of these warning signs. As long as the owner of a website is in full control of how to include these pop-ups the intended effects could be mitigated. The design of these pop-ups itself might influence their impact on the trust assigned to a website. Companies such as Facebook or Google have the resources to design pop-ups in a way that clicking on them is quick and effortless, compared to other sites. In the worst case, this could lead to sites that warrant higher trust to have badly designed pop-ups that lower trust significantly, but sites that warrant only lower trust to have perfectly engineered pop-ups without much of an effect on assigned trust. To counteract this issue the implementation of such pop-ups ought to be standardised – which perhaps moves the solution back towards the second option discussed.

Moreover, for these pop-ups to have the desired effect they require substantial background knowledge on what they actually indicate. Making personalisation salient does not do the trick if one has no idea about the effects of personalisation. However, this might be supplemented by a strategy on an individual level. The goal thereby is to improve the relevant cognitive abilities of users so that they are able to competently respond to available defeaters by lowering trust put into a website.[14] Heersmink (2018) suggests a version of this strategy with an emphasis on educating for online intellectual virtues, that is, an emphasis on teaching how to apply instances of general intellectual virtues in an online environment based on relevant background knowledge. Part of this educational goal is internet literacy skills, which then in turn allow an agent to apply their general intellectual virtues properly in the online environment. It might be a long shot to train us to not be victims to the psychological effects of processing fluency, but it is less of a long shot to teach us all we need to use institutionally mandated prompts as a way of making defeaters salient. Perhaps, it is even possible to acquire online intellectual virtues that by themselves decrease the default trust for websites, such that not even a salient defeater is necessary to compensate for fluency effects via user-friendly design. The challenge is to find concrete ways of teaching these intellectual virtues. Kotsonis (2020) argues that teaching for intellectual virtues in a social media environment is possible. Similarly, Heersmink (2018) remains hopeful that we can teach online intellectual virtues properly. However, the details of how such an education towards online intellectual virtues exactly looks like are still up in the air, which leaves plenty of work for future research.

Notes

1. An earlier version of this chapter was discussed in the "Manipulation Online" workshop series organised by Fleur Jongepier and Michael Klenk, a research meeting organised by Andreas Müller and a seminar at the University of Duisburg-Essen. Thank you to all participants. Further thanks to the editors of this volume for helpful suggestions.
2. This is the rough definition of "user friendly design" that I work with. User-friendly design has to be kept distinct from *persuasive design*. Persuasive design uses psychological and social means to change user behaviour (cf. Fogg 2009a, 2009b). In contrast, user-friendly design is solely focused on making it as easy as possible for the user to perform a task. It is not aimed at changing the task the user wants to perform. Some design choices that aim at speed and ease of use can also influence the tasks intended. Autocomplete features might fall into this category in a dangerous way (Noble 2018). I will bracket this issue.
3. Although the argument applies to some online systems other than websites I will limit myself to websites.
4. Perhaps not all of the trust is unwarranted, because sometimes aspects that play a role in user-friendly design and cognitive integration are also indicators for the care put into a website and hence plausibly play a role in justifying beliefs formed in relation to a website. For instance, correct spelling is no direct warrant for a claim but might be an indicator for care put into a website and provide higher-order warrant (Tollefsen 2009). However, I argue that at least some amount of trust lacks an epistemic ground because it is based on the effort required to engage with the website and not on any feature indicating truth-conduciveness.
5. The argument from cognitive integration is not the only one available. A similar argument can be provided if we take websites to be instances of testimony. Bracketing issues of who the trust would be directed at the argument would have roughly the following steps:

   1. Information written on and read of a website constitutes a form of testimony.
   2. Generally, user-friendly design of a website increases trust in the website.
   3. Trust based on user-friendly design is not fully epistemically warranted.

   C: Generally, user-friendly design promotes an overly trusting attitude towards a website as a source of testimony (From 1, 2 and 3).

   Thank you to Eva Schmidt for suggesting this version.
6. The alternative is virtue reliabilism. See Sosa (2007), Greco (2009), and Pritchard (2012).
7. And people, but for simplicity I focus on artefacts here.
8. This is not a complete contrast to Clark's work but rather a contrast to the early formulations of the extended mind thesis. See Wilson and Clark (2009).
9. Heersmink himself is not committed to this distinction.

10. For a discussion of similar non-obvious indicators for truth-conduciveness in websites, see Tollefsen (2009).
11. Again, there might be relations in some cases. However, in the following empirical examples it will be clear that we often trust because of speed and ease of processing information that has absolutely no relation to truth.
12. This does not rule out that lack of fluency also leads to a lack of trust. Perhaps, there is a particular point of fluency that helps us to neither overly or underly trust a source. I bracket this issue. All I require is that high fluency leads to an overly trusting attitude.
13. The approach can be broadly qualified as a version of what Lewandowsky, Ecker, and Cook (2017) label "technocognition". See also Kozyreva, Lewandowsky and Hertwig (2020).
14. This approach qualifies as a cognitive "boosting" strategy (cf. Hertwig and Grüne-Yanoff 2017; Kozyreva, Lewandowsky and Hertwig 2020).

7 References

Adams, Fred, and Ken Aizawa. 2010. "Defending the Bounds of Cognition." In Menary 2010, 67–80.

Alter, Adam L., and Daniel M. Oppenheimer. 2009. "Uniting the Tribes of Fluency to Form a Metacognitive Nation." *Personality and Social Psychology Review* 13 (3): 219–35. doi:10.1177/1088868309341564.

Baehr, J. 2015. "Cultivating Good Minds. Retrieved from: A Philosophical and Practical Guide to Educating for Intellectual Virtues." https://intellectualvirtues.org/why-should-we-educate-for-intellectual-virtues-2–2/.

Baier, Annette. 1986. "Trust and Antitrust." *Ethics* 96: 231–60.

Battaly, H. 2008. "Virtue Epistemology." *Philosophy Compass* 3 (4): 639–63.

Begg, I. M., Anas, A., and S. Farinacci. 1992. "Dissociation of Processes in Belief: Source Recollection, Statement Familiarity, and the Illusion of Truth." *Journal of Experimental Psychology* 121 (4): 446–58.

Cassam, Quassim. 2019. *Vices of the Mind: From the Intellectual to the Political.* Oxford: Oxford University Press.

Chatterjee, Samir, and Parvati Dev, eds. 2009. *Proceedings of the 4th International Conference on Persuasive Technology – Persuasive'09*. New York, NY: ACM Press.

Clark, Andy. 2008. *Supersizing the Mind: Embodiment, Action, and Cognitive Extension.* New York, NY: Oxford University Press.

Clark, Andy. 2010. "Memento's Revenge: The Extended Mind." In Menary 2010, 43–66.

Clark, Andy, and David Chalmers. 1998. "The Extended Mind." *Analysis* 58 (1): 7–19.

Fogg, B. J. 2009a. "A Behavior Model for Persuasive Design." In Chatterjee and Dev 2009.

Fogg, B. J. 2009b. "Creating Persuasive Technologies." In Chatterjee and Dev 2009.

Gertler, Brie. 2007. "Overextending the mind?" In Gertler, & Shapiro 2001, 192–206.

Greco, John. 2009. "Knowledge and Success from Ability." *Philosophical Studies* 142 (1): 17–26.

Hasher, L., Goldstein, D., and T. Toppino. 1977. "Frequency and the Conference of Referential Validity." *Journal of Verbal Learning and Verbal Behavior* 16 (1): 107–12.

Hawley, Katherine. 2019. *How to be Trustworthy*. Oxford: Oxford University Press.

Heersmink, Richard. 2015. "Dimensions of Integration in Embedded and Extended Cognitive Systems." *Phenomenology and the Cognitive Sciences* 13 (3): 577–98.

Heersmink, Richard. 2018. "A Virtue Epistemology of the Internet: Search Engines, Intellectual Virtues and Education." *Social Epistemology* 32 (1): 1–12. doi:10.1080/02691728.2017.1383530.

Hertwig, Ralph, and Till Grüne-Yanoff. 2017. "Nudging and Boosting: Steering or Empowering Good Decisions." *Perspectives on Psychological Science* 12 (6): 973–86. doi:10.1177/1745691617702496.

Hutchins, Edwin. 1995. *Cognition in the Wild*. Cambridge, MA: MIT Press.

Kirsh, David, and Paul Maglio. 1994. "On Distinguishing Epistemic from Pragmatic Action." *Cognitive Science* 18 (4): 513–50.

Kotsonis, Alkis. 2020. "Social Media as Inadvertent Educators." *Journal of Moral Education*, 1–14. doi:10.1080/03057240.2020.1838267.

Kozyreva, Anastasia, Stephan Lewandowsky, and Ralph Hertwig. 2020. "Citizens Versus the Internet: Confronting Digital Challenges with Cognitive Tools." *Psychological Science in the Public Interest: A Journal of the American Psychological Society* 21 (3): 103–56. doi:10.1177/1529100620946707.

Lewandowsky, S., U. K. Ecker, and J. Cook. 2017. "Beyond Misinformation: Understanding and Coping with the 'Post-Truth' Era." *Journal of Applied Research in Memory and Cognition* 6 (4): 353–69.

McGlone, M. S., and J. Tofighbakhsh. 2000. "Birds of a Feather Flock Conjointly (?): Rhyme as Reason in Aphorisms." *Psychological Science* 11 (5): 424–28. doi:10.1111/1467-9280.00282.

Menary, Richard, ed. 2010. *The Extended Mind*. Cambridge, MA: MIT Press.

Nguyen, C. T. forthcoming. "How Twitter Gamifies Communication." In *Applied Epistemology*, edited by Jennifer Lackey. Oxford: Oxford University Press.

Noble, S. U. 2018. *Algorithms of Oppression*. New York, NY: New York University Press.

Oppenheimer, Daniel M. 2006. "Consequences of Erudite Vernacular Utilized Irrespective of Necessity: Problems with Using Long Words Needlessly." *Applied Cognitive Psychology* 20 (2): 139–56.

Pritchard, Duncan. 2012. "Anti-Luck Virtue Epistemology." *Journal of Philosophy* 109 (3): 247–79.

Reber, R., and N. Schwarz. 1999. "Effects of Perceptual Fluency on Judgments of Truth." *Consciousness and Cognition* 8 (3): 338–42. doi:10.1006/ccog.1999.0386.

Rupert, Robert D. 2004. "Challenges to the Hypothesis of Extended Cognition." *Journal of Philosophy* 101 (8): 389–428.

Schwengerer, Lukas. 2021. "Online Intellectual Virtues and the Extended Mind." *Social Epistemology* 35 (3): 312–22.

Smart, P. 2018. "Emerging Digital Technologies: Implications for Extended Conceptions of Cognition and Knowledge." In *Extended Epistemology*, edited by J. A. Carter, Andy Clark, Jesper Kallestrup, and Duncan Pritchard, 266–304. Oxford: Oxford University Press.

Sosa, Ernest. 2007. *A Virtue Epistemology: Apt Belief and Reflective Knowledge*. Oxford: Oxford University Press.

Sterelny, Kim. 2004. "Externalism, Epistemic Artefacts and the Extended Mind." In *The Externalist Challenge*, edited by Richard Schantz, 239–54. Berlin: De Gruyter.

Sunstein, Cass R. 2016. "Fifty Shades of Manipulation." *Journal of Marketing Behavior* 1 (3–4): 214–44. doi:10.1561/107.00000014.

Sutton, J. 2006. "Distributed Cognitions: Domains and Dimensions." *Pragmatics and Cognition* 14 (2): 235–47.

Sutton, J. 2010. "Exograms and Interdisciplinarity: History, the Extended Mind and the Civilizing Process." In Menary 2010, 189–225.

Tollefsen, Deborah P. 2009. "'Wikipedia' and the Epistemology of Testimony." *Episteme* 6 (1): 8–24.

Wilson, R. A., & Clark, A. 2009. "How to situate cognition: Letting nature take its course." In M. Aydede, & P. Robbins (Eds.), *The Cambridge Handbook of Situated Cognition* (pp. 55–77). New York: Cambridge University Press.

Zagzebski, Linda T. 1996. *Virtues of the Mind: An Inquiry into the Nature of Virtue and the Ethical Foundations of Knowledge*. Cambridge: Cambridge University Press.

16 Online affective manipulation

Nathan Wildman, Natascha Rietdijk, and Alfred Archer

1 Introduction

In January 2012, members of Facebook's Data Science Team conducted an experiment: using algorithms, they tailored the feeds of 689,003 Facebook users. For one week, some of this group saw significantly fewer posts featuring negative emotional words and phrases while others saw fewer posts with positive words or phrases. The aim was to assess the possibility of *emotional contagion* – that is, whether individuals exposed to extremely negative or extremely positive online emotions would affectively respond in kind.[1] And, strikingly, the results suggested that they do: individuals exposed to fewer positive posts produced fewer positive posts, while those exposed to fewer negative posts made fewer negative posts (Kramer, Guillory, and Hancock 2014).

This experiment is a particularly explicit example of *online affective manipulation*: the affective states of Facebook users were manipulated by Kramer's team via online tools to serve the team's purposes. But there are a number of other, more subtle instances of this kind of manipulation. For example, in a recent interview, Jack Dorsey, CEO of Twitter, admitted that Twitter's algorithm highlights the "most salacious or controversial tweets".[2] This, being charitable, is an unintended outcome of the algorithm's designed function of spotlighting those tweets that garner the most responses, as prompting strong affective reactions is a good way to increase response numbers. Relatedly, a number of blogs and online newspapers now publish so-called *hate click* pieces: web-content angled to intentionally anger or infuriate an audience so much that they are driven to engage with the content, even if just to tell the author how stupid or wrong they are.[3]

The aim of this chapter is broadly exploratory. We want to better understand online affective manipulation and especially what, if anything, is morally problematic about it. To do so, we begin by pulling apart various forms of online affective manipulation. We then proceed to discuss why online affective manipulation is properly categorized as manipulative, as well as what is wrong with (online) manipulation more generally. Building on this, we next argue that, at its most extreme, online affective manipulation

DOI: 10.4324/9781003205425-19

constitutes a novel form of affective injustice that we call *affective powerlessness*. To demonstrate this, we introduce the notions of affective injustice and affective powerlessness and show how several forms of online affective manipulation leave users in this state. The upshot is that we now have a better grip on the nature of online affective manipulation, as well as some tools to help us understand when and why it is morally problematic.

Before moving to the body of our discussion, two quick points. First, we'd like to briefly address one potential general concern about affective manipulation, online or not. It is plausible to say that, unlike actions, affective states do not seem to be under our voluntary control. This can lead one to question whether emotions can be the proper target of manipulation. If, goes the objection, there is no autonomy in the affective domain, how can there be manipulation directed at them?

A full response to this worry likely requires a complete theory of emotions. That said, it is useful to note a few points in reply. First, one can readily deny the implicit assumption that only those things over which we normally have voluntary control *can* be manipulated. Beliefs are typically seen as potential targets of manipulation, and our voluntary control over them has been questioned (Williams 1973, 148; Rudinow 1978, 338). Further, even though, like beliefs, emotions may not *normally* be under an agent's direct voluntary control, she can influence them through various strategies – think of taking anger management classes or cheering herself up by watching a comedy show. Additionally, strong emotions can be our primary responses to kindness, injustice, or tragedy. But an individual may come to realize that her emotion is disproportional, ill fitting, or that it does not serve her. In that case she can try to reason with herself, offering arguments to temper or change her emotion.[4] Cognitive behavioural therapy can arguably be said to use this kind of strategy, as can many forms of stoicism. Alternatively, a more indirect approach to changing one's emotions may be taken, including using emotional regulation strategies like situation management, attentional deployment, cognitive reappraisal, and response modulation (Gross 2015). Since we generally do have this kind of control over our emotions, we think it is legitimate to speak of manipulation when that control is thwarted or undermined, or when others exert similar control over our affective states.

Second, while most of the examples that we discuss come from social media – Facebook, Twitter, and so on – the points that we make apply to any and all online platforms/spaces that are governed by or include algorithmic recommender systems (e.g., Spotify, Netflix, Amazon).[5]

2 Forms of affective manipulation

Given the myriad potential types of online interaction, it is no surprise that online affective manipulation comes in an equally dizzying number of different forms. In this section, we aim to pull apart some of these forms, to better clarify their morally problematic aspects.

To get a grip here, it is helpful to consider four key questions. The first concerns the nature of the manipulator: is it an individual (or a group of individuals) or an algorithmic system? For example, compare the Facebook study with the case of the Twitter algorithm. The former involves a group of individuals – Kramer's team of data scientists and psychologists – doing the affective manipulation while the latter is an adaptive algorithmic system.

Of course, this is an overly simplistic description of these situations. Workers at Twitter could, in theory, modify the algorithm to change its behaviour. In this way, they are (intuitively) partially responsible for its outputs and so could be counted as (partial) manipulators too. Conversely, in the Facebook case, Kramer's team did their manipulation via slightly modifying Facebook's algorithm, such that the algorithm was a means for their affective manipulation.

To accommodate this complication, it is useful to distinguish between *active* and *passive/indirect* affective manipulators, where the active manipulators are the primary actors, and the passive manipulators are either individuals/algorithms used as means to the primary manipulators' end (e.g., the algorithm in the Facebook emotional contagion case) or are potential regulators who do not stop the active manipulator (e.g., Twitter workers in the Twitter case). Keeping track of the role an individual/algorithmic system plays in a case of manipulation might prove helpful in distributing responsibility (compare how legal systems tend to be less harsh on those who were inadvertent accessories to a crime than those who are directly/actively responsible).

The second question is whether the affective manipulation was done *intentionally* or *unintentionally*. Or, to phrase the question more explicitly, were certain actions undertaken with the intention of drawing out specific kinds of affective responses from the audience? For example, in the Facebook case, the affective manipulation was intentional – the team modified users' Facebook feeds with the express purpose of bringing out certain affective reactions. Similarly, hate click articles are clear instances of intentional affective manipulation: they are written so as to make the audience affectively respond and, in so doing, engage with a particular web page (and thereby see some ads!). In contrast, the affective manipulation in the Twitter case is, at least according to @jack, an unintentional consequence of how the algorithm was designed. The algorithm is designed to promote tweets that prompt user engagement; the fact that user engagement goes up when users have certain affective responses to tweets is an unintended consequence.[6]

The third question concerns what *mechanism* was used to perform the affective manipulation. This roughly divides between *top-down* mechanisms, where the very framework or architecture of the relevant online space is presented or designed in such a way as to promote specific reactions/thoughts, and *bottom-up* mechanisms, where the content presented (and potentially the online space as well) is tailored to particular individuals over time (Alfano, Carter, and Cheong 2018). For example, a clear case of a top-down mechanism is when a website like Breibart News files articles about Hunter Biden under certain drop-down menu keywords – for example, "Democrat Party Corruption" – thereby using the architecture of their web page to prompt certain affective reactions. In contrast, a case of bottom-up manipulation is the so-called alt-right entryist content on platforms, like YouTube, which involve gaming algorithms in order to appear in a large number of people's recommended video suggestions.[7]

Notably, the aforementioned trio of questions arises whenever we consider online manipulation. But the fourth and final question is specific to *affective* manipulation and concerns the manipulation's "aim". Specifically, is the manipulation aimed at bringing about an affective state in the target or some behaviour that is an expected product of an affective state?[8] That is, is it just the affective state itself that is the goal, or is that affective state simply a step on the road towards a larger end? To label them, let's call the former *affect directed*, since it is all about the affective state itself, and the latter *effect directed*, since it is about the (expected) effects of inducing the affective state. Arguably, many internet trolls engage in *affect directed* manipulation – pissing people off is their sole aim. In contrast, the use of Facebook by Myanmar military personnel to stir up anger against the Rohingya is clearly *effect directed* manipulation, as the (eventual) aim of the manipulation was to build public support for the ongoing genocide.[9]

It is clear that certain combinations of answers to the earlier questions are fairly natural. For example, the aforementioned Facebook study involved top-down, intentional, *affect directed*[10] manipulation undertaken by a specific group of individuals (Kramer's team). Meanwhile, when Spotify suggests a series of soft, slow songs after you've just finished listening to the songs in playlists like *Top Most Sad Songs Ever* and *sad songs to cry to*, the manipulation is bottom-up, unintentional, *affect directed* manipulation by Spotify's recommendation algorithm. However, there can be mixed cases: for example, individuals might exploit their knowledge of relevant bottom-up algorithmic systems to bring about certain emotional responses (this looks like a natural way to categorize the alt-right entryist content) or design a platform that unintentionally promotes certain ways of affectively responding to the world (e.g., Twitter's promoting of tweets that generate more responses might indirectly push us towards producing more hot takes rather than measured responses).

Regardless, it is clear that answering these four questions helps specify the who, how, and why of instances of online affective manipulation. In turn, this clarifies what is (or is not) morally problematic about said case, as well as indicates how one might potentially respond to it. For example, if we know that there are some individuals responsible, then we have a (fairly) clear indication of who bears the blameworthiness burden for an instance of online affective manipulation. However, if the manipulator turns out to be an algorithmic system, then, due to familiar responsibility gap problems, it isn't entirely obvious who, if anyone, is blameworthy.[11] Relatedly, if the manipulation is being done via a bottom-up mechanism, then we can reassess the underlying algorithm to try and avoid the same issue emerging. Meanwhile, if it is top-down, then we can rethink the framework or architecture of the relevant online space. Finally, if the aim is to prompt action via an affective state, then we might combat the manipulation by offering information about the badness of the relevant action.

Of course, there is an elephant in the room here: there are many ways to influence emotions online (or otherwise) that most of us would agree are completely innocent, like confessing to being sad to a loved one via WhatsApp, tweeting about your happiness at having a paper accepted, or even posting a message of sympathy on a friend's Facebook wall after the death of a family member. These can be seen to be like cases of online affective manipulation in the sense that they involve actors doing things online which involve modifying another's affective state. But what sets these intuitively unproblematic cases that prompt affective responses apart from the intuitively problematic instances of online affective manipulation described above? To make sense of this distinction, it is helpful to turn to a different question: what exactly is *manipulative* about online affective manipulation?

3 What is *manipulative* about online affective manipulation?

Manipulation is typically regarded as a way to influence someone that is neither rational persuasion nor coercion (Rudinow 1978). But that condition in itself cannot be sufficient, as it would imply that any form of emotional appeal, rhetoric, or framing is automatically manipulative. So, various authors have introduced further conditions.[12] It is beyond the scope of this chapter to settle this debate. Instead of choosing and defending some particular position, we will briefly review three specific options to see what they can tell us about our examples of online affective manipulation. Our key concern here will be trying to ascertain (i) what, if anything, on these accounts, makes our example cases manipulative, and (ii) what this means for how we should morally evaluate them.

The first account, from Rudinow (1978), argues that the influence used must involve deception, pressure, or playing on the target's weaknesses. In this way, it identifies manipulation as a pluriform phenomenon. Often, it works in virtue of its covertness, but sometimes, it works precisely because of its transparency (Rudinow 1978, 340–41). The Facebook emotional contagion experiment is an example of the former: it got people to feel specific emotions because their feeds were altered to trigger precisely those emotions. The unwitting participants were deceived about what was going on in their online social environment (in that some messages they otherwise would have seen were filtered out) and also about being involved in an experiment. We can see how a personalization algorithm recommending only sad songs could work very similarly when a user is unaware of how it works. Even though such algorithms need not be designed for this purpose, they do tend to one-sidedness and feedback loops. On the other hand, Rudinow's account also allows for instances of manipulation that are much more blatant – for example, think of online public shaming or peer pressuring. Even just the way in which social media make emotions such a central feature of online communication can make users feel pressured to conform and express the same emotion that is already dominant in their social network. Finally, a play on the target's weaknesses can clearly be seen in the example of the hate click articles, which work precisely because they tempt users to engage by invoking anger and indignation. Often, these articles' headlines will be more or less misleading, thus also ticking the box for deception.

On the second account, influence is manipulative when it does not sufficiently engage the target's reflective and deliberative capacities. Instead, it operates on other, more "peripheral routes" of decision-making – for example, affective states (Fischer, in this volume). As the main proponent of this account, Cass Sunstein (2016), emphasizes, engagement with reflective and deliberative capacities is not always necessary. Some forms of affective influence, like cheering someone up by smiling, are not (at least not normally) manipulative. This account neatly distinguishes between the innocent forms of online affective influence listed at the end of the previous section and the problematic forms of online affective manipulation discussed before. It may also explain why we think of Instagram asking users "Are you sure you want to post this?" when they are about to post a message containing hateful language as an unproblematic form of affective influence: it prompts the user to reflect and possibly reconsider a problematic action. However, this account is also question-begging in that it still requires a criterion to determine what level of reflection and deliberation is necessary in a given situation. Presumably, reflection is called for when an emotion or an action is likely to result in significant harm, either to the target or to others. Yet, this broadly consequentialist line can't be the whole story, as it seems problematic to cheer someone up by misleading them, even if we assume that it won't harm anyone. Intuitively, there is something off about an unfounded joyfulness or a false happiness, especially if we knowingly instill it in others. This either pushes us back to the first account, or alternatively, to the third.

The third account identifies manipulation with the attempt to get someone's beliefs, desires, or emotions to fall short of the ideals that, in the view of the influencer, govern beliefs, desires, and emotions (Noggle 1996). Here, the problem is not so much with the means not being rational enough but about the intended outcome not being ideal. The advantage of this account is that it deals explicitly with emotions, not just as a vehicle for, but also as the aimed target of, manipulation. What the relevant ideals for emotions are will be determined by the psychology of the influencer, but likely candidates would be aptness in quality and in degree. If there is grave political injustice being done in my country, my appropriate emotion would be indignation, not joy. If the injustice was only minor and short-lived, and the responsible parties have since atoned and taken steps to prevent it in the future, then my enduring outrage would be disproportionate and not ideal. Having an unfitting emotion, or having the right one excessively or possibly not enough, would not be ideal. If someone is sad because the Facebook team filtered out all positive messages from their feed, their emotion may be appropriate to the information they have, but it still falls short of being ideal given that it constitutes a response to a misrepresentation. A similar thing applies to being angry about what has unjustly been labelled "Democratic Party Corruption" by Breitbart *because* of this framing.

These are three different but not wholly unrelated ways to think of certain forms of online affective influences as manipulative: because they involve deception, pressure, or playing on weaknesses, because they do not sufficiently engage reflective and deliberative capacities, or because they make the target fall short of ideals for practical reasoning. Each of these accounts also gives an indication for why we might think of manipulation – including the kind of manipulation involved in our opening cases – as at least prima facie morally wrong. On Rudinow's (1978) account, there is always something illegitimate about the influence manipulation exerts on the target: it works because the target is misled, because they were put under such pressure that they were not properly free to choose, or because their vulnerability was exploited. These means are all, albeit in different ways, undermining of autonomous agency or at least a violation of recognition respect. In Sunstein's and Noggle's accounts, the wrongness of manipulation is stipulated in its definition: manipulation does not engage deliberation where it should, or it prohibits someone from ideal practical reasoning. There is then automatically a moral wrong involved. This is not to say that there cannot be extenuating circumstances permitting the use of manipulative means, for instance, if this is in the target's own best interest or if it can prevent some graver moral harm. We will not go into the exact conditions under which manipulation is warranted, but we do want to hold open the theoretical possibility that there are such conditions.

Finally, to come back to the issue highlighted at the end of Section 1, these accounts help explain why other intuitively innocuous forms of online affective influence, like sharing our sadness, happiness, or sympathy through social media, are, at least under normal circumstances, not manipulative. They are neither deceptive nor forms of pressure, nor do they prey on weaknesses. Engagement of deliberative capacities is not normally necessary in these instances, and they do not inhibit the receiver from attaining any ideals of practical reasoning. Thus, these accounts give us a means to demarcate genuinely problematic affective manipulation from unproblematic affective engagement.

To conclude, instances of online affective influence can be considered manipulative either because they involve deception, pressure, or playing on weaknesses because they do not sufficiently engage deliberative and reflective capacities or because they make the target fall short of ideals for practical reasoning. Manipulation is prima facie morally problematic because by definition it violates certain norms of agency and interaction, be they related to practical reasoning and deliberation or recognition respect in the means we use to influence others; these considerations apply also to online affective manipulation.

4 Online affective manipulation as affective injustice

Characterizing certain ways in which people's emotions are influenced online as manipulation may be enough to persuade many that online affective manipulation is wrong. However, we want to highlight the social and structural problems that arise from this special form of affective manipulation. Manipulation can take place between social equals who may be involved in manipulating each other. By itself then, analyzing online affective influence in terms of manipulation does not highlight the extent of this problem nor the fact that this manipulation is one-directional. We will now extend our analysis by arguing that online affective influence often constitutes a form of affective *injustice* – that is, an injustice someone faces as an affective being.

Srinivasan (2018) uses "affective injustice" to describe the situation victims of oppression face when they are put in contexts in which anger would be a *fitting* but *counterproductive* response. In other words, this anger accurately evaluates the situation as one in which they have been wronged but where expressing this anger would damage their attempts to combat their oppression.[13] For example, when black Americans express their anger at racism it may lead members of the white majority to lose sympathy with their cause and for racist oppression to worsen as a result. In cases like this, victims of oppression face two kinds of injustice: the original injustice of oppression and the second injustice of having to manage the normative and psychological conflict between their justified emotional response and their need to combat the original injustice.[14] It is this second, emotional form of injustice that Srinivasan calls an *affective* injustice (Srinivasan 2018, 135).

Whitney (2018) also talks about affective injustice, though her focus is different from Srinivasan's. Specifically, Whitney is interested in cases where the affective responses of oppressed people do not receive uptake. For example, a man who responds to a woman's anger by classing it as hysterical takes this anger as a sign that there is something wrong with the woman rather than something wrong with the world.[15] Based on cases like these, Whitney outlines three different forms of affective injustice. *Affective marginalization* occurs when oppressed people's affective responses are pushed to the margins of the shared world of affect circulation. *Affective exploitation* occurs when powerful people extract the affective labour from those with less power in an exploitative way. *Affective violence* is a uniquely affective form of violence that Whitney argues results from the combination of the two other forms of injustice. Building on this, Archer and Matheson (2022) argue that enforced participation in certain commemorative practices can constitute a form of affective injustice they call *emotional imperialism*, which occurs when a powerful group imposes its emotional norms and standards on another less powerful group.

We understand these different accounts as outlining different forms of affective injustice, rather than competing accounts of what affective injustice is. Following Archer and Matheson (2022), we understand the general concept of affective injustice as an injustice done to someone "specifically in their capacity as an affective being".[16]

The various specific forms of affective injustice outlined earlier offer important insights into the different ways in which people may be wronged as affective beings and may provide some tools for analyzing what is wrong with the forms of online affective manipulation we have been considering. For example, the way Twitter and hate click articles promote engagement through eliciting strong emotional responses may be viewed as a form of affective exploitation. Similarly, eliciting anger in this way about situations people may be powerless to change may lead to an increase in apt but counterproductive anger. However, we wish to focus on a novel form of affective injustice that is also present in these examples.

These cases are examples of an as yet unrecognized form of affective injustice, *affective powerlessness*. In the examples of online affective manipulation we have considered, someone (or, in cases where an algorithmic system is the active manipulator, some*thing*) is wielding a great deal of power over the emotions of users. To see why this should be considered a form of affective injustice, consider Young's (1990) analysis of powerlessness as one of five faces of oppression. The powerless, according to Young (1990, 56), are "those over whom power is exercised without their exercising it; the powerless are situated so that they must take orders and rarely have the right to give them". For example, in a typical workplace in an advanced capitalist country, the majority of workers have little power to influence the decisions that are made about their work. They must follow the order of others rather than making their own decisions concerning their work. Of course, some in such a workplace both follow other people's decisions *and* impose their own decisions on other people, such as line managers and supervisors. These workers are not powerless. Rather, the powerless are those who must follow orders and obey the decisions made by others without being able to make decisions that others must follow. Young (1990, 57) takes this lack of autonomy and decision-making power over one's own life to be oppressive both in itself and due to the lack of status it provides.

Using this view of powerlessness as a form of oppression, we propose that *affective powerlessness* is being subject to affective control by those with affective power. Those experiencing affective powerlessness will have their affective lives governed by the decisions or actions of others, and similarly have little to no control over other's affective experiences. Affective powerlessness arises, then, when power relations are asymmetrical. This means that although two lovers who are involved in mutually shaping each other's emotions will have a great deal of affective influence; neither lover would be in a situation of affective powerlessness. When affective powerlessness extends across significant portions of an individual's life, there is good reason to think that this constitutes a form of affective injustice.

One immediate application for this notion of affective powerlessness is that it allows us to see what specifically is wrong with the cases of online affective manipulation we are considering. Consider again the Facebook case. The key element of this study was the manipulation of the Facebook feeds of a number of test subjects by Kramer's team. This feed manipulation consists of a one-sided form of affective power: the Team made decisions that effectively controlled whether selected users would have positive or negative affective responses. In this way, we have a group of individuals – Kramer's team – actively, intentionally manipulating the emotional states of a number of Facebook users, where the users having certain states is the "end goal" of said manipulation. And, at least partially because this manipulation was done via a top-down mechanism, these users could do little to respond to or counteract this affective manipulation.

Similarly, Twitter's promotion of "salacious" tweets that are deemed likely to generate strong affective responses also involves controlling the affective responses of its users. However, in this case, we have an algorithm unintentionally manipulating users' affective states by a bottom-up generative process. In this way, Twitter users are going to be exposed to the most affectively stimulating tweets in their network – even if posted by someone they don't follow (e.g., because someone they do follow liked or responded to it). And users can do next to nothing to respond to this affective manipulation, short of simply not engaging with the platform.

A natural objection to the aforementioned is that the possibility of disengagement shows that these are not really cases of *powerlessness*. After all, users of social media have the power to decide not to use the platform and so cannot really be said to lack any kind of power here. There are, though, several responses that can be made to this objection. First, powerlessness may be something one experiences in some areas of one's life but not others. In Young's discussion of powerlessness, she focuses on people who lack power in the workplace that feeds into how they are viewed in other public settings. These people though, may still experience significant power in the private sphere of their family. Similarly, people whose emotions are subject to high levels of external control when they engage with social media may have significant degrees of affective power in other areas of their life. This, though, does not undermine the idea that they may lack power when they are using these platforms or in these online spaces.

Alternatively, one might object that, if this powerlessness influences people only when they are engaging with certain online platforms, then it does not constitute a sufficiently *significant* portion of that person's life to be considered a form of powerlessness. This reply seems unpersuasive when we consider the amount of time people spend engaging with online spaces and social media in particular. According to a report by Global Web Index (2020, 7), digital consumers spend an average of two hours and 24 minutes on social media each day. This rises to over three hours a day for 16–24 year olds in the UK, Spain, and Portugal and over four hours for Russians in the same age group Global Web Index (2020, 8). For at least some users of social media then, the time they spend engaging with social media seems significant enough to justify classing this as a form of powerlessness. Moreover, some internet users have little choice but to engage with social media. Myanmar offers an extreme example of this. Facebook entered the country in 2010, just as internet access was becoming widespread after decades of censorship, and came preloaded on the majority of available mobile phones. As Silvia Venier (2019, 233) describes it, "Facebook soon acquired a position of complete monopoly, as the only online portal supporting Burmese text and often the only internet tool available for the vast majority of people".[17] It is hard to see how cases like this shouldn't be described in terms of powerlessness.

More generally, the specific reasons to worry about powerlessness will vary depending on the details of the case. For example, in cases where some person or group is controlling users' emotions, like the Facebook case, we are likely to have a clear case of domination. According to republican conceptions of liberty, freedom is a matter of not being dominated by others. Someone is dominated, according to Lovett's (2010, 20) articulation of the view, "whenever they are dependent on a social relationship in which some other person or group wields arbitrary power over them". Republicans offer different accounts of what it takes for power to be wielded arbitrarily. According to procedural views, power is wielded arbitrarily when it is not constrained by widely known legitimate procedures (Lovett 2010). The democratic view, on the other hand, holds that power is arbitrary when those who are subject to it do not have any control over how that power is exercised (Pettit 2012). Finally, according to the well-being view, power is arbitrary when it is not used to promote the well-being or interests of those who are subject to it (Pettit 2012). The cases of online affective manipulation that we are considering are arbitrary in all three senses. The manipulation of Facebook feeds was not constrained by widely known procedures, as they did not inform users that they would be conducting such research.[18] Nor were the users involved given any control over how that power was exercised. Finally, the research was not conducted in order to promote the well-being of the users but rather to investigate whether "emotional states can be transferred to others via emotional contagion, leading people to experience the same emotions without their awareness" (Kramer, Guillory, and Hancock 2014). Importantly, concerns about domination do not only occur when this power is actually wielded – the fact that someone wields arbitrary power over another counts as domination, whether that power is wielded or not.

Similarly, when algorithms are open to being gamed and controlled, we have reason to worry that the tech-savvy will effectively gain the power to manipulate the emotions of others. For example, consider the so-called "radicalization pipeline" on YouTube: viewers of even just slightly right-leaning/conservative content are, via the YouTube recommendation algorithm, presented with content that becomes ever more extreme, subtly driving them to adopt radically right-wing, hate-filled political stances (see, e.g., Ribeiro et al. 2020 and Munger and Phillips 2020 for discussion of the pipeline). This is an example of affective manipulation because a key part of this radicalization is an effort to try and trigger strong negative affective reactions toward certain individuals or groups. It is achieved by a careful curation/tagging of videos in order to (equally carefully) manipulate the recommendation algorithm into recommending the video to more people. In this way, as Kaiser and Rauchfleisch (2018) put it, a user is "only one or two clicks away from extremely far-right channels, conspiracy theories, and radicalizing content". And, importantly, *it doesn't even matter if the user watches the video*; often, simply being exposed to a video title or preview image is enough to prompt at least the beginnings of certain affective responses. So, by gaming the algorithm, alt-righters can (and it seems do)[19] subtly manipulate the affective states of YouTube users.

The reasons to worry are a little different when these affective responses are being controlled only by an algorithm. Here, there is no person or group who has the ability to control what users feel (though they might have some indirect control), so it does not fit clearly in republican accounts of domination. Nevertheless, we have reason to worry about how these algorithms may be reflecting and reinforcing existing forms of oppression. For example, Noble (2018) argues that the algorithms driving major internet search engines replicate and help reinforce existing forms of racist and sexist oppression. If she is correct, then there is good reason to think that online affective manipulation governed by algorithms may have a similar impact.

5 Conclusion

In the chapter, we have considered a number of points about online affective manipulation in order to clarify the nature of this phenomenon. More specifically, we began by pulling apart some of the different forms that online affective manipulation can take. We tried to distinguish these forms by focusing on four questions:

i. Who (or what) is doing the manipulation?
ii. Is the manipulation intentional or unintentional?
iii. What mechanism is used to do the manipulating?
iv. Is the manipulation *affect* or *effect directed*?

How we answer these questions helps clarify the particular nature of the affective manipulation in specifics cases. We then turned to the broader question of what is *manipulative* about online affective manipulation. Here, we argued that instances of online affective influence can be considered manipulative either because they involve deception, pressure, or playing on weaknesses, because they do not sufficiently engage deliberative and reflective capacities, or because they make the target fall short of ideals for practical reasoning. Further, online manipulation is prima facie morally problematic because by definition it violates certain norms of agency and interaction. By itself though, analyzing online affective influence in terms of manipulation does not highlight the extent of this problem nor the fact that this manipulation is one-directional. In the final section, we extended our analysis by sketching a new form of affective injustice, *affective powerlessness*. We argued that many cases of online affective manipulation are instances of it because they feature a kind of affective domination over users by someone (or something, in cases where algorithmic systems are the active manipulator).

The upshot of all this is that we now have a better grip on the nature of online affective manipulation, as well as some tools to help us understand when and why it is morally problematic. Of course, this naturally leads to a number of questions. Perhaps, the most pressing two are, first, what can be done to address existing cases of online affective manipulation? And, second, how can we design future online platforms/systems so as to avoid supporting online affective manipulation? These we hope to address in future discussion.

Notes

1. Here, we use "emotion" and "affect" interchangeably, though we recognize that some in the literature treat the latter as a broader category involving other feelings (e.g., moods, sensations, etc.).
2. Jack Dorsey, CEO of Twitter, as quoted in Jackson and Ibekwe (2020).
3. Hate clicks are so widespread that there is arguably a growing backlash against sites that use them. Still, they regularly generate a brief but huge spike in viewers. For example, Jonelle La Foucade's 2016 article, "Beyonce is Overrated AF" on *The Edit* generated "more views and more clicks than any [other *The Edit*] articles before" (La Foucade, quoted in Way 2019) – so many that, as thousands of angry comments (and hence clicks!) came in, editors decided to pull down the article a day after it was published.
4. In some cases where the initial affective response is not fitting, mere acknowledgment of this lack of fit may succeed in resolving the mismatch.
5. In fact, one can easily extend our claims to *all* online platforms/spaces by simply dropping talk of algorithmic manipulators and bottom-up mechanisms.
6. This was arguably a foreseeable (and obvious) consequence of the algorithm's design, such that Twitter employees could (and probably should) have intervened as it developed. Still, we will here be charitable to @jack and grant that they were unaware of this potentially problematic outcome.
7. For more, see, for example, Institute for Research and Education on Human Rights (2020).
8. For present purposes, "behaviour" can be understood very broadly, so as to include actions, forming beliefs, or even being in subsequent affective states. So, for example, manipulating in order to prompt anger in a target so that they later

feel guilty about this anger could be seen as an instance of a manipulation aimed at some "behaviour".

9. For more, see Mozur (2018).
10. Notably, the study actually looked at produced *e*ffects (i.e., posting actions), but these were of course merely an indicator of the relevant affective states, which is what Kramer et al. were really interested in.
11. For more on the idea of a responsibility gap, see, for example, Matthias (2004) and Sparrow (2007).
12. For a more in-depth discussion of these various positions, see Jongepier and Klenk (in this volume) and Barnhill (in this volume). Here, we highlight these three accounts because their definitions of manipulation are, in our view, nicely equipped to deal with instances where emotions are the target of the manipulation and hence capable of capturing both answers to our fourth question.
13. Though, as Srinivasan accepts, there will also be cases where expressing the anger would have overall positive effects for combatting oppression.
14. For further discussion, see Archer and Mills (2019).
15. This example comes from Frye (1983).
16. For a more substantive account of what makes affective injustices unjust, see Gallegos (forthcoming).
17. Facebook's near monopoly on internet use, combined with the use of the platform to incite violence, was criticized by UN human rights investigators in 2018 (United Nations Human Rights Council 2018).
18. Indeed, this led to an editorial expression of concern, which said that "the collection of the data by Facebook may have involved practices that were not fully consistent with the principles of obtaining informed consent and allowing participants to opt out" (Verma 2014, 10779).
19. Ledwich and Zaitsev (2020) argue that "data shows that YouTube does the exact opposite of the radicalization claims". However, it is not clear whether their results are generalizable to users who are logged-in to their accounts, as arguably the whole point of the recommendation algorithm is personalizing recommendation to individual users.

6 References

Alfano, Mark, J. A. Carter, and Marc Cheong. 2018. "Technological Seduction and Self-Radicalization." *Journal of the American Philosophical Association* 4 (3): 298–322. doi:10.1017/apa.2018.27.

Archer, A., and B. Matheson. 2022. "Commemoration and Emotional Imperialism." *Journal of Applied Philosophy*. doi: https://doi.org/10.1111/japp.12428

Archer, Alfred, and G. Mills. 2019. "Anger, Affective Injustice and Emotion Regulation." *Philosophical Topics* 46 (2): 75–94.

Barnhill, Anne. 2022. "How philosophy might contribute to the practical ethics of online manipulation."In *The Philosophy of Online Manipulation*, edited by Jongepier, F. and Klenk, M., 49–71. New York: Routledge.

Fischer, Alexander. 2022. "Manipulation and the Affective Realm of Social Media." In *The Philosophy of Online Manipulation*, edited by Jongepier, F. and Klenk, M., 327–352. New York: Routledge.

Frye, M. 1983. "A Note on Anger." In *The Politics of Reality: Essays in Feminist Theory*, edited by M. Frye, 84–94. Berkeley, CA: Crossing Press.

Gallegos, F. "Affective Injustice and Fundamental Affective Goods." *Journal of Social Philosophy* (forthcoming).

Global Web Index. 2020. "Social: GlobalWebIndex's Flagship Report on the Latest Trends in Social Media." www.gwi.com/reports/socia.

Gross, James J. 2015. "Emotion Regulation: Current Status and Future Prospects." *Psychological Inquiry* 26 (1): 1. doi:10.1080/1047840X.2014.940781.

Institute for Research and Education on Human Rights. 2020. "From Alt-Right To Groyper." Accessed September 03, 2021. www.irehr.org/reports/alt-right-to-groyper/.

Jackson, Lauren, and Desiree Ibekwe. 2020. "Jack Dorsey on Twitter's Mistakes." *The New York Times*, July 8. Accessed March 09, 2021. www.nytimes.com/2020/08/07/podcasts/the-daily/Jack-dorsey-twitter-trump.html.

Jongepier, Fleur, and Michael Klenk. 2022a. "Manipulation Online: Charting the field." In *The Philosophy of Online Manipulation*, edited by Jongepier, F. and Klenk, M., 15–48. New York: Routledge.

Jongepier, Fleur, and Michael Klenk, eds. 2022b. *The Philosophy of Online Manipulation*. New York, NY: Routledge.

Kaiser, J., and A. Rauchfleisch. 2018. "Unite the Right? How YouTube's Recommendation Algorithm Connects The U.S. Far-right." https://medium.com/@MediaManipulation/unite-the-righthow-youtubes-recommendation-algorithm-connects-the-u-s-far-right-9f1387ccfabd.

Kramer, A. D. I., J. E. Guillory, and Jeff Hancock. 2014. "Experimental Evidence of Massive-scale Emotional Contagion Through Social Networks." *Proceedings of the National Academy of Sciences* 111: 8788–90.

Ledwich, Mark, and Anna Zaitsev. 2020. "Algorithmic Extremism: Examining YouTube's Rabbit Hole of Radicalization." *FirstMonday*. doi:10.5210/fm.v25i3.10419.

Lovett, Frank. 2010. *A General Theory of Domination and Justice*. Oxford: Oxford University Press.

Matthias, Andreas. 2004. "The Responsibility Gap: Ascribing Responsibility for the Actions of Learning Automata." *Ethics and Information Technology* (3): 175–84.

Mozur, Paul. 2018. "A Genocide Incited on Facebook, With Posts From Myanmar's Military." *The New York Times*, October 15. Accessed August 23, 2021. www.nytimes.com/2018/10/15/technology/myanmar-facebook-genocide.html.

Munger, Kevin, and Joseph Phillips. 2020. "Right-Wing YouTube: A Supply and Demand Perspective." *The International Journal of Press/Politics*, 194016122096476. doi:10.1177/1940161220964767.

Noble, Safiya Umoja. 2018. *Algorithms of Oppression: How Search Engines Reinforce Racism*. New York: NYU Press.

Noggle, Robert. 1996. "Manipulative Actions: A Conceptual and Moral Analysis." *American Philosophical Quarterly* 33 (1): 43–55.

Pettit, Philip. 2012. *On the People's Terms: A Republican Theory and Model of Democracy*. Cambridge: Cambridge University Press.

Ribeiro, Manoel H., Raphael Ottoni, Robert West, Virgílio A. F. Almeida, and Wagner Meira. 2020. "Auditing Radicalization Pathways on YouTube." In *Proceedings of the 2020 Conference on Fairness, Accountability, and Transparency*, edited by Mireille Hildebrandt, Carlos Castillo, Elisa Celis, Salvatore Ruggieri, Linnet Taylor, and Gabriela Zanfir-Fortuna, 131–41. New York, NY, USA: ACM Press.

Rudinow, Joel. 1978. "Manipulation." *Ethics* 88 (4): 338–47. doi:10.1086/292086.

Sparrow, Robert. 2007. "Killer Robots." *Journal of Applied Philosophy* 24 (1): 62–77.

Srinivasan, A. 2018. "The Aptness of Anger." *Journal of Political Philosophy* 26 (2): 123–44.

Sunstein, Cass R. 2016. "Fifty Shades of Manipulation." *Journal of Marketing Behavior* 1 (3–4): 214–44. doi:10.1561/107.00000014.

United Nations Human Rights Council. 2018. "Report of the Independent International Fact-finding Mission on Myanmar." www.ohchr.org/Documents/HRBodies/HRCouncil/FFM-Myanmar/A_HRC_39_64.pd.

Venier, S. 2019. "The Role of Facebook in the Persecution of the Rohingya Minority in Myanmar: Issues of Accountability Under International Law." *The Italian Yearbook of International Law Online* 28 (1): 231–48.

Verma, I. M. 2014. "Editorial Expression of Concern: Experimental Evidence of Massivescale Emotional Contagion Through Social Networks." *Proceedings of the National Academy of Sciences* 111 (29): 10779. doi:10.1073/pnas.1412469111.

Way, Katie. 2019. "Hate Clicks Are the New Clickbait." https://contently.com/2019/02/20/hate-clicks/.

Whitney, Shiloh. 2018. "Affective Intentionality and Affective Injustice: Merleau-Ponty and Fanon on the Body Schema as a Theory of Affect." *Southern Journal of Philosophy* 56 (4): 488–515.

Williams, Bernard. 1973. *Problems of the Self: Philosophical Papers, 1956–1972.* Cambridge: Cambridge University Press.

Young, Iris M. 1990. *Justice and the Politics of Difference.* Princeton, NJ: Princeton University Press. www.loc.gov/catdir/description/prin021/90036988.html.

17 Manipulation and the Affective Realm of Social Media

Alexander Fischer

1 Introduction: The #StopTheSteal Manipulation

When the former republican US President Donald J. Trump used Twitter (especially in the last months of the year 2020) to vent accusations of an election being stolen by the Democrats, the goal was not to present rationally accessible proof concerning widespread voter fraud or dysfunctional voting machines. Instead, he used his favorite social media channel as a cogwheel in a broad strategy[1] to create a destructive fictitious pseudo-environment[2] interweaved with controversial, even outrageous accusations to stir up people's affectivity in order to spread mistrust in the democratic electoral process of the Unites States and get supporters and "believers" moving in opposition to what was supposedly happening. This ultimately culminated in the violent riot on Capitol Hill in Washington, D.C. on January 6, 2021.[3] Furthermore, even a few days *before* the lost election in November 2020, the Trump campaign reached out to supporters via email and various social media channels with what became known as the "#StopTheSteal" campaign. This campaign was supposedly curated by the well-known, infamous strategic communication specialist Roger Stone in 2016 to be used whenever elections might not turn out victorious. Regarding the presidential election in 2020 said campaign was revitalized (Atlantic Council's DRFLab 2021; Spring 2020). As a fundraising effort (or rather scam) and, yet another tactical cogwheel, part of a thoroughly planned multifaceted manipulative strategy, it swept across inboxes, flooded Facebook and Twitter timelines spreading not only misinformation but also using destructive language and imagery to, then again, stir up people's affectivity and muster up motivation to act on these affects. Multiple agents joined in supporting the #StopTheSteal campaign, putting it in a broad frame, thus making it visible widely. On Facebook, a group with the same name ("Stop the Steal") and effort formed which was banned for the same misinformation and attempt to emotionalize quickly after its emergence. Furthermore, numerous Trump allies appeared on national TV trying to support the stolen-election-narrative that was pushed online all the more. Besides the carefully planned roll-out of the campaign, truly bizarre and dilettantish moments could be seen as well, not only showing us tools for the attempt to manipulate in a blatant way as they were handled poorly

DOI: 10.4324/9781003205425-20

in some cases but also how badly Trump supporters wanted this attempt to manipulate to be effective. One example: Trump's private lawyer Rudy Giuliani holding a rushed press conference erroneously in front of a landscaping business – instead of a famous hotel chain – with the name *Four Seasons Total Landscaping*. Then again, venting the same baseless claims of a stolen election while trying to suggest a (non-existent) credibility usually supported by means such as context, authority and the right timing to curate certain content. Trump's allies, in other words, tried to use an arsenal of symbols at the right time, at the right place, and by authorities that act personally worried, stating lies, formulating grave threats and painting a dark picture of the future to manufacture a public opinion and, once again, stir up people's affectivity. They provided a visually powerful blueprint of what to feel (and consequently also think) about an event no one has actually experienced for real, as "[t]he only feeling that anyone can have about an event he does not experience is the feeling aroused by his mental image of that event"(Lippmann 2008, 13).

For this extensive multi-channel effort to misinform and, amalgamated with that misinformation, *manipulate*, we can even suppose that Trump's strategists and willing allies had specific affective states in mind that they planned to stir up, such as (a) the feelings of frustration or indignation after the election night, (b) acute or persistent emotions such as anger and fear, and (c) moods of mistrust in democratic voting and the political system of democracy on the whole.[4] Of course, it was paramount that Trump seemed to be the one able to slay the monster they build up before. To target these affective states via controversial and destructive messaging is especially effective to bundle *attention* – the main 'online currency' – and motivate individuals to act in a possibly destructive manner intended by a manipulator (think again of January 6, 2021). Furthermore, the systematic spreading of false information and the constant attacks on and disdain for reliable truth sources created (and still creates, as we saw yet again during the COVID-19 pandemic) a bedrock of disorientation – and thus an environment where affective states become primary tools for (self-)orientation and steering individuals.

So, the whole campaign was not at all about sound arguments and it was not based on tenable facts but instead it was about inciting all of the aforementioned affective states as a foundation to create, reassure and motivate believers. This ultimately gave ground for the suspension of Trump's mainstream social media accounts and the attempt for a second impeachment just shortly after the insurrection of the US Capitol. Once again, Trump impressively harvested the fruits of frustration, indignation, fear, anger, and mistrust in the political system mixed with faith in him in an at least partly disoriented, riven society. What we saw by Trump and his allies was a systematic attempt to *manipulate* the American people and undermine an essential democratic process by orchestrating false information and unsubstantiated accusations to muster up and connect to affective states that help motivate individuals to act in a certain direction.[5] In this attempt to manipulate, social media played a vital role as a digital realm of affectivity where a low threshold for publication makes it possible to anonymously reach a myriad of individuals, where their affectivity can be targeted in specifically tailored content on the grounds of big data, where our own affects are tools for orientation in the mass of stimuli, where algorithms favor controversial and destructive content, and where reviewing the substance of this content is, at times, tricky.

As social networks[6] possess "phenomenological effects significant to the actions we take and the decisions we make . . . [and] are of no small consequence, raising questions about how and under what circumstances we are shaped by social media" (Nelson 2018, 3), my argument in this chapter will be the following: as digital realms of affectivity in which rational persuasion often plays a secondary role for many users, social networks offer potentials for manipulating users, modulating their feeling, thinking, and acting, which result from its very design. Political, economic, and also private agents try to use these potentials in *not necessarily always but often* questionable ways. Trump and his campaign can function as a crass example for a questionable use.[7] From this starting point I, first, want to focus on the phenomenon of manipulation which is an important part of how our feeling, thinking, and acting comes about. I will offer a sketch of the phenomenon from the perspective of action theory and provide a conceptualization of what I see as the primary mechanism of manipulation which consists of rendering certain ends as pleasurable/unpleasurable, motivating us to act in a certain way without a coercing us but also without primarily using our capacities to rationally deliberate. Still, our rationality plays an important role in the context of manipulation as 1) a secondary rationalization of what and why we are feeling something and 2) the capacity to deduce reasons on the grounds of our affective states to act in a certain way.[8] Thus, the manipulated remains, at least in a minimal sense, free to act in this manner or not. The manipulation I am going to be concerned with are instances of manipulation of affective states *suggesting* a certain direction for our decision and acting.[9] This, in my opinion, marks the core of manipulation per se, in opposition to other types of influence like the – usually ethically esteemed – rational discourse where every decision is ideally based on rational deliberation (instead of affective states) or the – usually ethically debatable – use of coercion where there might be a lot of affectivity involved but which lacks an option to really decide freely in the end. The development of this manipulation model will be my primary focus. In a second step, I want to elaborate briefly on a assumed potentials of manipulation in social media which are based on a design of affective messaging and also on the interface's design. My hope is to offer a fundamental perspective with an account of how specifically we are manipulated and, in more general terms, how social media supports this, thus laying the groundwork for more specific case studies regarding different phenomena like big data, interaction patterns, tailored advertising, fake reviews, and influencers as new manipulative tools are being developed constantly.

In my understanding, social media is constructive of human behavior and not with an unidirectional influence but as technology that acts on us as well as we act with it (cf. Nelson 2018, 4): "technological artifacts are not neutral intermediaries but actively coshape people's being in the world; their perceptions and actions, experience and existence" (Verbeek 2011, 8). My focus here will primarily regard the interpersonal communication in the online realm whereas, for example, social bots will only be mentioned but not analyzed in depth.[10] My proposed concept of manipulation will be applicable to the intentions of communicators who use certain interface designs and algorithmic effects to manipulate. Whereas I will claim that there *is* the potential to manipulate efficiently and effectively via social media it is neither denied that there are rational discourses and the exchange of sound arguments on social media platforms, nor am I adopting the claim that every attempt to manipulate is undermining our ability to decide freely, or that manipulation is by default morally problematic. To build this argument, I will now at look at manipulation as a certain type of influence in the landscape of influences attempting to cause an agent to feel, think, and act, differentiating it from other types of influence in order to gain a neutral account of manipulation with focus on its impact on affective states, called the *Pleasurable-Ends-Model* of manipulation (PEM) (cf. Fischer 2017, Fischer/Illies 2018, Fischer 2022).

2 What Is Manipulation, and How Does It Make Us Act? Conceptualizing a Type of Influence Offline and Online

The internet as a whole and social media in particular offers an environment that targets the affective side of our agency. Thus, it provides several instruments that use our peripheral routes of decision-making and also help create an impactful pseudo-environment.[11] Consequently, now there is talk of "online manipulation" (mostly, so far, outside of the discipline of philosophy), a term that simply aims at the exercise of manipulative influence in the digital realm of the internet (cf. Abramowitz 2017; Susser, Roessler, and Nissenbaum 2019b, 3). Here, manipulative influence is intended by human agents and then performed by means of a programmed, digital architecture that aims at various aspects of our affectivity (ranging from barely tangible qualitative reactions like us being prone to be comfortable to intense affective states like frustration, fear or anger). This architecture helps to lay the groundwork for shaping a certain action. Some of these structural intricacies are often called "dark patterns" which rely heavily on our slothfulness and avidity for convenient, functional default options and shortcuts. These effects apply to the basic design of an interface and your own profile settings (Facebook handles this masterfully): it is often used here that it is very easy, even in one click, to achieve something that is beneficial for the interface provider, whereas it just seems too hard to click through the depths of the settings to check certain privacy options or get rid of a product. Obstacles that make it hard to do something the providers do not want, are – as they get on our nerves – effective and often found (like nagging questioning or laborious clicking). We also find social pressure via fake reviews or because of a large number of likes and shares, time pressure (in the economic context), or obscured ads all aiming at making us feel instead of thinking too much.

But social networks also serve as a potentially manipulative vehicle for private, political, economic, and other agents using a presorting algorithm logic that helps agents with an agenda to manipulate to feed specifically designed attention-seeking affective content into the realm of social media. Various algorithms support this as they are, at least partly, designed to engage us even more by showing supposedly relevant content. This content in turn has to be designed in a certain way, e.g. as controversial and destructive, in order to be considered by the algorithm. The rule of thumb here is that attention-seeking content has a better chance to reach many users. So, evoking an acute intense affective state like making someone angry counts for a lot as it helps engaging that someone to do something with the content (liking/disliking, commenting, sharing). This is at the center of my considerations. Before we dive deeper into this, let us try and understand how manipulation functions as a cogwheel in the workings of us being feeling, thinking, and acting agents.

In general, manipulation can be understood as an omnipresent form of influence of human agency which in its many costumes aims at shaping our affective states and with it our thinking and acting. It can be qualified as a form of influence on our capacities as acting agents alongside others. On a rough map of influences, rational persuasion and coercion are well explored forms whereas manipulation has been less well researched. This is not surprising as (a) its position seems to be somewhere in the messy in-between of these two, which can be seen as poles in a continuum of influence (cf. Rudinow 1978, 338; Beauchamp 1984; Coons and Weber 2014a; Fischer 2017, 53). These poles are not understood as a strict dichotomy. They are rather like brightness and darkness between which we can find many different shades. And (b) these many shades make the dissection of manipulation harder as things get opaque and more difficult to describe analytically the closer we get to our affectivity, the suggestive aspects of our communication and even our unconscious.[12]

Let us make the two mentioned poles clearer to gain a first understanding of manipulation *ex negativo*. In the context of coercion, an agent usually does not have the opportunity to choose between alternatives – or at least preferable alternatives – and thus to act freely in an extensive sense.[13] Coercion can even be forceful so that agents have to reckon potential personal damage which often issues basic automatic behavior patterns like fight, flight, or freeze reactions as a threatening fear of consequences is evoked. In contrast, in the case of rational persuasion agents can reach a free decision on the basis of the correct and relevant information by forming good reasons without heavy pressure but with the help of their rational capacities. While coercion in extreme cases marks the absence of free decision-making, rational persuasion is generally regarded as an ideal since it accounts for our autonomy and freedom and is not "contaminated" by the unnerving prospect of cruel consequences or factors that make free deliberation more difficult such as underhandedness or deceptive information. In general, something else is supposedly missing from rational persuasion, at least ideally conceptualized: our affectivity. The capacity to be guided by our affectivity plays an ambivalent role in the debate about the nature of free decision-making and acting, ranging from being an important part of it as our affectivity is thought of as intertwined with and a vital part of our rationality to being a grave threat to free decision-making. The truth, as it often does, might lie somewhere in between. Our affectivity seems to play an important role for a reasonable decision-making process as it shapes a meaningful perception of the world, helps us to judge, identify values, and be motivated; at the same time a decision that is worth to be called "free" also accounts to being able to prescind ourselves from what we are feeling to ultimately gain an integral decision as the foundation of our action.[14]

In contrast to an idealistic account of rational persuasion, manipulation is supposed to be a threat – at least in an everyday understanding. Here, underhandedness, deception, negative consequences, and selfish manipulators play a vital role. However, I want to oppose most of these characteristics as necessary conditions of manipulation. But there is one that does not seem to be possible to reason away: the "contamination" of our decision-making process by our affectivity, consisting of our feelings, emotions, and moods. In contrast, underhandedness, deception, and negative consequences can be seen as *amplifying* conditions even though they are not necessary ones. Thus, manipulation in my understanding neither accounts for a purely rational decision-making process nor does it force agents to do something specific like coercion does. It is not so much good reasons and the presentation of all the relevant information (or a gun to our head as a brutal form of coercion) that lead us to act in a certain way within the framework of manipulation but rather the curation of our affective states and the peripheral routes of decision-making.

In the literature on manipulation, we can find various suggestions on how exactly to understand manipulation and differentiate it from other forms of influence. I have already hinted to an understanding which nonetheless needs elaboration. In order to provide this elaboration, I will very briefly summarize the general discussion on manipulation and its different definitions. Following this, I will propose a new, integrative definition of the term "manipulation", inspired by previous philosophical attempts to define it, with regard to action theory, while trying to avoid some of the problems of earlier concepts of manipulation. In the light of all this, manipulation is understood as a type of influence where a manipulator actively leads the manipulated to choose a certain end (e.g., an action or a product), but the manipulated stays at least in a minimal sense free to choose this end or not.[15]

Manipulation consists of intentionally modulating the affective attraction of certain ends or their realization by rendering them as pleasurable/unpleasurable, thus making some options more (or even extremely, whereas others not at all) appealing to the manipulated and consequently more likely to be chosen (cf. Fischer 2017, 2018, 2020, 2022; Fischer and Illies 2018). Our affective states are actively modulated so that the evaluation of a certain end can change, resulting in often complex affective experiences which eventually will boil down to a desire or an aversion to act accordingly to the manipulator's goal (or not).

This understanding rests on the premise that an action (in contrast to mere behavior) is a realization of a pro-attitude toward an end, selected as being more fitting than alternative ends (this is basically how Aristotle understood it). A chosen purpose then leads to an action, if we conclude that an action does not conflict with other ends that we have and if there are no limiting conditions to fulfilling that end. Choice-worthy ends for our actions are then manifold: both good and bad, objective and subjective and maybe because we just like them. To gain a better understanding of why we act it is helpful to turn to Aquinas's more general application of Aristotle's practical syllogism where he distinguishes three types of choice-worthy ends: those we (1) desire *for their own sake* as ultimate ends (such as truth), those we (2) have because they are *useful* and serve, in direct or indirect ways, other ends we have (such as healthy food that makes us healthy), or we (3) desire because they are *pleasant* (such as appreciation or, more mundane, chocolate) (for a more detailed account see Fischer and Illies 2018, 35–39).[16] In the case of manipulation, our affective evaluation of an end is modulated by presenting an end as pleasurable/unpleasurable (instead of presenting something as useful or choice-worthy for its own sake). A manipulative stimulus thus is used to trigger an immediate qualitative reaction, an affective response, to a pleasurable/unpleasurable end which might create a desire or aversion to do or not do something. This desire/aversion aims at the alteration of the reality so that reality accords. (Sure, we often cannot directly act on these grounds and consequently have to cope and see where and when we can really act in accordance to our affective response.)

In order to achieve this, underhandedness, deception, negative consequences and even careless, selfish manipulators are not necessarily needed but can function as amplifying conditions. This may seem like an atypical definition, as it partly leads away from our everyday understanding of the phenomenon, which usually degrades "manipulation" to a fighting word (even though the term was long used neutrally but, in a neo-Marxist tradition, manipulation was tinted as unethical), thus often blocking a clear view of what is happening in detail as we are so convinced it is something devilish.[17] So let us take a quick look at the steps that lead me to this definition and thereby summarize existing accounts (for a detailed discussion, see Fischer 2017, 26–78).

3 The Different "Schools" of How to Understand Manipulation

The first perspective on manipulation emphasizes the character of manipulation as an intentionally underhand influence that unfolds beyond our consciousness and is therefore almost not at all controllable for the manipulated (cf. Baron 2003; van Dijk 1998; Goodin 1980; Ware 1981; Noggle 2018). Daniel Susser et al. also identify the essential feature of online manipulation "as the use of information technology to covertly influence another person's decision-making, by targeting and exploiting decision-making vulnerabilities" (Susser, Roessler, and Nissenbaum 2019a, 6; for an argument against this see: Klenk 2021). This applies to many dark patterns where something is, for example, secretly placed in our shopping cart, where tricky questions lead us to answers that we did not intend, where additional costs remain hidden or where advertising is disguised so that we click on it because we simply think it is something else than advertising. However, highlighting "covertly" does not help much to differentiate manipulation. Almost everything that happens in a hidden manner, trickily or secretively might then be understood as manipulative – such as lies, cheating (e.g., in a game), or even magic tricks. While there are many examples in which underhandedness is part of an attempt to manipulate and might be considered an amplifying condition, it is not a necessary or sufficient feature of manipulation. This can be seen not only in the personal context (e.g., when relatives blatantly induce guilt) but also in the affective realm of social media: it is widely known that advertising tries to grab us by our affectivity and that this is done in a specifically personalized manner or that right-wing trolls want to stir up an acute emotion like anger (often as a part of a long-term strategy). It still works.

Closely related to this notion of manipulation as a hidden and secretive tactic to influence is the understanding of manipulation as an encroachment of an individual's perception of reality, in other words: as a form of deception or trickery (cf. Scanlon 1998, 298; Noggle 1996, 44; Cave 2007). Very often the terms "deception" and "manipulation" are even used synonymously.[18] For the specific online context, Susser, Roessler, and Nissenbaum (2019b) refer to the weaknesses of agents in the decision-making process which are exploited, for example, the difficulty of being able to directly falsify every piece of deceptive information. This association with deception explains the often-made (and not false) affiliation of "fake news" with manipulation.[19] But also dark patterns can be associated with deception, for example, involuntarily sharing more information than wanted without knowing it (ironically called "privacy Zuckering" – aiming at Facebook's CEO Zuckerberg). However, these attempts to influence an individual in a certain direction are not necessarily manipulative as they are simply obscure paths and/or false information, which represent a form of deceptive rationality but do not always aim at influencing our affective states (but *can* be used for that as #StopTheSteal illustrates). This characterization of manipulation also seems to be over-inclusive since it, for example, includes any form of marketing as manipulation that goes beyond the presentation of factual statements about the product offered.[20] Emphasizing deception as inherently manipulative also narrows our vision of manipulation down to the presentation of incorrect information and false reasons. This, again, can be seen as an amplifying condition of manipulation by specifically using false information to affect and to arrange the perception of reality of an agent. That helps us to understand how editorially curated and preselected content via algorithms can be part of an elaborate manipulation scheme. Nevertheless, deception does not have to be understood either as a necessary or as a sufficient condition of manipulation because it is also quite possible to use actual facts to manipulate. An adequate concept of manipulation should consider both cases. Those where underhandedness and deception play a vital role and those where they do not.

Another characterization of manipulation associates it with the manipulator's pursuit of egoistic purposes which yield negative consequences for the manipulated. Here, two components are addressed: first, that manipulation is tied to corrupt, selfish characters, who often carelessly use manipulation as a means for a clearly selfish end; second, that they pursue ends that are useful and pleasurable to *them* and harm the manipulated (Green and Pawlak 1983, 35–37). Against the background of egoism Marcia Baron describes manipulation as a condemnable form of harmful and selfish behavior, even as a "vice" involving "arrogance" (Baron 2003, 37, 49). However, this also seems to distort the perspective on manipulation (just think of Shakespeare's Iago). Ultimately, there is no doubt that there are many cases where manipulation involves careless, harmful and/or selfish intentions on the part of a manipulator and where it consequently does damage to the manipulated. However, at the same time it is neither reasonable to claim that any kind of manipulation is careless and harmful, nor that it serves only negative purposes of a selfish manipulator. In general, the same mechanism that leads people to bad actions can also lead them to good actions (Fischer and Illies 2018, 31). Manipulation is also often very thoroughly planned (and not careless at all) and can even be regardful. In such cases manipulation might even do something good for the manipulated as they are nudged, for example, to a healthier or more environmentally friendly life (cf. Noggle 2018). It is also far reaching to say that anyone who manipulates is careless, has a corrupt character, or that the benefit of the manipulator always constitutes the direction of the manipulation. We just have to think of romantic relationships in which one person is selflessly concerned about the welfare of the other and yet does not try to convince rationally (e.g., because they know the other one will stubbornly reject a good argument or else). In consequence, negative consequences and careless, selfish manipulators should not be considered as necessary or sufficient conditions of manipulation. It is, in fact, interesting why we seem to cling to a negative understanding of manipulation (more on this in Fischer 2022).

The most important characterization of manipulation, which is often implicitly contained in the previous ones (because the manipulated do not notice being manipulated, are misinformed, or have fewer abilities, etc.), conceptualizes manipulation as a form of influence that at least partially bypasses our rationality and possibly even undermines it completely (cf. e.g., Wood 2014; Gorin 2014b). Robert Noggle suggests that manipulation is an act where the manipulator controls someone by using their "psychological levers" (Noggle 1996, 44). He suggests that manipulation leads the manipulated "astray from certain paths toward certain *ideals*" (Noggle 1996, 44) by the already mentioned deception (changing a belief), changing situations or conditioning (changing desires) or inducing emotional states like guilt (changing feelings). Noggle seems to assume that the ideal way of decision-making lies in the rational deliberation on the basis of good reasons. Susser et al. assume with regard to online manipulation that ideal decision-making processes are prevented by the use of weaknesses. Noggle's view of the circumvention of rationality points in an important direction: as assumed previously, manipulation does, on the one hand, not primarily use rationality and good reasons or, on the other hand, coercion; it at least circumvents our rationality to a certain extent and strengthens the role of our affectivity in decision-making. The view that manipulation disconnects the links between good reasons and our decisions is still very popular (Fischer 2004; Wood 2014; see also Barnhill 2014; Gorin 2014a). This is because of a threat to autonomy that manipulation supposedly entails. Ethical concerns usually take over by that point.[21] But, in the case of manipulation, even though it uses the biological and bounded rational side of our being, there still seems to be room for a rational and free decision, not always following the path of a modulated affectivity that motivates us to act in a certain way. Free agency remains robust but at least might be challenged. This is probably one of the reasons why manipulation is a particularly interesting type of influence in liberal societies as we are still able to act on the basis of *our own* affectivity and can usually decide for or against its suggested direction – even if this is not always easy. Manipulation can make it difficult to act in a way we would rationally choose, it can lead us in a certain direction (although hardly generate completely new feelings, emotions or moods) while flying beyond our rational radar, but it does not establish a one-way street of decision-making (a manipulative influence can be very weak when, e.g., default options use our slothfulness[22]). Otherwise, we should speak of "coercion" and not "manipulation". Consequently, at least partially circumventing the rationality of an agent is a necessary condition for manipulation. We are finite, boundedly rational beings with a talent for rational deliberation and a colorful affectivity – both constantly interacting with each other. To overstress our rational capacities and dodging our complex affectivity seems to be one of the standard moves or even the "life-style" of Western society (Gellner 1992, 136) – something Martha Nussbaum once called (borrowing a term by Frans de Waal) "anthropodenial" (Nussbaum 2008; see also Fischer 2017, 91–103).[23]

4 An Integrative Understanding of Manipulation: The *Pleasurable-Ends-Model*

Even if it is not reasonable to simply define manipulation as the vitiation of human rationality (and our freedom and autonomy) and as underhand, deceptive, or a harmful means to achieve selfish ends as a manipulator (and all of these features together), all of the mentioned approaches point to items worth discussing. What they lack, though, is a detailed description of the mechanism as a foundation of manipulation that allows us to include or not include the mentioned (and at times amplifying) features. Generally speaking, manipulation introduces an influence into the development of our thoughts, decisions, and acting by modulating our affectivity. It is especially the evaluation that is strongly suggested by our affectivity that seems to be the target of a manipulation (and where all kinds of measures are used for it); in other words, the affective significance of an end is tried to be changed. In order to focus on the description of manipulation and its mechanism (without simultaneously involving an ethical assessment), I suggest understanding the phenomenon as follows.

The *how* of manipulation includes three steps. (1) An attempt is made to actively change the affective attraction of certain ends or their realization, in the sense that the realization of the respective end is more pleasant or unpleasant than the felt status quo. This is usually done by depicting a change that reaches us by our affectivity through effectively contrasting what is and what could be (cf. Ben Ze'ev 2001, 15) – for example, thieves stealing an election and the dark future after that. The prospect of a pleasant or unpleasant change in the status quo then makes (2) an option more attractive (or even extremely attractive) for the manipulated, whereas others not at all and thus (3) more or less likely that this option is chosen (Fischer and Illies 2018, 27; Fischer 2017, chapter 1). Attractive is everything that is connected to a sense of well-being (or vice versa in regard to unattractiveness). Well-being is founded in our interests, dispositions, and artificial and natural needs. As these things want to be satisfied, they can be used manipulatively. The evaluation of a certain end changes, resulting in often (but not always) complex affective responses which eventually boil down to a desire or an aversion regarding a certain end. This motivates the manipulated to act in a manner according to the manipulator's goal (Fischer 2022). Thus, manipulation can be seen as a form of influence where a manipulator leads the manipulated person to choose an end (an action, a product, etc.) but where the manipulated remains free at least in a minimal sense to decide whether she or he adapts this end or not. This is where the necessary and sufficient prerequisites for manipulation lie.[24]

By shedding more light on the mechanism an integrative and neutral understanding is achieved that leaves space for the various aspects discussed earlier and thus also upholds a connection to the everyday usage of the term "manipulation", while refining it. The *Pleasurable-Ends-Model* of manipulation can, but does not have to, include deception, underhandedness, negative consequences and selfish characters as possibly amplifying conditions of a manipulation since it focusses on the mechanism, which can be supported with certain more or less effective and applaudable means. According to this definition, it is necessary to push rationality aside (at least to a certain extent and as a primary mode of judging) as well as modulating our affectivity to change the evaluation of an end along the lines described previously. Just bypassing rational capacities is not sufficient, as this can also be done in ways other than manipulation (e.g., through underhandedness or deception). The active modulation of our affectivity, more precisely: our feelings, emotions, and moods with regard to the attraction of an end, is necessary (cf. for the properties of our various affects see Ben Ze'ev 2018, 112–137).

These three spheres of our affectivity make clear why there are many different suitable ways and instruments of manipulating as they have to be addressed specifically. Feelings are qualitative inner triggers that seem to be primarily responsible for making us act in a short-termed manner and can be used manipulatively by triggering impulses. Emotions, acute or persistent, are more complicated as they make us feel, think, evaluate, believe, and ultimately decide and motivated to act; they consist of cognitive and affective states simultaneously.[25] They can be used manipulatively for acute purposes or in broader, long-termed schemes. Moods tint certain things in life in a long-term manner, they often "belong" to us, can sometimes become firm dispositions, and establish certain manipulation-relevant triggers that can reach us more effectively (you can seldom catch the melancholic with outright fun).

With this characterization, we gave manipulation a place on the map of influences by determining its mechanism which enables us to distinguish it from other forms such as rational persuasion and coercion. If and how manipulation can count as a morally legitimate type of influence is so far only hinted at: if selfish manipulators, deception, underhandedness, and negative consequences are involved, manipulation tends to become morally problematic. However, there is more differentiation needed to dissect the difficult question about the morality of manipulation (for more see, e.g., Fischer 2017, chapter 3, 2018; Noggle 2018; Wood 2014).

But back to social media: due to the hinted at many shades in the evolution of an affective state/thought/decision/act, we are faced with a conceptual problem that becomes particularly clear in the online world, where technology stands between users and an interest pursuing beneficiary who uses certain technology, e.g., Trump and his campaign. It is already just not always clear how exactly an affective state/thought/decision/act comes about in the analogue world. The online world maybe makes this even messier. Here, when trying to get a better grasp of online manipulation, we can make a difference between (a) built-in structures of an interface that are leaning towards coercion and some that are outright manipulative and (b) interpersonal communication and the message design online.[26] This kind of online interpersonal communication via digital surfaces seems to make, in comparison to offline attempts, a pointed version of manipulative communication possible.[27] Sure, social media is used in various ways, e.g., as a major news source, for social interaction and self-presentation, but exactly the mixture of these features makes it so interesting for manipulation. Not just because people create a vital part of their construct of reality there (when using it as a news channel and collecting basin for perspectives) but also because individuals can be reached easily and are generally interested in interacting in the realm of social media (from clicking the "like" button to sharing, commenting, and posting). All the more, relatively few agents can reach large, interacting groups creating certain dynamics. But how then does social media support to actively change an attraction of an end or its realization? I will make a few remarks about this toward the end of this chapter.

5 Social Media as an Affective Realm Providing an Environment and Tools for Efficient and Effective Manipulation

Now that we have conceptualized manipulation, it is important to outline the characteristics of social media that render an efficient and effective manipulation possible. Let us come back to the example from the beginning for a moment. #StopTheSteal by Trump and his allies counted on misinformation. But it did more: it used affectively loaded, destructive language and imagery prone to set peoples' affectivity on fire with a certain narrative rendered salient, counting on controversy to stir up affective responses like frustration, indignation, fear, anger, mistrust, and a belief in Trump; it provided a memorable phrase for this to simplify a very complex issue (as, e.g., hashtags on Twitter often do) and used video clips and pictures reminiscing a dark and possibly violent future (if the election stays "stolen") whereas a "heroic leader" like Trump could slay the threatening monster – of course with the help of the recipients. By means of all this, #StopTheSteal literally tweaked algorithms and created widespread attention. So, there are two aspects to be differentiated: first, there is the designed controversial messaging trying to effectively carry content – how is it trying to influence (by using controversy, hope, . . .)? In what way does it primarily aim at our affectivity? Is it trying to depict an end as pleasurable/unpleasurable (in a blatant or subtle sense[28])? Which means does it use do that? Does it stand alone or is it framed in a bigger context? And second, there is the interface that serves as a vehicle for that affective messaging and is designed for this purpose – is the platform aiming at affectively engaging its users, making them prone to affective messaging? How does the affective message benefit from certain interaction possibilities, platform rules and algorithms?

Regarding #StopTheSteal, social media served as an ideal surface for a campaign like this that was a multichannel purpose using a multitude of communication devices where heavily affective messaging with controversial and destructive dark and heroic imagery, was pushed and constantly repeated to call for action. It aimed at user's affectivity, presented an end state as pleasurable (Trump staying president)/unpleasurable (the election staying "stolen"). The design of social media platforms like Facebook is a useful vehicle for such affective messaging as the interface itself aims at affectively engaging us. Involvement is a key concern for social media that thus try to provide a convenient, surprising, fun and informative platform with its algorithmic news feeds, videos, pictures, written messages provided by various agents calling us to react by consuming, commenting, liking, sharing. Publishers want the attention of users, which is created by constant repetition and flashy messages. Reaching the user's affectivity guarantees the biggest success: more followers, more likes and more shares – all of which broaden the distribution of a message (which is of course supported by programmed social bots which create even more likes and shares). Rational arguments and the confirmability arguments need are drowned out, as primarily affective content goes viral more effectively. Thus, it is only logical to design messages this way for successfully being recognized and achieving a goal. This fundamental construction of social media creates a manipulative potential as the affective realm can ultimately function as an affectivity catalyst, modifying the attraction of certain ends and thus making it more or less likely for them to be chosen. Let us look at the different bricks that provide the walls of this realm and connect them to how they help manipulative messages that render an end as pleasurable/unpleasurable to be successful and widely recognized as well as pointing out specific interface features that rely on, invite, and reward affective interaction, thus making it a useful tool for efficient and effective manipulation.

1. Social media interfaces are designed in a manner that tries to make users stay. Nir Eyal offers the thesis that it is a discomfort (feeling bored, lonely, confused, fearful, lost, or indecisive) that brings us online to find (often very short-termed) relief in interactions that distract us, even make us feel good and thus offer relief (Eyal and Hoover 2013). So, we are using social media (at least partly) to fulfill affective needs. If our own content is recognized and actually evaluated positively we like to come back. Recognition feels good (at least often). The short-high that comes with it seems to be close to what we are feeling when we shop. The interface design is in a simple way focused on basic conditioning, as rewards bring us back now and again. But, usually rewards do not carry us too far; keeping someone at it in regard to a certain direction and, on this foundation, the development of habits and at last manufacturing an

inner cognitive and affective connection need more than simple reward systems. Thus, e.g., Facebook offers far more than quick rewards: a news feed which is pleasurable in itself by being convenient, informative but all the more affectively stimulating by being sometimes surprising, the positive amazement or the outrage conveyed through affective messaging and the attention through receiving likes and shares. This is the multifaceted foundation social media is built on, creating lust and a routine to use it. As the interface already aims at engaging us primarily on an affective level, users are, at worst, ultimately made to be manipulable as we are kept in 'affective mode' when using social media; thus rationality is potentially put on the back burner.

2. Being an informative platform plays an important role in modifying an end in the context of manipulation. As soon as we dive into the world of social media, it becomes clear how much a significant part of our reality consists of using the shimmering bluish screens of our computers, smartphones, and tablets. Here, it is the representation of controversial messages suggesting drastic (often depicted as destructive) changes of personal relevance that grasp our attention and stimulate our affectivity but not so much a rational discourse. On the message side controversial and destructive content like #StopTheSteal becomes effective in creating a pseudo-environment that users turn to and use as a basis for their feeling (including the evaluation of a certain attraction of ends), thinking (i.e., their beliefs), and ultimately acting. To communicate affectively, this is hardly surprising, pictures, videos, and rather short messages are often more effective than long texts (Döveling 2015; Sachs-Hombach 2003). The evolution of advertising in the twentieth century gives proof to this tendency: words are less and less important (if, at all, they are important in the form of slogans), and the focus on pictorial messages is of growing importance. The increasing significance of Instagram and TikTok seems to show exactly this. The convenient presentation due to algorithms helps to manifest a pseudo-environment as it connects well to our slothfulness whereas the multitude of stimuli supports affective heuristics to sort through all the content. Difficult, potentially blinding, but intense states of affectivity (that can stand out), like fear and anger, on the grounds of the formula "excitement instead of information" are thus guiding principles for designing social media content in order to reach high visibility and, for example, unravel users.[29] This bears not only the danger of a rational discourse being drowned out in certain contexts and regarding certain topics but also to provide a distorted affectively loaded perspective onto reality. Shortcutting the ways to create a construct of reality opens up potential for misinformation. This also helps create a foundation on which manipulation is advantaged by the possibility of a presentation of certain ends as pleasurable/unpleasurable so that an effort would be needed for users to differentiate or distance themselves from the

attempt to modulate our affectivity by this presentation and, in broader strategies, a whole pseudo-environment.

3. With regard to the interface design, where algorithms conveniently provide you with related posts, groups, and sites "you might like", a strong temptation might occur to design your social media space in the form of an echo chamber in which existing convictions and affective states are reinforced within a relatively hermetic system. So, this is not only done by algorithms themselves but also supported by conveniently being able to subscribe in one click without checking certain agendas before. This invitation to create echo chambers can support social and political polarizations and the normalization of problematic opinions, since outside influences are hardly able to penetrate this bubble. Again, to be worthy of entering a user's echo chamber there is a need to be visible with which affective messaging helps. Also, it helps to drown out the need for rational checking as convenience and social pressure might just make a user give in. The phenomenon of being presented, searching for, and interpreting information according to one's own expectations is also known under the term "confirmation bias" (Pohl 2004, 93). In addition, claims might turn into felt truth (something Trump liked to legitimize constantly while talking about the "stolen election") if they are often and constantly repeated, ultimately manifesting an end as pleasurable/unpleasurable and supporting the realization of an end. So, sharing and liking of posts, which thus receive greater distribution and attention through algorithms, helps produce this phenomenon of felt (not known) "truth" solely through the widespread attention and an accompanying principle of repetition (Heath 2015, 191) rendering certain narratives and their images salient, highlighting certain ends as especially pleasurable/unpleasurable.[30]
4. The design of social media platforms with its masses of stimuli also invites using shortcuts to evaluate something. In our modern societies we find a steady high frequency of information stimuli. The media has always played an important role in structuring these stimuli and the environment they stem from for us: "For the real environment is altogether too big, too complex, and too fleeting for direct acquaintance. We are not equipped to deal with so much subtlety, so much variety, so many permutations and combinations" (Lippmann 2008, 16). Since much more content is being produced than we can look at, affective orientation is a useful method. Again, algorithms support this form of orientation by showing posts and related ones with which users interact the most. Also, we receive more and more sometimes crude suggestions that possibly cement an individual's echo chamber and the pseudo-environment that comes with it. Whatever most effectively appeals to our affectivity, thus successfully creating attention, can go viral. After all, what counts in the realm of social media is evaluation: "like" (on Facebook there is more options like "love", "haha", "wow", "sad",

"angry"); there is no "needs fact-checking" button. Like this, we are constantly engaged affectively and less so rationally, giving way for effective affective messaging.

5. Such affective messaging design seldom causes a problem in regard to publishing rules in social media. Anyone can say or show something about almost anything at any time, often without having to give up anonymity. Trump and his allies did not have to go through *any* process of certification to forgo their manipulation, they just published and the only obstacles they had to face (very late in the process) were deleted groups on Facebook or tweets marked with a little flag telling us such and such claim "is debated" (in the end this campaign seems to be one of the main reasons why Trump and some allies got banned on social media platforms for years – a historic intervention). Other than that, they could count on reaching millions of recipients within seconds, distributing their attempt to manipulate even more with the help of affective-design-rewarding algorithms.[31] With the interface design disregarding identities, a manipulative attempt becomes intransparent whereas at the same time the efficient distribution of affective messaging provides users with a suggestion of something being true and heart-felt. Selfish manipulators are hidden, motives unclear, and self-regulation by authors as well as recipients seems in some case to be massively weakened (Trump is a perfect example for this). On a sidenote: the lack of efficient social and legal control supports the virtue of "temperance" to crumble in digital communication (Vallor 2010).[32]

Let us sum up: by using the logics of seeking attention, social media tries to engage its users by means of its interactive design and strongly algorithmically selected affective content. Attention is created by flashy affective messaging depicting controversial, often destructive and drastic changes, trying to touch users; the visibility of these messages is often supported artificially by social bots and/or trolls, thus blowing up specific topics. The content itself mainly contains pictures, videos, mostly (very) short texts, if any, rich with simple messages and/or symbolism, so that the interpretation of this specific content is left to a mixture of confiding in friends who shared it, confiding in friends of the friends who are engaged in groups, making up the foundation of an echo chamber, and an affective evaluation of what is shown. Data analyses helps tailoring (economic but also political) ads groupwise and even individually, making it possible to target specific feelings, emotions, and moods.[33] Here lies a big potential for online manipulation; as well as with convenient dark patterns using our slothfulness, fake reviews and other means creating social pressure, and influencers providing a parasocial interaction where ends are rendered pleasurable/unpleasurable efficiently and effectively.[34]

Christopher Wylie, the whistleblower in the 2018 *Cambridge-Analytica*-scandal, once said: "We exploited Facebook to harvest millions of people's profiles and built models to exploit what we knew about them and target their inner demons" (Cadwalladr and Graham-Harrison 2018). What Wylie stated to the journalists of the *Guardian* describes the core of Trump's #StopTheSteal campaign: to target inner demons and to stir up possibly problematic affective states to make certain individuals become a part in what could be said was a tried – and luckily unsuccessful – *coup d'etat*. What Trump and his campaign did is clearly not ethically applaudable. They tried to create a pseudo-environment that had nothing to do with the facts; they targeted affective states in a multi-channel effort that are especially motivating while possibly blinding us for being able to see the complexity of things: frustration, indignation, fear, anger, mistrust, and an affection for Trump. They well knew the affectively engaging attention-focused design of social media and instrumentalized it for a harmful purpose on the grounds of selfish ends with deceptive (whereas not so much underhanded) and, most of all, heavily affective controversial content, suggesting that a *status quo* would result in an unpleasant reality, thus making democracy-undermining actions a pleasurable end. They were influenced in a way that can serve as a destructive example of an illegitimate (online) manipulation.

These basic structures certainly do not seduce everyone to move affectively guided in the social media world, to be manipulated, or use these structures to manipulate. However, social media has a strong potential for this in the sense of an almost optimally designed affective realm with an efficient interface using our human condition (especially by patterns), parasocial interaction and our need to be connected, as well as the possibility to retrieve data about users via tracking their behavior online which can make an attempt to manipulate even more efficient and effective.

Notes

I want to thank Klaus Sachs-Hombach (University of Tübingen), Damian Cox (Bond University), Christian Illies (University of Bamberg), Fabian Geier and Sebastian Krebs (CODE University Berlin) for comments on parts of this chapter which I presented at various occasions. Also, I want to thank the organizers and participants of the workshop preceding this volume for their suggestions.

1. Trump – who was not the only but certainly the most prominent one – first began tweeting allegations of fraud in April 2020. Since then, he made these allegations occasionally here and there, soon tweeting more often and regularly about it, ultimately leading to the systematic attempt we saw in the last weeks before, then during (establishing the hashtag #StopTheSteal and videos of suggested election fraud going viral), and after the election. Trump and his allies sure did not just tweet. They used every outlet possible to weave in the fraud allegations into the public discourse and the heads of Americans. What we witnessed here was a classic build-up of fear, frustration, and anger by using a fictitious scenario to create images in the heads of US citizens that should ultimately make them act in certain manner: to fight a "rigged" election, possibly making it possible for Trump and his allies to stage a coup and try to subvert democracy. Social media played a vital role in this attempt which means that

there comes a whole lot of responsibility with it when designing and curating this online realm.

2. This is a term I am borrowing from Walter Lippmann, meaning fictions, images in our heads, that shape our perception of the reality of the "world out there": "We shall assume that what each man does is based not on direct and certain knowledge, but on pictures made by himself or given to him" (2008, 25).
3. Lippmann wrote, that "it is clear enough that under certain conditions men respond as powerfully to fictions as they do to realities" (Lippmann 2008, 14). This has become painfully clear once again when looking at how events unfolded over the course of Trump's presidency, especially in regard to the presidential election of 2020, but also during post-election times where the republican party continued to keep the narrative of a stolen election alive, leading to severe changes in the appearance of the party and its political direction which in part seems to approach a fascist posture.
4. In my understanding we can divide our affectivity as a whole into these three categories which are all related to one another but not quite the same. *Feelings* are qualitative bodily impulses that can be very basic just like pain. Pain is clearly a feeling but not yet an emotion. *Emotions* are more complex because they contain not only feelings but also other components of a cognitive, evaluative, and motivational nature. They are intentionally related to an object in the environment and usually acute (like anger) or extensive and persistent (like a very complex emotion like love). Love, on the other hand, despite its persistence is not a *mood* like melancholia (which can also be constant), because it is more specifically related to an object, that is, the loved one, while a mood has a generalized scope and colors our lives in many areas (cf. Ben Ze'ev 2001).
5. This stands as a warning for every other nation grappling with right-wing populists and others with a disrespect for reason, truth, and democracy as the latter rests in large parts on exactly this: reason and the will to adhere to a factual basis and solid measures of how to count something as evidently true or false.
6. I am going to use "social media" and "social network" synonymously.
7. Also briefly consider this positive example: in the beginning of the COVID-19 pandemic Vietnam's Health Ministry produced an upbeat song (called "Ghen Cô Vy"), inspired by a precursory dance challenge on TikTok, teaching the necessary measures like handwashing, how to correctly wear a face mask, and so on. The video went viral all over social media and reached millions of people with its slightly kitschy animation which includes demonstrations of handwashing, warnings about face masks and public gatherings, and a gloved hand flicking away an angry-looking green coronavirus particle. This song (there are many others in the Asian music world) as fun and upbeat it is (while also being informative), grabbed lots of individuals by their affectivity manipulatively nudging them to act in the safe way necessary.
8. I developed a detailed account in my book *Manipulation. Zur Theorie und Ethik einer Form der Beeinflussung* (published in late 2017). Some thoughts from it are published in English; see, for example, Fischer and Illies (2018) and Fischer (2022).
9. I take it that affective states always help to shape our thinking and thus consequently our decision-making process.
10. At least as long as they are not trying to imitate interpersonal communication (which they, as of yet, often don't do very well). If that is the case, many of my theses should be transferable.
11. See again endnote 2.
12. Literature often offers even more insightful accounts, for example, on specific emotions than philosophical analysis or empirical psychological studies.

Our affectivity does not seem to be measurable to the full extent in a scientific sense or to be broken down into clear propositions. We can conceptualize the outlines of it and describe its main workings in cold, technical terms, but by this we will not grasp them to the full extent. Literature masterfully fills this gap and warmly plasticizes our affectivity and shows itself as a vehicle of non-propositional knowledge, an essential foundation for our practical and phenomenological perception of the world that can function as a completion of the propositional kind of knowledge we can find in science and most areas of philosophy (cf. Fischer 2018). See with a slightly different focus than here Olivia Sudjic's novel *Sympathy* (2017).

13. Of course, you can, for example, act against a law (which counts as a form of coercion) but to understand that as freedom is, at least in judicial systems where fairness is fundamental, understanding freedom in a distorted I-do-everything-as-I-think-is-best-way. Sure, coercion also leaves an option to just abide by the law but not to decide out of unencumbered alternatives.
14. Even though our affectivity seems to have a bad reputation, we can find accounts that attest our affectivity a share of an integrated rationality. Aristotle famously thought that our affects are rational when they show themselves at the right time, for the right amount, and the right reason, for example, grief when a loved one died. Robert C. Solomon, as an example for a modern-day emotion theorist, also conceptualizes our affectivity as a vital part of rationality as it provides meaningful judgments, discerns value, is trainable, functions as an engine of our actions, is strategic, and creates meaning (Solomon 2001). In consequence, this also means that our affectivity does not in every case and necessarily undermine our autonomy.
15. I will look at active, intentional attempts of manipulation which does not mean that manipulation does not happen unconsciously or by just careless individuals. Nonetheless, many attempts of manipulation are thoroughly planned and it is not enough to assume that every manipulating agent is just careless. It is something else that seems to be at the core of the phenomenon: trying to use affectivity and our peripheral routes of decision-making to bring about a certain action.
16. These three types of ends are, of course, in many cases mixed with one another. Often, too, an agent is not fully aware of her (unacknowledged) ends.
17. Suggestions to not speak of "manipulation" anymore open up the question if it is possible to meaningfully speak about the phenomenon without the term "manipulation". Listeners just would not know anymore what we are talking about. There seems to be no everyday language term that marks a positive emotional influence anyhow. "Emotional influence" itself may be a neutral candidate which is less pejoratively connotated than "manipulation" (a connotation that can be closely linked to neo-Marxist thinking and the grim times of national socialism in Germany and capitalism worldwide). But the negative connotation seems to have more to do with a critique of challenges to our rationality than with the actual literal sense of the word "manipulation" (which is nonetheless understandable against the background of Nazi propaganda and the rise of capitalism). Since we are all finite, boundedly rational beings, it seems that we must admit that manipulation can count as a normal mode of communication. I am voting for keeping the term "manipulation" instead of erasing it because it marks an influence that goes beyond our rational radar and can still be connected to the everyday use of it, even if I try to carve out its characteristics, the when and how a bit more (cf. Fischer 2022).
18. In 2020, Sven Feurer of the Bern University of Applied Sciences (Switzerland) and I conducted an empirical study with a representative sample (in regard to

gender, age, and education) of 1000 German consumers to research the perception of marketing as manipulative. We asked the sample what they perceive as manipulation and ethically problematic with regard to new marketing measures that heavily rely on the internet and social media platforms like influencer marketing, fake reviews, targeted ads, and so on. The study confirmed that manipulation is not necessarily associated with deception but stronger with an attempt to affectively involve consumers to buy a product. Whereas deception was seen as very morally problematic, manipulation was perceived with a general skepticism but not necessarily seen as ethically problematic in *every* case. Targeted ads were seen as much less problematic than sentiment analysis or fake reviews, for example. For more, see Feurer and Fischer (2022).

19. This association is not necessarily wrong due to the usually strong emotionalization of fake news which plays a vital role besides the objectively wrong content. Manipulation, in the end, always also influences our thinking even when it primarily tries to modulate our affective states.
20. The empirical study Sven Feurer and I conducted shows that most people perceive marketing strategies usually as manipulative as they are aiming at our affectivity instead of presenting a product in a non-affective kind of way (Feurer and Fischer 2022).
21. Moral concerns arise, since an unencumbered rationality is seen as a necessary condition for autonomy. Manipulation, however, does not "*sufficiently* engage or appeal to [agents'] capacities for reflective and deliberative choice" (Sunstein 2016, 443; my highlighting) or even "*perverts* the way that [a] person reaches decisions, forms preferences or adapts goals" (Raz 1986, 377, my highlighting). Not being able to dive deeper into this particular discussion, I want to put forward that a manipulatively induced behavior does not automatically yield a degradation of an agent to an object as it is claimed in a Kantian tradition (e.g., Wood 2014). See for the link of our affectivity and our autonomy again endnote 14.
22. This is almost not tangible but still works with an affective state: that of not wanting to invest anything but instead staying comfortable. This points us to a certain problem in conceptualizing manipulation: some attempts at influencing an agent are, due to the use of many different factors, not 100% or 0% manipulative, but of a more or less manipulative quality and certainly tangible to almost not tangible. So there could be rational devices like arguments involved in manipulatively influencing an agent while certain forms of contextualization (e.g., an incident like the US election 2020), framing (e.g., as a "fraudulent election" of "corrupt individuals"), or presentation (e.g., by supposedly trustworthy authorities, with effective images etc.) add an affective and maybe manipulative character to what tries to reach an agent by argument cursorily (even though it might be objectively wrong). So, it is not always easy to distinguish how we influence each other between the mentioned poles in the analogue but also digital realm as it is not easy to determine what exactly makes an agent think, act, or decide in this or that direction. After all, this is not only based on a more or less rational *and* affective basis but always shaped by concrete situations, specific contexts, habits, and individual character traits. Burrhus F. Skinner's operant condition might perhaps be the simplest example of manipulation through its use of rewards that are pleasurable and motivate us to do something again and punishments that make us avoid repeating certain behavior. But there are many more ways of using our affective states to influence us.
23. It becomes clear that all three ways of characterizing manipulation tend to blur the boundaries between descriptively and normatively defining said phenomenon by usually seeing manipulation as a negative type of influence. This

coincides with the everyday use of the word as an encumbered term which usually is intended to highlight that an outrageous type of influence has been used. Regarding the case of manipulation it makes sense to separate the question of how it works from the question of whether it is ethical or unethical (cf. Wood 2014, 19; Coons and Weber 2014a, 6–8). First, this can be explained by the fact that the term was once used neutrally and received its negative connotation only in the course of the twentieth century – which may not come as a surprise if one considers, for example, the horrors of Nazi propaganda (cf. Fischer 2017). However, this may at the same time stimulate us to look carefully and try to understand what exactly happens in the context of manipulation instead of leaving it blurry by immediately rejecting it as something evil. Stripped from this tinted looking glass, it becomes clearer that manipulation is constantly present in our social life and might even be qualified as a rather normal mode of communication between individuals that is not just malicious (although it can be) because, after all, nobody communicates purely rationally throughout. Additionally, a normatively loaded definition from the outset threatens to block a differentiated ethical debate because it supposedly seems clear from the get-go that manipulation is devilish. If we turn to the history of rationality we can quickly learn how rationality became the sun that supposedly helped grow the bulk of the grass of our humanity and that, especially by the discipline of philosophy, became an even dazzling light that might perhaps have blinded us for being at peace and with trust in regard to our affectivity. For more thoughts on our attitude regarding manipulation, see Fischer 2022.

24. Defining manipulation this way finds precursors in the concepts of Baron (2014, 109), that manipulation plays upon emotions, uses pressure to acquiescence (which is not yet coercion) or weaknesses of character, as well as in Noggle's (1996) and Barnhill's (2014) examples using guilt or Marcuse's observation that manipulation works via systematically inducing libidinal needs (Marcuse 1969, 31).
25. The relationship of our cognitive and affective states in the case of emotions is often complicated. Just think of jealousy where a strong feeling component contaminates our thoughts drastically, even creating tunnel vision, while we feel bad, evaluate harshly, and are motivated to act in an often destructive manner. It sure would be interesting to fan out a phenomenology of other difficult affects like anger, indignation, fear, and so on. But this is not the place for that.
26. Recommender systems, for example, can merge a) and b); for more detail, see Klenk, in this volume.
27. This is applicable especially for cases of, e.g., strategic political communication and marketing, whereas this is not necessarily true for every case of interpersonal communication where body language, facial expression, pitch of voice, and various other factors of nonverbal communication can intensify an attempt to manipulate more effectively than online.
28. In Sven Feurer's and my empirical study, we found that consumers themselves believe that manipulation in marketing has become much more subtle over the last two decades.
29. Sure, excitement can also be gained by other affective states than a feeling of indignation. I have also mentioned the acute emotions fear and anger or the persistent mood of mistrust in regard to the #StopTheSteal campaign. However, we can also be, for example, humorously affected or enthused in a positive manner and be guided by this through the affective online realm. See again endnote 7.
30. On December 10, 2020, Trump tweeted: "78% of the people feel (know!) the Election was RIGGED". With this tweet we get a small affidavit of means as it shows the (often) undifferentiated and in consequence dangerous equalization of feeling and knowing something is true. For Trump, it is convenient to stylize

feelings to equal truth as his only goal is to stir up the affectivity of his recipients without providing proper evidence that can actually be verified or falsified and thus *known* in a rational sense.

31. The suspension of Trump's social media accounts happened extremely late. After the U.S. Capitol was insurrected, Twitter and Facebook were able to argue that violence is actively incited (which breaches their rules – not an actual (and maybe needed) law, which regulates what is allowed to be done online). Before this specific, huge outbreak of violence both social networks held back, pointing to Trump's status as the president of the United States and thus a person of interest, when, of course, inciting violence played a big role for all of the years of Trump's presidency just on a different scale.
32. This often is because of a crooked understanding of freedom of expression. Trump tried to depict himself as a victim of censorship after his accounts were closed down. But objectively considered he was not at all a victim. He had a press room in the White House where he could address the nation and answer questions. Consequently, he was neither censored nor was his freedom of expression destroyed.
33. Because of its growing importance for the future the phenomenon called "microtargeting" should be kept in mind. Here, psychographic profiles are built to let content creators decide which advertisement or campaign design can make the biggest impression to which group of individuals – something that is heavily used by campaign strategists to efficiently grab voters by their affectivity. Interestingly enough the empirical study Sven Feurer and I conducted with regard to e-commerce marketing strategies and their manipulativeness showed that targeted ads are perceived as manipulative in general but not too morally problematic (in comparison to influencer advertising, sentiment analysis or fake reviews). Of course, this evaluation concerns product advertisement and not politically used targeted advertising where one can expect a different answer (cf. Feurer and Fischer 2022).
34. Cf. Kramer, Guillory, and Hancock 2014 for an empirical work that has investigated affective influences on Facebook via the modification of hundreds of thousands of news feeds, without the knowledge of the users. The authors come to the (admittedly quite general) conclusion: "Online messages influence our experience of emotions, which may affect a variety of offline behaviors" (Kramer, Guillory, and Hancock 2014, 8788). Kramer himself is a data analyst at Facebook. This gives the work an interesting and a bit dazzling component: it is not only that users were experimented with without their consent. But Facebook also has an interest to show its advertising customers that manipulation works on the platform, which is why a scientific output can help to substantiate this claim (at the same time, users must be told that they are not easily manipulated on the platform). The cautious wording cited might result from ultimately not excessively strong experimental effects observed by the authors. However, they rightly point out that against the background of the size of social networks even small effects are based on a large number of people. It is also interesting to know that the paper had to take a lot of criticism. For an overview of this criticism see Grohol 2018.

6 References

Abramowitz, Michael J. 2017. "Opinion: Stop the Manipulation of Democracy Online." *The New York Times*, November 12. Accessed August 20, 2021. www.nytimes.com/2017/12/11/opinion/fake-news-russia-kenya.html.

Atlantic Council's DRFLab. 2021. "#StopTheSteal: Timeline of Social Media and Extremist Activities Leading to 1/6 Insurrection." Accessed August 20, 2021. www.justsecurity.org/74622/stopthesteal-timeline-of-social-media-and-extremist-activities-leading-to-1–6-insurrection/.

Barnhill, Anne. 2014. "What is Manipulation?" In Coons and Weber 2014, 51–72.

Baron, Marcia. 2003. "Manipulativeness." *Proceedings and Addresses of the American Philosophical Association* 77 (2): 37. doi:10.2307/3219740.

Baron, Marcia. 2014. "The Mens Rea and Moral Status of Manipulation." In Coons and Weber 2014, 98–109.

Beauchamp, Tom L. 1984. "Manipulative Advertising." *Business and Professional Ethics Journal* 3 (3/4): 1–30.

Ben Ze'ev, Aaron. 2001. *The Subtlety of Emotions*. Cambridge, MA: Bradford Books.

Ben Ze'ev, Aaron. 2018. "The Thing Called Emotion: A Subtle Perspective." In *Philosophy of Emotion*, edited by Aaron Ben Ze'ev and Angelika Krebs, 112–37. London: Routledge.

Cadwalladr, Carole, and Emma Graham-Harrison. 2018. "Revealed: 50 Million Facebook Profiles Harvested for Cambridge Analytica in Major Data Breach." *The Guardian*, March 17. Accessed August 20, 2021. www.theguardian.com/news/2018/mar/17/cambridge-analytica-facebook-influence-us-election.

Cave, Eric M. 2007. "What's Wrong with Motive Manipulation?" *Ethical Theory and Moral Practice* 10 (2): 129–44.

Coons, Christian, and Michael Weber. 2014a. "Manipulation: Investigating the Core Concept and its Moral Status." In Coons and Weber 2014, 1–16.

Coons, Christian, and Michael Weber, eds. 2014b. *Manipulation: Theory and Practice*. Oxford: Oxford University Press.

Döveling, Katrin. 2015. "Emotion Regulation in Bereavement: Searching and Finding Emotional Support in Social Net Platforms." *New Review of Hypermedia and Multimedia* 21 (1–2): 106–22.

Eyal, Nir, and Ryan Hoover. 2013. *Hooked: How to Build Habit-Forming Products*. London: Penguin Books.

Feurer, Sven, and Alexander Fischer. 2022. *Exploring the Ethical Limits of Manipulation in Marketing: A Discussion Based on Consumer Perceptions*. under review.

Fischer, Alexander. 2017. *Manipulation: Zur Theorie und Ethik einer Form der Beeinflussung*. Berlin: Suhrkamp.

Fischer, Alexander. 2018. "Die Moral durch die Geschicht': Erzählen als vernünftige Integration von Verstand und Affekten im moralischen Handeln." *Internationale Zeitschrift für Philosophie und Psychosomatik* 10 (1): 1–15.

Fischer, Alexander. 2020. "Im Schraubstock der Angst: Manipulation und unsere Disposition zur Ängstlichkeit." *Hermeneutische Blätter* 26 (1): 20–37.

Fischer, Alexander. 2022. "Then Again, What is Manipulation? A Broader View of a Much-Maligned Concept." *Philosophical Explorations*. doi: 10.1080/13869795.2022.2042586

Fischer, Alexander, and Christian Illies. 2018. "Modulated Feelings: The Pleasurable-Ends-Model of Manipulation." *Philosophical Inquiries* 1 (2): 25–44. Accessed August 06, 2020.

Fischer, John M. 2004. "Responsibility and Manipulation." *Journal of Ethics* 8 (2): 145–77.

Gellner, Ernest. 1992. *Reason and Culture: The Historic Role of Rationality and Rationalism*. Oxford: Oxford University Press.

Goodin, Robert E. 1980. *Manipulatory Politics*. New Haven, CT: Yale University Press.

Gorin, Moti. 2014a. "Do Manipulators Always Threaten Rationality?" *American Philosophical Quarterly* 51 (1): 51–61. Accessed June 04, 2019.

Gorin, Moti. 2014b. "Towards a Theory of Interpersonal Manipulation." In Coons and Weber 2014, 73–97.

Green, Ronald K., and Edward J. Pawlak. 1983. "Ethics and Manipulation in Organizations." *Social Service Review* 57 (1): 35–43.

Grohol, John M. 2018. "How Facebook's Squishy Ethics Got Them into Trouble." *PsychCentral*, July 8. https://psychcentral.com/blog/how-facebooks-squishy-ethics-got-them-into-trouble.

Heath, Joseph. 2015. *Enlightenment 2.0: Restoring Sanity to Our Politics, Our Economy, and Our Lives*. Toronto: Harper.

Klenk, Michael. 2022 "Manipulation, Injustice, and Technology." In *The Philosophy of Online Manipulation*, edited by Jongepier, F. and Klenk, M., 108–132. New York, NY: Routledge.

Klenk, Michael. 2021. "Manipulation (Online): Sometimes Hidden, Always Careless." *Review of Social Economy*. 80: 1, 85–105. doi:10.1080/00346764.2021.1894350.

Kramer, A. D. I., J. E. Guillory, and Jeff Hancock. 2014. "Experimental Evidence of Massive-scale Emotional Contagion Through Social Networks." *Proceedings of the National Academy of Sciences* 111: 8788–90.

Lippmann, Walter. 2008. *Public Opinion*. Miami: BN Publishing.

Marcuse, Herbert. 1969. "Aggressivität in der gegenwärtigen Industriegesellschaft." In *Aggression und Anpassung in der Industriegesellschaft*, edited by Herbert Marcuse, 7–29. Frankfurt a.M.: Suhrkamp.

Nelson, Lisa S. 2018. *Social Media and Morality: Losing Our Self Control*. Cambridge: Cambridge University Press.

Noggle, Robert. 1996. "Manipulative Actions: A Conceptual and Moral Analysis." *American Philosophical Quarterly* 33 (1): 43–55.

Noggle, Robert. 2018. "The Ethics of Manipulation." In *Stanford Encyclopedia of Philosophy: Summer 2018*, edited by Edward N. Zalta.

Nussbaum, Martha. 2008. "Compassion: Human and Animal." http://idsk.edu.in/wp-content/uploads/2015/07/SL-4.pd. IDSK Special Lecture 4.

Pohl, Rüdiger. 2004. *Cognitive Illusions: A Handbook on Fallacies and Biases in Thinking, Judgement and Memory*. New York, NY: Psychology Press.

Raz, Joseph. 1986. *The Morality of Freedom*. Oxford: Oxford University Press.

Rudinow, Joel. 1978. "Manipulation." *Ethics* 88 (4): 338–47. doi:10.1086/292086.

Sachs-Hombach, Klaus. 2003. *Das Bild als kommunikatives Medium: Elemente einer allgemeinen Bildwissenschaft*. Köln: Halem.

Scanlon, Thomas M. 1998. *What We Owe to Each Other*. Cambridge, MA: Harvard University Press.

Solomon, Robert C. 2001. *True To Our Feelings: What Our Emotions Are Really Telling us*. Oxford: Oxford University Press.

Spring, Marianne. 2020. "'Stop the Steal': The Deep Roots of Trump's 'Voter Fraud' Strategy." *BBC News*, November 23. www.bbc.com/news/blogs-trending-5500995.

Sunstein, Cass R. 2016. *The Ethics of Influence: Government in the Age of Behavioral Science*. Cambridge: Cambridge University Press.
Susser, Daniel, Beate Roessler, and Helen Nissenbaum. 2019a. "Online Manipulation: Hidden Influences in A Digital World." *Georgetown Law Technology Review* 4 (1): 1–45. Accessed February 27, 2020.
Susser, Daniel, Beate Roessler, and Helen Nissenbaum. 2019b. "Technology, Autonomy, and Manipulation." *Internet Policy Review* 8 (2): 1–22. doi:10.14763/2019.2.1410.
Vallor, Shannon. 2010. "Social Networking Technology and the Virtues." *Ethics and Information Technology* 12: 157–70.
van Dijk, Teun A. 1998. *Ideology: A Multidisciplinary Approach*. London: SAGE Publications.
Verbeek, Peter-Paul. 2011. *Moralising Technology: Understanding and Designing the Morality of Things*. Chicago, IL: University of Chicago.
Ware, Alan. 1981. "The Concept of Manipulation: Its Relation to Democracy and Power." *British Journal of Political Science* 11 (2): 163–81.
Wood, Allen W. 2014. "Coercion, Manipulation, Exploitation." In Coons and Weber 2014, 17–50.

18 Social media, emergent manipulation, and political legitimacy

Adam Pham, Alan Rubel, and Clinton Castro

1 Introduction

Psychometrics firms such as Cambridge Analytica[1] (CA) and troll factories such as the Internet Research Agency (IRA) have had a significant effect on democratic politics, through narrow targeting of political advertising (CA) and concerted disinformation campaigns on social media (IRA) (U.S. Department of Justice 2019; Select Committee on Intelligence, United States Senate 2019; DiResta et al. 2019). It is natural to think that such activities manipulate individuals and, hence, are wrong. Yet, as some recent cases illustrate, the moral concerns with these activities cannot be reduced simply to the effects they have on individuals. Rather, we will argue, the wrongness of these activities relates to the threats they present to the legitimacy of political orders. This occurs primarily through a mechanism we call "emergent manipulation," rather than through the sort of manipulation that involves specific individuals.

We begin by examining two cases. The first is the 2010 Cambridge Analytica "Do So!" campaign, which aimed to tip the balance of a closely contested election by promoting youth apathy in the (ethnically split) Trinidad and Tobago elections. The second is a suite of campaigns by the IRA, which involved the organization franchising its activities to evade detection (Channel 4 2018; Alba 2020). Next, we develop and discuss the concept of emergent manipulation, explaining how it differs from other scholarly accounts. Then, we argue that the presence of this sort of manipulation in electoral politics threatens the legitimacy of the elections themselves. Legitimacy, we argue, requires that a citizenry be unmanipulated in a holistic way, independently of whether individuals are manipulated and have their autonomy undermined.

2 Manipulation campaigns around the world

2.1 Cambridge Analytica and the Do So! campaign

Cambridge Analytica has become infamous for its involvement in the 2016 US elections and the Brexit referendum, but more recent reports have revealed

DOI: 10.4324/9781003205425-21

that the reach of the political consulting and marketing firm has extended far beyond the United States and the UK. Alexander Nix, the former CEO of CA, was caught undercover bragging about extortion operations in Sri Lanka (Channel 4 2018). Though Nix's penchant for exaggeration is well known,[2] a brochure obtained by the BBC revealed schemes in Nigeria, Latvia, and several Caribbean nations, among them Trinidad and Tobago (*BBC News* 2018).

Like many postcolonial societies, Trinidad and Tobago has faced deep ethnic divides since the departure of its colonial government. Although interethnic relations are cordial in public, cultural differences and weak institutions have led to professional segregation and a clientelist political system. The primary divide is between Indo-Caribbeans – who tend to support the United National Congress party – and Afro-Caribbeans – who tend instead to support the People's National Movement party – with neither ethnic group owning a majority allowing it to claim durable political control (Premdas 2019). In such a political climate, where elections are always bound to be closely contested, the sort of manipulative practices associated with CA can be not just influential but decisive.

Trinidad and Tobago's 2010 elections, which were highlighted in detail by the 2019 Netflix documentary *The Great Hack*, provide an illuminating case study of CA's techniques (Noujaim and Amer 2019). The crux of CA's intervention into the elections involved capitalizing on an opposition movement called "Do So." The movement began when a disaffected pensioner, Percy Villafana, refused to allow the then-prime minister to traverse his property during a political walkabout, with Villafana's arms crossed in defiance of the stunt. The movement, which came to be branded by an emblem of crossed arms, went viral on Facebook and soon attracted the attention of CA, which began to bolster the movement via astroturfing efforts in the form of an "ambitious campaign of political graffiti" that "ostensibly came from the youth (*BBC News* 2018).

CA's own promotional web materials painted their influence in that election as decisive, arguing that "the employment of CA's research-based differential campaigns and establishment of consistent policy and variegated communications contributed to the [United National Congress]" (CA Political 2018). Their strategy, more plainly, involved increasing political apathy among *all* young people in Trinidad and Tobago, while anticipating that this would differentially depress voter turnout among Afro-Caribbean youth relative to their Indo-Caribbean peers. In audio from a sales presentation, Nix himself is strikingly candid about the strategy:

> There are two main political parties, one for the blacks and one for the Indians. And, you know, they screw each other. So, we were working for the Indians. We went to the client and we said, "We want to target the youth." And we try and increase apathy. The campaign had to be non-political because the kids don't care about politics. It had to

> be reactive because they're lazy. So, we came up with this campaign which was all about: be part of the gang. Do something cool. Be part of a movement. And it was called the "Do So!" campaign. It means "I'm not going to vote." "Do so! Don't vote." It's a sign of resistance against, not the government, [but] against politics and voting. . . . We knew that when it came to voting, all the Afro-Caribbean kids wouldn't vote, because they Do So! But all the Indian kids would do what their parents told them to do, which is go out and vote. They had a lot of fun doing this, but they're not going to go against their parents' will. . . . And the difference in the 18- to 35-year-old turnout was like 40%. And that swung the election by about 6%, which was all we needed in an election that was very close.
>
> (Noujaim and Amer 2019)

Following the release of *The Great Hack*, officials in the People's National Movement called the legitimacy of the election into question (George 2019).

Whatever threat to legitimacy the campaign might have caused, the threat did not appear to operate through direct affronts to anyone's autonomy or quality of agency. This, in turn, provides the grounds for deniability. To this end, Nix offered a statement in response to allegations of election manipulation, in which he claimed that "[t]he objective of this campaign was to highlight and protest against political corruption," that "[t]here is nothing unlawful or illegal about assisting with this activity," and that "[CA] has never undertaken voter suppression and there is no evidence to the contrary" (Hilder 2020). Taken at face value, his argument is surprisingly difficult to resist. Since the Do So! campaign *did* begin in a grassroots fashion and *was* furthermore supported by a broad coalition of youth voters, CA's activities cannot be viewed as involving the outright fabrication of a social movement. Rather, we must view these activities as a distorted amplification of an existing movement.

Regardless of its relation to other movements, the Do So! case has several interesting features. First, no individual person or persons were targeted for behavior modification by CA; second, no one's autonomy was necessarily undermined (though someone's might have been); third, there was no publicly disclosed source of central influence. What matters here is that the kind of manipulation we are addressing need not turn on any individual being affected enough for them to lose autonomy.

2.2 *The Internet Research Agency and "active measures"*

Cambridge Analytica is best known for its connections to the 2016 US presidential election and the UK Brexit referendum. Christopher Wylie (and similar accounts) describes various interactions between CA and Russia (e.g., testing social media messaging about Vladimir Putin, campaigns for Lukoil, and relationships with pro-Russia factions in the Russia-Ukraine conflict). As a result, CA's actions in 2016 are often conflated with direct Russian involvement. CA denies any connection to Russian state actors, admitting only to working for private interests in Russia.

The most notable example of Russian "active measures" in US presidential politics is by the Internet Research Agency. The IRA is a Russian state-supported influence operation, described by DiResta et al. as a "sophisticated marketing agency" (2019, 6). It has trained and employed "over a thousand people to engage in round-the-clock influence operations" to influence citizens, social organizations, and political processes in a range of countries, including Russia, Ukraine, and the United States. In February 2018, the US Department of Justice indicted the IRA and several of its principals (all of whom are Russian nationals) based on the results of the investigation into Russian interference in the 2016 election conducted by Special Counsel Robert Mueller. The charges in the indictment include conspiracy to commit fraud, wire fraud, and bank fraud (U.S. District Court for the District of DC 2018).

The activities underlying the charges are social media disinformation campaigns, known as "active measures." The IRA and its agents engaged in a years-long operation to understand US politics and its points of conflict (including agent visits to the United States under false pretenses in order to better understand political culture). They created interwoven networks of ersatz social media profiles and groups that appeared to have a large and "organic," unplanned presence. The IRA purchased ads on social media sites that were targeted at users likely to follow the fake profiles and join the fake groups (U.S. Department of Justice 2019). The IRA then used these networks to seed and promote inflammatory, divisive content. Notably, the IRA did not focus on any particular ideology or political affiliation. Rather, it sought to engage and enrage social media users from a broad swath of US political positions. The IRA did focus particular attention on Black Americans, targeting this group with ads, creating groups that appeared affiliated with racial and social justice, targeting ads toward places with large African American populations, and focusing on issues that divide Americans along racial and ethnic lines. The Senate Select Committee on Intelligence writes that "[b]y far, race and related issues were the preferred target of the information warfare campaign designed to divide the country in 2016" (Select Committee on Intelligence, United States Senate 2019, 6). So, for example, the IRA created social media pages and groups such as "Blacktivist" and posted to social media comments about Colin Kaepernick and other athletes' kneeling protests and about police shootings of Black people (Select Committee on Intelligence, United States Senate 2019, 6–7).

The pattern of finding groups receptive to provocative, negative rhetoric extended across a broad range of social, cultural, and political affiliations. Some efforts appealed to nativism ("Stop All Immigrants," "Secured Borders"), others targeted messages toward racial and ethnic minorities ("Black Matters," "United Muslims of America"), and some aimed to exploit other cultural and political divides ("Tea Party News," "Don't Shoot Us," "LGBT United") (U.S. Department of Justice 2019, 24–25). It is difficult to determine the magnitude of effects these efforts had. However, US Department of Justice reports that the IRA's accounts "reached tens of millions of U.S. persons" and had "hundreds of thousands of followers" (U.S. Department of Justice 2019, 26).

Moreover, the IRA's social media accounts' effects went beyond online viewing. They were the basis for organizing rallies in person, for recruiting political activists to engage in organizing, and for promoting content promulgated by the IRA (U.S. Department of Justice 2019, 31–32). In a study of social media and misinformation, members of the Oxford Internet Institute found that in 2016, prior to the US presidential election, "Twitter users got more misinformation, polarizing, and conspiratorial content than professionally produced news" (Howard et al. 2017, 1).

The IRA's activities during the 2016 election cycle ranged across social media platforms, including Twitter, Facebook, Instagram, and YouTube. It did not limit its targets to particular political or social orientations but instead aimed to influence a broad range of views. And while its precise aims remain unclear, its tactics include influencing users to refrain from voting, to support and vote for third parties, to diminish overall voter participation, to undermine support of political leaders generally, and to build support for "Brexit-style" movements for states (e.g., Texas and California) to secede (DiResta et al. 2019, 8–10). Similarly, the IRA sowed distrust in traditional news media by seeding Russian disinformation stories in news media (DiResta et al. 2019, 65–66).

There is no definitive information connecting the IRA to Cambridge Analytica, but that is not crucial for our argument here. What matters for our purposes is that several things occur in close, mutually reinforcing order. First is massive data collection based on lack of privacy protections in social media environments (and in particular on Facebook), the increasing power of data analytics that can use the data collected to better target influence campaigns, and automated systems that recommend how clients can target advertising and which promote content to social media users. The precise relationship between CA and the IRA may be important for determining responsibility or legal liability, but it is not key in understanding manipulation in the sense we are addressing here.

In addition to the connection between CA and the IRA being unclear, the efficacy of their efforts (individually or collectively) is unclear. Election and policy outcomes are complex phenomena and it is impracticable to identify a single set of events as their cause. And even so, it is unclear whether tactics like those of CA and the IRA are effective at all. According to Kogan, media accounts exaggerate the effectiveness of data analytics and social media campaigns generally, and in particular "[w]hat Cambridge has tried to sell is magic" (Weaver 2018). During the 2016 Republican party primary, the Ted Cruz campaign maintained that its data-driven tactics drove its victory in the Iowa caucus (Hamburger 2015). That view changed as the primary campaign unfolded, with the Cruz campaign growing skeptical and eliminating its use of psychological profiling after it lost the South Carolina primary (Detrow 2016).

Yet, there is a growing body of evidence for the effectiveness of psychological targeting. In particular, a team of psychologists has recently argued that the CA case "illustrates clearly how psychological mass persuasion could be abused to manipulate people to behave in ways that are neither in their best interest nor in the best interest of society" (Matz et al. 2017). At the same time, Nix's cynical argument looms large: there is nothing unlawful, illegal, morally objectionable, or necessarily even manipulative about directing people's attention to information about corruption. To understand how and why such activities could threaten the political legitimacy of otherwise legitimate governments, we must first understand how the activities are manipulative.

The actions of CA and the IRA surrounding Do So!, Brexit, and the 2016 US presidential election are in some sense old news. The 2020 presidential election has seen more homegrown misinformation campaigns. Among the most successful of these has been the false claims that states had voting irregularities. These claims have been extensively litigated, and the political pressure for election officials to throw out vote tallies were ultimately unsuccessful. However, a surprisingly large portion of the population took the claims seriously. And this campaign led directly to a violent assault on the US Capitol building that sought to prevent the US Congress from accepting the electoral votes from the states. Indeed, the misinformation campaign has convinced many Americans that the election was illegitimate and is underwriting a number of actions to restrict voting access in many US states. The 2020–21 campaigns are still unfolding, and analyzing them in depth now is premature. However, we can note here that the same kinds of emergent processes we discuss in this chapter are present in 2020–21.

3 The forms of manipulation

3.1 Disputes about manipulation

The philosophical literature on manipulation is rife with scholarly debates about its nature, its extent, and what, if anything, makes it wrong. Is manipulation an effect, an act, or an event? Is manipulation constitutively wrong – applying only to morally unjustifiable conduct – or is it merely usually wrong? How can manipulation be distinguished from similar, possibly overlapping practices such as coercion and persuasion? Which specific activities – online or offline – count as manipulative? Finally, precisely what values are undermined by manipulative conduct? These are important debates, but we are not going to take a determinate position on most of them. A range of conceptions of manipulation is compatible with the arguments we make in the following. Whether we conceive of it as overlapping with coercion, or whether we demarcate it from coercion in terms of a distinctive sort of harm, trickery, or carelessness that sets it apart from coercion, the downstream implications of emergent manipulation on issues of legitimacy remain largely the same.

The literature on manipulation most often links its wrongness (if and when it is wrong) to impingements on autonomy, which we will here understand in terms of a capacity for self-government (Coons and Weber 2014; Yeung 2017; Susser, Roessler, and Nissenbaum 2019).[3] One way to understand the relationship between manipulation and legitimacy is grounded in the close link between manipulation, the loss of individuals' autonomy, and the implications of this loss on the possibility of democracy. Such an argument works in the following way: if the citizens of a community face a sufficiently strong affront to their capacities for autonomy, they will be left unable to live up to an important civic responsibility, which involves being an informed, conscientious citizen genuinely capable of holding the government democratically accountable. Each of them must be able to critically assess the government's activity and then mobilize accordingly – either in support of the good or in rejection of the bad – or the community will lack a crucial mechanism of democratic accountability. No government can act efficiently unless its citizens can carry out this responsibility, rising to the challenge of holding a government responsible. So, the effects of the IRA and CA's activities at scale is a weakened civil society, rendering effective and responsible government more difficult to achieve (if not impossible altogether). In short, since carrying out one's responsibilities to support civil society requires exercising one's capacity for autonomy, diminishing people's autonomy undermines their ability to underwrite democratic legitimacy to laws, policies, and government actions. Manipulation of this sort makes legitimacy impossible.

Yet, strictly speaking, this argument does not neatly apply to most cases of interest. Not all manipulation has the effect of undermining autonomy or is even harmful. Consider, for instance, apps such as StayFocusd, which allow users to restrict or control their own access to sites and platforms (Klenk and Hancock 2019). To be sure, examples of extreme destruction of autonomy can be found (and appear to be gaining prominence in some online communities see Kang and Frenkel 2020), but this model is, in our view, incomplete. Most election-oriented manipulation is not best understood as deeply affecting the autonomy of any one individual social media user, and the degree to which the IRA and CA campaigns affected any one individual person's autonomy was almost always low. 126 million people – the number exposed to IRA-backed content on Facebook – were not epistemologically incapacitated simply in virtue of having seen IRA-backed content. Even if some of the disaffected youth voters in the Do So! case were simply manipulated, this would not explain the drag on legitimacy posed by the Do So! campaign. This is because the manipulation involved was independent of whether some youth voters were individually manipulated or had their individual autonomy undermined.

Several authors in this volume discuss aggravating factors which appear to make online manipulation more pernicious than manipulation in its more traditional, offline form (cf. Jongepier and Klenk in this volume). It is finely targeted, it exploits dark patterns, and so on. In this chapter, we add another: the practices we discussed in the previous section are examples of what we call "emergent manipulation," which occurs (and matters morally) primarily at the population level.

Here, we adapt the "careless influence" account of individual-level manipulation from Michael Klenk to provide an account of group-level manipulation (Klenk 2021). Specifically, a manipulator (*M*) aims to manipulate a group (*G*) when:

1. *M* aims for *G* to perform some act (φ) through the use of some tactic (t), and
2. *M* disregards whether t reveals eventually existing reasons for *G* to φ.

Klenk's focus is on the manipulation of individuals, and he claims that a key feature of manipulation is carelessness: manipulators are not appropriately sensitive to the reasons of those they manipulate. Our focus is different in two ways. First, we are interested in group-level manipulation. Second, and more importantly, we are interested in a particular *type* of group-level manipulation, viz., emergent manipulation, which involves three additional features. One is that it is *holistic*: it cannot be reduced to the manipulation of individuals. A second is that it is *multiply realizable*: it does not depend on the identities of any specific individuals within the group but can be instantiated by many distinct combinations of those individuals. And third, it involves *distinctive group-level powers and regularities* which do not appear at the level of the individual, such as the mobilization of a social group.

Next, we will distinguish two types of emergent manipulation, and we will discuss each in turn.

3.2 *Stochastic manipulation*

One type of emergent manipulation, we will call "stochastic manipulation." This involves interventions in which no individual is specifically targeted for intervention, and no individual is (or few individuals are) affected so much that their autonomy is undermined. Such practices do, of course, affect some individuals, but they do not affect (or intend to affect) any individual very much, because the intended effect is at the population level. As we see it, stochastic manipulation has two essential features:

1. The approach to the intervention is *dragnet*; it makes initial contact with *many* people but is predicated on the assumption that the behavior of only a *few* will be modified.

2. The aim of the intervention is *marginal*; only relatively few people's behavior needs to be modified to obtain the desired effect.

In addition to these essential features is an additional feature that bolsters the effectiveness of the intervention:

3. The content of the intervention is seductive; those who receive it might already be inclined to agree with it.

3.3 Fragmented manipulation

Another form of emergent manipulation, we will call "fragmented manipulation." This involves interventions in which there is no openly centralized source of influence, and the manipulation is distributed through more localized (and perhaps unwitting) third parties, such as social media influencers. The features of fragmented manipulation are:

1. The approach to the intervention is *distributed*; those who receive it do not receive it from its actual originator but receive it through a more localized trusted source.
2. The appearance of the intervention is *misleading*; the intervention appears to be associated with a genuine social movement but has in fact been produced by a centralized group with a disguised agenda, redirecting support from the genuine movement to an ersatz movement.

Though the two forms of emergent manipulation are different (and they can occur at the same time), what makes them morally significant in this context is their intended effect, which is to increase mistrust. Those who receive emergently manipulative interventions are nudged to lose trust either in their fellow citizens or in prevailing institutions. As we will see next, the effect that these sorts of interventions have on social trust can, under the right conditions, play a delegitimizing effect on governments themselves.

4 Emergent manipulation and drags on legitimacy

In this section, we address some of the moral considerations surrounding emergent manipulation. We argue that the phenomenon can, in some cases, serve as a drag on the legitimacy of a political order (regardless of whether that order would otherwise be legitimate).

Following Fabienne Peter (forthcoming), we see two possible sources of legitimacy for political authorities. One possible source of legitimacy flows from the assent of the democratic will, meaning, as Peter puts it, "how well [the authority] can adjudicate between the potentially conflicting wills of the citizens." We will call this sort of legitimacy "democratic legitimacy." Some theorists describe this criterion of legitimacy in terms of public reason (Rawls 2005, ch. 4), while others describe it in terms of civic participation

(Pettit 2014, ch. 5), but in general, this sort of legitimacy is premised on Rawls's idea of citizens as "self-originating sources of valid claims" (Rawls 1985, 242), whose claims carry moral weight simply in virtue of having been issued from an autonomous will.

A second possible source of legitimacy, Peter argues, involves a higher sort of normative authority to make binding decisions. On this "epistemic" understanding of legitimacy, legitimate policies are those that are "appropriately responsive" to justified beliefs about what should be done (Peter forthcoming). Joseph Raz's "service conception" of authority exemplifies this epistemic source of legitimacy: on this view, duties to comply with authorities can arise when a subject "is likely better to comply with reasons which apply to him" by "accept[ing] the directives of the alleged authority as authoritatively binding and tr[ying] to follow them, rather than by trying to follow the reasons which apply to him directly" (Raz 1986, 53). The exercise of authority over someone, in other words, is justified when that authority is exercised in service of the reasons that person already has. This second way of understanding legitimacy allows some space between what is dictated by the democratic will and what can be regarded as politically legitimate.

In this section, we will argue that emergent forms of manipulation drag on both democratic and epistemic sources of legitimacy.

4.1 Affronts to democratic legitimacy

There is considerable disagreement among scholars of democracy, both about what genuine democracy *is* and about what the value of achieving it might be. We might formulate democracy in direct terms – that is, in terms of majority rule or unanimous consent – or indirectly – in terms of satisfying certain deliberative mechanisms. And we might regard the value of democratic decision-making as instrumental – that is, democracy is useful insofar as it facilitates good outcomes – or we might think that certain procedural features of democratic politics inherently confer legitimacy on the decisions it produces. In any case, democratic politics always has the same basic aim: to adjudicate the conflicting wills of the citizens in service of promoting the common good. Achieving this aim is the key to democratic legitimacy. The challenge, then, is that – contrary to Rawls – it is not plausible to think that people are, in general, self-originating sources of valid claims. Rather, people are often manifestly ignorant, irrational, or unreasonable, and it is difficult to avoid the conclusion that this ignorance, irrationality, and unreasonableness can extend into the political domain.

There is more than one way of viewing the source of democratic legitimacy. One way, often associated with Rousseau, involves the idea of a holistic "general will." According to this view, the common good – which is revealed by but not constituted by deliberative processes – is taken to be distinct from the interests of any individual citizen. Another way of viewing the legitimacy-conferring character of democracy focuses on the structure of the deliberation itself. Josh Cohen, for instance, regards a deliberative procedure as offering legitimacy when the procedure satisfies certain conditions: when it constitutes an ongoing and independent association with final authority, characterized by mutual respect, transparency, and value pluralism, with no suggestion that the results of this process somehow lie apart from the wills of individual citizens (Cohen 1989).

Regardless of whether we view the citizenry holistically or as merely aggregative, successfully executing the deliberative processes of the sort outlined by Cohen still requires a citizenry that has achieved a kind of collective autonomy that stands apart from the interests, preferences, desires, or values of any one citizen. Several of Cohen's conditions refer not to the capacities of any one individual within the democracy but to an irreducibly population-level property: its degree of social trust. The way to understand this property, in turn, is in terms of collective autonomy.

Scholars, of course, have long disagreed about the nature of *individual* autonomy. Some, such as John Christman, understand autonomy as, at bottom, a matter of how individuals' internal motives relate to their history and psychology, while others, such as Marina Oshana, understand autonomy as fundamentally relational (Christman 2011, 154; Oshana 2006, 21–49). Setting aside issues related to collective competence and collective relations for a moment, we can see that the crux of collective autonomy involves what we might think of as "collective authenticity." This is the extent to which a collective would not be, in Christman's terms, "alienated" from a given decision "upon (historically sensitive, adequate) self-reflection." To satisfy this nonalienation condition is to feel and judge that the decision could "be sustained as part of an acceptable autobiographical narrative" (Christman 2011, 155).

Groups, or collectives, can be alienated from their decision-making just as individuals can. To illustrate this notion of collective alienation, we might imagine an assembly of individually well-informed, rational, and reasonable citizens, who all share an agenda of supporting some sort of public good, such as the construction of a public school, park, or health care clinic. However, suppose that the collective lacks adequate social trust, at least in the sense that vague rumors abound throughout the community about "free-riders," leading each of the assembly members to reasonably question the motives of the others, and thus, to question the ultimate practicality of the agenda itself. The failure here involves a lack of common knowledge within the collective, rather than a shortcoming on the part of any individual. This is because although everyone can (by hypothesis) be counted on to contribute to the good (or at least to behave according to some norm of reciprocity) even in the absence of external enforcement, none of the citizens are in a position to reasonably believe that they can count on their fellow citizens in this way. Whatever its merits might be, the policy lacks democratic legitimacy.

There is more than one way to interpret this collective failure. We might interpret it robustly; in terms of, say, a failure to form the "joint intention" implied by each of their individual views (List and Pettit 2013). Or we might maintain a more individualist outlook, arguing that the assembly doesn't "really have any moral status" but that the "distinctively collective interests of individuals mean we must, in some respects, act as if" it does (Lovett and Riedener forthcoming). The key point is that on either interpretation of the failure, the moral of the story is stark: since it is (individually) rational for each member to contribute nothing to the (presumed to be hopeless) public good, everyone voting their own individual interests is a highly stable equilibrium, meaning that no single assembly member would have an incentive to change their voting. It is difficult to imagine a collective that is more alienated: the assembly will not be able to support its own stated agenda despite the unanimous support of that policy from its members.

For an assembly in a complex democratic society to function appropriately – or even get off the ground – it must holistically embody some degree of mutual trust. Within a group, a collective lack of trust functions as a drag on the democratic legitimacy of any group proposal they might consider together: it would be reasonable for any of the assembly members to vote down the proposal. As we will see in Section 4.3, one of the primary effects of CA and the IRA campaigns is to undermine the basis of that trust without violating anyone's individual autonomy.

4.2 *Affronts to epistemic legitimacy*

At first, it might not be evident that there could ever be any source of normative authority apart from that which flows from the will of the people (at least indirectly). How, in a genuine democracy, could there ever be "sufficiently justified beliefs about what should be done" that depart substantively from what the governed themselves have consented to? What kinds of parties could have the standing to interfere with a genuinely democratic decision? And what kinds of issues could be at stake in such cases?

Peter, for instance, offers "[p]olitical decisions that sanction unnecessary harms to small children, that promote slavery, call for genocide, or incite rape and other forms of violence" as clear examples of cases where normative authority can be justifiably exercised against the democratic will (Peter forthcoming). Yet, even in these "clear" cases, it is difficult to decisively justify what should be done and by whom. Any political decision involving guns in schools, for instance, can be expected to raise complex, quasi-empirical issues related to the welfare of children (and others), and a great many decisions involving labor regulations will raise subtle questions about which status inequalities are morally tolerable. As Peter acknowledges, "the epistemic circumstances of politics tend to be such that [epistemically grounded] normative authority is often difficult to establish" (Peter forthcoming). In such an uncertain, risky, and contentious social environment, how could it ever be possible to establish normative authority?

Just as in the context of democratic legitimacy, the linchpin of epistemic legitimacy is social trust and collective autonomy. However, regarding the sort of higher normative authority that is characteristic of epistemic legitimacy, the critical component of collective autonomy is not (collective) authenticity but competence, which is in essence the "ability to effectively form intentions to act, [] along with the various skills that this requires" (Christman 2011, 154). In most cases, assessing an individual person's competence is usually straightforward: is the person minimally rational, self-controlled, and capable of forming intentions that, under normal circumstances, would be effective? Assessing the competence of a collective, in contrast, is much less straightforward. What would it mean to say that a collective is rational, self-controlled, or capable of forming intentions at all?

The key to understanding collective competence involves seeing that when people act collectively, they often do so through public institutions, formal or otherwise. These institutions can be viewed as population-level tools, whose primary function is to stabilize and govern certain kinds of large-scale civic activity. In the United States, the most effective institutional agencies, such as the National Institutes of Health (NIH) and the Federal Reserve, embody forms of bureaucratic competence that allow the population as a whole to respond quickly and flexibly to large-scale problems that do not lend themselves to either political or market-based solutions. But, as the history of economic and political development has shown, these institutions cannot be created overnight or imported from elsewhere. To be effective, they must be grown organically over a long period of time, while exhibiting a proven track record of competence. To be credibly viewed as trustworthy, meanwhile, they must be given a degree of independence from mechanisms of direct democratic accountability – such as electoral politics – that is well-matched to their capacities. Under favorable conditions, and only under such conditions, can these institutions serve as truly self-sustaining sources of trust, and insofar as such institutions can manifest forms of collective competence that cannot be obtained otherwise, we will regard them (where they appear) as collectively good in themselves. So, when bad actors sow misinformation to undermine trust in these institutions, without regard to whether they serve a critical role in supporting public infrastructure or providing any sort of alternative, they serve as a drag on a source of epistemic legitimacy.

While collectively aligned democratic assemblies embody democratic legitimacy, effective autonomous bureaucracies embody epistemic legitimacy. As we have argued, both depend crucially on the presence of adequate social trust to function properly. As we will see next, in addition to undermining collective alignment of democratic will, emergent forms of manipulation can also undermine the effectiveness of self-sustaining trustworthy institutions.

4.3 Emergent manipulation and the sources of legitimacy

The practices of CA and the IRA conflict with both democratic and epistemic sources of legitimacy and without seeming to involve impingements of the autonomy of any particular person.

CA's Do So! campaign bears the hallmarks of emergent manipulation. First, it was stochastic; it did not involve targeting any particular voter for intervention, by getting those specific individuals to behave in any specific way. Rather, the campaign targeted an entire class of voters – youth voters – with the aim of achieving a certain predictable effect only at scale, under specific environmental conditions. Moreover, the campaign did not seek to seriously undermine any one individual's autonomy; that is, exploiting the specific weakness of those who might be highly sensitive to such operations was not the primary goal, and was (in the majority of cases) plausibly not achieved. Rather, the goal was only to persuade a small number of potential voters – recall that Nix described the change as involving only 6% of voters – to feel sufficiently disenfranchised to abstain. Second, the Do So! campaign was fragmented; it did not consist in the open and transparent sponsorship of a political operation. Rather, it involved surreptitious amplification of an existing grassroots movement, paying contributors to propagate the graffiti campaign. Thus, it illicitly borrowed on the populist credentials of that preexisting movement to achieve its goals unencumbered by the mechanisms of accountability that govern political activity.

So, the Do So! campaign falls under the rubric of emergent manipulation. But what – if anything – raises a moral concern with CA's practices in that case? The key threat relates to democratic legitimacy: the practices prevented the political process from reflecting democratic will in the way necessary to avoid collective alienation. While the individual youth voters who abstained from voting might have been able to genuinely affirm their abstention as part of an acceptable autobiographical narrative, the youth voters considered as a group could not have. Indeed, the fact that the Do So! campaign was indifferent to the group's interest in voting, and also depressed that voting, is what makes the campaign manipulative on the definition we articulated in Section 2.1.

The IRA campaigns also involve emergent manipulation. Their main mode of operation includes elements of both stochastic and fragmented manipulation. The goal of the active measures was not necessarily to influence any particular individual not to vote (or alternatively, to essentially spoil one's ballot by voting for a third party) but to mix influences with disenfranchising effects into a media ecosystem in which they have the appearance of organically generated content. And as with CA and the Do So! campaign, the primary mechanism by which the IRA exerted its influence was not by wholly disabling the autonomous capacities of any voter but rather by weakening those capacities or misdirecting them in a subtler fashion. Yet, there is an important difference between the Do So! campaign and the

IRA's "active measures" operations, in terms of their effects on legitimacy. The IRA's practices do, of course, threaten democratic legitimacy in many of the same ways as the Do So! campaign did, but the IRA's operations also threaten epistemic legitimacy. They do not aim simply to manipulate persons, either individually or at scale, but they also aim to undermine the legitimacy of institutions that might otherwise serve as self-sustaining sources of trust (and thus, normative authority), such as the independent news media (DiResta et al. 2019, 65–66). Without a media that enjoys this sort of trust, a government will not be able to implement and publicly justify policies that are appropriately responsive to reasonable beliefs about what should be done. This problem arises regardless of whether the IRA's operations impinge on individual autonomy, because what is required to avoid this problem is not simply an assembly of individually rational and reasonable citizens but a citizenry that is holistically unmanipulated, and that shares common knowledge, understanding, and trust.

5 Conclusion

Within any democratic polity, there will inevitably be individuals whose values are unsatisfied, and there will be others who are treated in ways that are alienating. Such individual-level phenomena may threaten legitimacy, but they are not the only threats to legitimacy. And in this chapter, we considered several examples of "emergent" manipulation that operate at the group level and not necessarily at the individual level. This sort of manipulation, we argued, threatens legitimacy where it is present. In CA's Do So! campaign, the manipulation was stochastic and situated within a polarized and narrowly balanced electoral system where small marginal changes can have a decisive impact. In the IRA's active measures campaigns, the manipulation was stochastic *and* fragmented in the sense that the interventions were not presented as coming from the IRA but were distributed to users through multiple, more localized sources of influence. The presence of these forms of manipulation in electoral politics threatens legitimacy. Understanding these forms of emergent manipulation, and avoiding the temptation to understand manipulation and legitimacy as strictly operating at the individual level, we can better understand the range of threats it can present.

Notes

1. For brevity, we include Cambridge Analytica's former parent company, the SCL Group, under the simple heading of "Cambridge Analytica."
2. Sources from the *Wall Street Journal* described "Mr. Nix's penchant for exaggerating the company's capabilities and work, sometimes to its own detriment." See Ballhaus (2018) and Wylie (2019).
3. In Rubel, Castro, and Pham (2021), we argue for an ecumenical account of autonomy, encompassing both agents' relationship to their wills and social structures within which agents' values, understandings, and preferences develop. Autonomy

requires both a degree of non-alienation and social structures that foster the ability to develop values, understandings, and preferences within reasonable alternatives. See Rubel, Castro, and Pham (2021, 21–42).

6 References

Alba, Davey. 2020. "How Russia's Troll Farm Is Changing Tactics Before the Fall Election." *The New York Times*, March 29. Accessed August 23, 2021. www.nytimes.com/2020/03/29/technology/russia-troll-farm-election.html.

Ballhaus, Rebecca. 2018. "Cambridge Analytica Suspends CEO Alexander Nix After Video's Release." *The Wall Street Journal*, March 20. Accessed August 23, 2021. www.wsj.com/articles/cambridge-analytica-suspends-ceo-alexander-nix-amid-facebook-data-uproar-1521572446.

BBC News. 2018. "Cambridge Analytica-linked Firm 'Boasted of Poll Interference'." March 25. Accessed August 23, 2021. www.bbc.com/news/uk-43528219.

CA Political. 2018. "Do So: Trinidad and Tobago." https://ca-political.com/casestudies/casestudytrinidadandtobago2009.

Channel 4.2018. "Exposed: Undercover Secrets of Trump's Data Firm." www.channel4.com/news/exposed-undercover-secrets-of-donald-trump-data-firm-cambridge-analytica.

Christman, John. 2011. *The Politics of Persons: Individual Autonomy and Socio-Historical Selves*. Cambridge: Cambridge University Press.

Cohen, Joshua. 1989. "Deliberation and Democratic Legitimacy." In *The Good polity: Normative Analysis of the State*, edited by Alan P. Hamlin and Philip Pettit, 15–31. Oxford: Blackwell.

Coons, Christian, and Michael Weber, eds. 2014. *Manipulation: Theory and Practice*. Oxford: Oxford University Press.

Detrow, Scott. 2016. "What Did Cambridge Analytica Do During The 2016 Election [audio file]?" Accessed February 21, 2022.

https://www.npr.org/2018/03/20/595338116/what-did-cambridge-analytica-do-during-the-2016-election

DiResta, Renee, Kris Shaffer, Becky Ruppel, David Sullivan, Robert Matney, Ryan Fox, Jonathan Albright, and Ben Johnson. 2019. *The Tactics & Tropes of the Internet Research Agency* 6, 8–10, 65–6.

George, Kinnesha. 2019. "Different Political Beast." *Trinidad and Tobago Newsday*, August 3. https://newsday.co.tt/2019/08/02/d/.

Hamburger, Tom. 2015. "Cruz Campaign Credits Psychological Data and Analytics for Its Rising Success, sec. Politics." *Washington Post*, December 13.

Hilder, Paul. 2020. " 'They Were Planning on Stealing the Election': Explosive New Tapes Reveal Cambridge Analytica CEO's Boasts of Voter Suppression, Manipulation and Bribery." www.opendemocracy.net/en/dark-money-investigations/they-were-planning-on-stealing-election-explosive-new-tapes-reveal-cambridg/.

Howard, Philip N., Bence Kollanyi, Samantha Bradshaw, and Lisa-Maria Neudert. 2017. *Social media, news and political information during the US election: Was polarizing content concentrated in swing states?* (Computational Propaganda Research Project Data Memo 2017.8/September 28 2017), 1. Oxford Internet Institute.

Jongepier, Fleur, and Michael Klenk. 2022. "Online Manipulation: Charting the Field." In *The Philosophy of Online Manipulation*, edited by Jongepier, F. and Klenk, M., 15–48. New York, NY: Routledge.

Kang, Cecilia, and Sheera Frenkel. 2020. "'PizzaGate' Conspiracy Theory Thrives Anew in the TikTok Era." *The New York Times*, June 27. Accessed August 23, 2021. www.nytimes.com/2020/06/27/technology/pizzagate-justin-bieber-qanon-tiktok.html.

Klenk, Michael. 2021. "Manipulation (Online): Sometimes Hidden, Always Careless." *Review of Social Economy*. 80: 1, 85–105. doi:10.1080/00346764.2021.1894350.

Klenk, Michael, and Jeff Hancock. 2019. "Autonomy and Online Manipulation." *Internet Policy Review*. Accessed February 28, 2020. https://policyreview.info/articles/news/autonomy-and-online-manipulation/1431.

List, Christian, and Philip Pettit. 2013. *Group Agency: The Possibility, Design, and Status of Corporate Agents*. Oxford: Oxford University Press.

Lovett, Adam, and Stefan Riedener. "Group Agents and Moral Status: What Can We Owe to Organizations?" *Canadian Journal of Philosophy* (forthcoming).

Matz, S. C., M. Kosinski, G. Nave, and D. J. Stillwell. 2017. "Psychological Targeting as an Effective Approach to Digital Mass Persuasion." *Proceedings of the National Academy of Sciences* 114 (48): 12714–19. doi:10.1073/pnas.1710966114.

Noujaim, Jehane, and Karim Amer. 2019. *The Great Hack*. Netflix. Documentary.

Oshana, M. 2006. *Autonomy in Society*. Hampshire: Ashgate.

Peter, Fabienne. "The Grounds of Political Legitimacy." *Journal of the American Philosophical Association* (forthcoming).

Pettit, Philip. 2014. *Just Freedom: A Moral Compass for a Complex World*. New York, NY: Norton.

Premdas, Ralph. 2019. "Ethno-nationalism and Ethnic Dynamics in Trinidad and Tobago: Toward Designing an Inclusivist Form of Governance." In *The Palgrave Handbook of Ethnicity*, edited by Steven Ratuva, 809–24. Singapore: Springer.

Rawls, John. 1985. "Justice as Fairness: Political Not Metaphysical." *Philosophy and Public Affairs* 14: 223–51.

Rawls, John. 2005. *Political Liberalism*. New York, NY: Columbia University Press.

Raz, Joseph. 1986. *The Morality of Freedom*. Oxford: Oxford University Press.

Rubel, Alan, Clinton Castro, and Adam Pham. 2021. *Algorithms and Autonomy*. Cambridge: Cambridge University Press.

Select Committee on Intelligence, United States Senate. (2019). *Report of the Select Committee on Intelligence, United States Senate, on Russian active measures campaigns and interference in the 2016 U.S. Election, volume 2: Russia's use of social media and additional views* (Report 116-XX), 6–7.

Susser, Daniel, Beate Roessler, and Helen Nissenbaum. 2019. "Online Manipulation: Hidden Influences in A Digital World." *Georgetown Law Technology Review* 4 (1): 1–45. Accessed February 27, 2020.

U.S. Department of Justice. (2019). *Report on the investigation into Russian interference in the 2016 Presidential Election, Volume II of II*, 24-6, 31-2.

U.S. District Court for the District of DC. 2018. U.S. v. Internet Research Agency, LLC, No. 1:18-cr-00032.

Weaver, Matthew. 2018. "Facebook Scandal: I am Being Used as Scapegoat – Academic Who Mined Data." *The Guardian*, March 21. Accessed August 23, 2021. www.theguardian.com/uk-news/2018/mar/21/facebook-row-i-am-being-used-as-scapegoat-says-academic-aleksandr-kogan-cambridge-analytica.

Wylie, Christopher. 2019. *Mindf*ck: Cambridge Analytica and the Plot to Break America*. New York, NY: Random House.

Yeung, Karen. 2017. "Hypernudge: Big Data as a Mode of Regulation by Design." *Information, Communication & Society* 20 (1): 118–36.

Part IV

Legal and regulatory perspectives

19 Regulating online defaults

Kalle Grill

1 Introduction

Our lives are increasingly lived online. We shop online, get informed online, get entertained online, date online, and stay in touch with our friends and family online.[1] The online environment where we do these things is largely created by private corporations that shape it according to their own aims. Meanwhile, the digital spreads into the physical, with increasingly intelligent machines tracking and influencing also our offline behavior (see Zuboff 2019, esp. ch. 7, 10, 12–16). Yet there are relatively few laws that regulate how our online environment may be shaped. This is probably due, at least in part, to the internet's global reach, its relatively young age, its rapid development and change, and to some extent its history as an arena for the free exchange of information (see van Dijk 2020, esp. ch. 9).

In this chapter, I focus on one aspect of corporate design – default-setting. Default-setting is a form of behavioral influence – a way to shape behavior, which is often quite effective.[2] While default-setting has been much discussed as a form of so-called nudging, it deserves more focused attention. In the nudging debate, the focus has primarily been on benevolent influence, but the main concern online is, arguably, default-setting for profit. By paying more attention to defaults, we can identify paths to regulation that promote and protect the social good and counteract corporate manipulation.

In the following, I will describe what I take defaults to be and how defaults online fail to align with users' interests, either as individuals or collectively. I will argue that defaults are instead often harmful and manipulative and so should be regulated. Specifically, I propose that we impose quality constraints, such that products whose default settings do not sufficiently promote users' interests should not be legal. Regulation of default quality must to a large extent be a piecemeal process, but this process can be guided by general principles. I end the chapter by proposing four such principles: *non-consumption*, *minimal collection of data*, *minimal opt-out costs*, and *truth*.

DOI: 10.4324/9781003205425-23

2 Seemingly innocuous defaults

A *default* option is that option that is realized if one *remains passive* in some sense and from which one can *opt out* by taking action. *Default-setting* is the design or shaping of some choice situation such that a particular option becomes the default. For example, it may be the design of a product (or service) such that some form of use becomes the default. One example is that operating systems or graphical user interfaces come with default browsers that are activated when users click on a link in a non-browser program. For Windows, the default browser is Explorer, for OS X, it is Safari, and for the Unix-based interface Debian GNOME, it is GNOME Web.

When default-setting is discussed in the behavioral economics and psychology literature, as well as by ethicists, it is typically taken for granted that the option that a designer perceives to be the default will also be perceived as the default for the agent in the choice situation. This need not be the case. For example, social pressure may cause it to be a subjective default to change one's default browser to Chrome upon installing an operating system. From an ethical perspective, it is arguably the subjective default that is most important. However, I will go along with the assumption of convergence between the designer and the agent perspective, since in the online environment, I believe the two typically converge.[3]

People often stick with the default, for several reasons: opting out requires more effort, which carries a cost; knowing when to opt out and what to opt for requires attention, deliberation, and decision-making, which often carries a higher cost; opting out entails personal responsibility and sometimes social accountability, both of which are nice to avoid; the default is often perceived as a reference point and so opting out entails a perceived loss; and we are often overcome by irrational inertia. In addition, for consumer products and in other contexts where agents are defaulted by design, the default is often perceived as implicit advice (see e.g., Choi et al. 2003; Sunstein 2015, 34–39).

While defaults are, for these reasons, "sticky", they are also seemingly innocuous, for two reasons. First, a situation lacks a default only if there is no way to remain passive, so that any behavior is an active choice. Not only are such default-free situations rare, they are also very difficult to produce by design, especially for groups or populations, since different people may perceive different options to be the default. Second, even when default-free situations can be created, this brings no obvious benefit. To the contrary, absence of defaults means that an active choice is required, which carries costs, as just explained. While agents are often willing to assume these costs, it is rarely a benefit to be forced to assume them.

While it is, for these reasons, innocuous to design products or choice situations with defaults rather than without them, it is *not* innocuous to implement some particular defaults rather than others. Designers often have some agenda for what outcomes are desirable from their perspective. If they are sufficiently acquainted with behavioral psychology, they can design products so that default use leads to those outcomes. When the methods and the aim of this influence are covert and do not engage an agent's rational agency, the influence is arguably manipulative. When the outcome goes against an agent's interests and desires, it is likely to cause harm and frustration.

3 Online defaults that track, distract, and misinform

Online defaults are particularly powerful because they leave few means of escape. In the physical world, default-setting is constrained by material circumstances and people can often be innovative and creative in how they relate to defaults. A landscape designer may decide the route of a footpath, but unless the path is surrounded by high and sturdy fences, people can choose to take shortcuts and detours. In the online environment, designers are less constrained by physical circumstances and users are more susceptible to the resulting design (unless they are capable hackers).

In this section, I will survey some prominent examples of online defaults and note how they are problematic for users. Consider first the collection of personal information. Until recently, most commercial websites routinely collected and stored information about our online behavior without giving us a choice in the matter. After the introduction of the EU General Data Protection Regulation (GDPR), many websites offer us a choice of what data to share (European Commission 2016). This choice, like most choices, typically comes with a default option.

While some web pages make it as easy to "reject all" as to "accept all" tracking, avoiding tracking is often the less salient and more cumbersome alternative, making tracking the default outcome. One common design approach is to present two options – either to accept the default of comprehensive tracking or else choose "more options" or some such, which leads to a list of different kinds of cookies, with more or less transparent names and descriptions, to be considered separately.[4] This is an instance of what Cass Sunstein calls "simplified active choosing" (Sunstein 2015, 115). By this, Sunstein means a primary choice between on the one hand a default option and on the other hand a richer set of options, which enables a more active choice. The point of designing for simplified choice is to harness the power of clear defaults, while also offering ambitious choosers many options. Of course, the richness of the options presented on the active choice path often makes this path quite cumbersome and time-consuming. In the case of online tracking, many users cannot be bothered with this exercise and so give up their privacy, despite reporting that they value it highly.[5]

Consider next social media engagement. As is now well known, social media platforms are built to catch and keep our attention.[6] Social media users are not the corporation's customers but rather their resources or products, since revenue comes from advertisers who pay for user attention. One downside of this business model is that many users become addicted, with associated depression and social anxiety (Foroughi et al. 2019). Another downside is that people simply spend a disproportionate amount of time and energy online, at the expense of more worthwhile pursuits. The long-term consequence of design for engagement in combination with collection of personal information is that information (or "surveillance") capitalists control more and more of our lives and induce us toward complacency and consumption (e.g., Frischmann and Selinger 2018; Zuboff 2019, esp. ch. 18).

The methods social media corporations use to keep us engaged include default-setting in no small part. Consider Facebook's News Feed and similar feeds on other social media – a scrollable column of information, centrally placed. While the content of the feed is partly based on each user's previous behavior, the resulting stream of content is a default selection from the vast amount of information available on the platform. Though it is technically possible to seek out other information while avoiding the feed, it is psychologically very difficult. As a result, social media users are defaulted into content.

Not only are social media users defaulted into particular content, but the way this content is presented is also itself a default. So called *infinite scrolling* (that we can scroll down for more content, without end or interruption) is an invention intended to promote prolonged engagement or addiction (Andersson 2018). Though it is often technically possible to avoid scrolling, it is psychologically very difficult.

Both infinite scrolling and content selection algorithms for feeds are of course contingent design features. Social media platforms could instead display a list of friends, group, events, and so on with only their stable properties – names and perhaps pictures or short descriptions, from which users could actively choose to get the latest updates (this is indeed what early versions of Facebook looked like). Such a design would be more driven by user activity and curiosity and less by automatic behavior. It would not, however, as effectively serve social media corporations' aim to maximize use (it might also be less interesting or less fun to use).

Recent political events like the storming of the US Capitol in January 2021 and related conspiracy theories about election fraud and more have helped shift the main concerns with social media away from addiction, scattered attention and time waste to a perhaps even more worrisome concern with false information, in particular "fake news" (Lazer et al. 2018). Facebook and other media companies routinely distribute "information" that is misleading or outright false. The problem is aggravated by personalization of content, which is essentially defaulting users into content selected for them based on their previous behavior and the aims of the platform provider. Personalization means that some users are consistently exposed to similar falsities, making them more vulnerable to integrating fundamentally false viewpoints on everything from political elections to the side-effects of vaccination.

As argued by Eli Pariser (2011, 9–10), the digital "filter bubble" we occupy online differs from conventional selective media consumption in three ways: 1) you do not choose to enter the bubble – in other words, it is a default, 2) the bubble is opaque or "invisible" – services do not typically tell you how or even that they are defaulting you into particular content, which can easily make you assume that the information you receive is true and unbiased, and 3) the bubble is adjusted to you as an individual, not to a group to which you belong, which makes it more effective (and divisive). Even for those of us who are aware that we are being defaulted into personalized content, the psychological tendency to interpret defaults as implicit advice means that we are inclined to suppose that information selected for us should be particularly good (personal confession: I believe that my social media flow is a great source of information because my connections are very wise and well informed).

Consider finally a different sort of default for prolonged engagement, which is more obviously designed to induce consumption: the autoplay function that has become standard for video streaming services, including public service platforms in some countries (such as Sweden). This standard was pioneered by Netflix. As a default, end titles are quickly interrupted by the start of the next episode in the series or else by a trailer for another movie or series. As of February 2020, Netflix users can opt out of this form of autoplay, as well as from the previews that otherwise autoplay when users hoover over some content. Still, both types of autoplay are operative by default, and many users are not aware that there is an opt-out. The opt-out is to some extent hidden away in the settings section, which many users have no (other) reason to access. Obviously, autoplay induces extended television watching, which contributes to a sedentary lifestyle, which is known to contribute substantially to poor health.[7]

4 Online defaults that harm and manipulate

As we saw in the previous section, online corporations employ sophisticated default-setting design features in order to induce particular user behavior. These behaviors are often harmful to users, in the sense that they cause loss of privacy, excessive consumption of social and other media, and distorted worldviews. Such outcomes are bad for the individual user and often bad for their societies and fellow citizens. Personalized marketing for consumer products helps cement a culture of consumerism that arguably causes decreased life satisfaction and definitely causes accelerated environmental degradation. The undermining of respect for privacy and the culture of exposure induced by social media may make us all more vulnerable to microtargeting designed to sway elections, undermine social trust or advance other dubious agendas (see Véliz 2020, ch. 4). Other people's warped worldviews may cause them to harm us as individuals as well as our liberal democratic institutions, where they exist.

The main way in which default-setting causes harm is by inducing users to accept suboptimal outcomes. However, default-setting can also harm users by imposing opt-out costs on the options that best align with their interests, so that even users who are attentive and proactive enough to opt out of poor defaults have to pay some cost in order to do so. Though this cost can be small on any one occasion, it can be excessive in the aggregate, because default-setting is so prevalent. For example, in order to minimize online tracking, a frequent internet user must assume the excessive cost of carefully checking the privacy settings on all websites visited and make sure to click the right boxes. Since websites typically do not remember earlier opt-out choices but rather revert to default tracking, the aggregate cost remains high over time.

Harmful as it is, what I have described may be seen as typical and expected market interactions. Markets are indeed potential arenas for ruthless exploitation and manipulation, and the associated deterioration of human relationships is perhaps a cost to be accepted for the benefits of efficiency. However, some factors make the online market especially worrisome. First, the internet is a global market and so the personal relationships and the sense of community that can be a counterweight to impersonal market interactions are weak. Second, online services can be provided without any personal interaction, unlike most physical contexts, where end users must at some point be faced. Third, the fact that users are oftentimes not the corporation's costumers, but rather their product, means that corporations have less reason to be responsive to user interests and possible grievances. Fourth, the online environment is relatively new and changes relatively fast, which means users are less prepared and less able to notice and counteract sophisticated corporate influence.

Though markets can be ruthless and manipulative, they need not be, but can instead be arenas for rational cooperation for mutual benefit (see, e.g., Maitland 1997 for this perspective on markets). Such cooperation, however, presupposes honesty and transparency. Much online default-setting is characterized by quite the opposite – deceit and nonrational influence.

Some scholars hold that deceit or covertness is at the core of manipulative behavior.[8] Of course, there is no moral requirement to always disclose one's intentions. Covertness may be morally problematic, however, when circumstances are such that without transparency, others will be induced to form false beliefs, rather than just stay uninformed (cf. Scanlon 1998, ch. 7). There are at least three distinct aspects of a behavioral influence that may be more or less covert: the very fact that there is an influence, the method of influence, and what motivates the influence.[9] The practices of online media corporations are arguably covert in all these three ways. There is lack of transparency regarding the basic fact that when we use online services, we are influenced to spend as much time as possible online, and that our behavior is tracked as we do so. There is also lack of transparency regarding the methods used to keep us online, for example how our personal information is stored and processed, and how we are induced to share content with friends and acquaintances to make them feel that they are socializing and conversing rather than receiving advertising. There is, finally, a lack of transparency regarding the reason these services are offered, for free, that is, the very business model of selling personal data and user attention.

More specifically for our current purposes, there is lack of transparency regarding the fact that online services use default-setting, in what way online defaults influence behavior, and for what purposes. While there is hopefully growing awareness about the business model of selling user data and user attention to advertisers, this is not due to any efforts by the corporations. More transparency could be easily achieved. There are global approaches, such as informing users upon signing up for a service how the service is financed and that it is designed to maximize use. There are also local approaches, such as making individual defaults more salient, by, for example, displaying some appropriate message before the auto play of another episode. Some problematic defaults, such as defaults into false information, are unintended side-effects of the business model. However, corporations are aware of these effects, and most do very little to either stop them or make users aware of them.

Another feature often taken to be central to manipulation is influence by means other than rational argument and information, which may be disrespectful of our rational agency and contrary to moral market interactions.[10] That it bypasses rationality is perhaps the most common complaint against benevolent nudging (e.g., Bovens 2009; Grüne-Yanoff 2012). Defenders of benevolent nudging have pointed out that nudges can induce people to deliberate more rationally and better respond to reasons that apply to them (e.g., Hanna 2015; Schmidt 2019). In the case of most online corporate default-setting, however, users are clearly not induced to be more rational. To the contrary, the defaults are arguably designed to overcome many users' rational desire to protect their privacy and limit their consumption. These defaults likely have negative impact both on user's process rationality – the degree to which they choose rationally, and their outcome rationality – the degree to which they realize their own long-term goals (e.g., Engelen 2019).

5 A potential countermeasure: minimal requirements

The problem we face as a society is that much of our online environment is harmful and manipulative. More specifically, nonrational behavioral influence is used covertly to induce behavior that does not align with our best interests. One central method of nonrational influence is default-setting, that is intentional design of online environments that makes some options the paths of least resistance. Because we are prone to stick with the default, default-setting is an effective means of influence. Because opting out of the default always requires some effort and often has other associated costs, setting a poor default is harmful also to users who opt out. Online default-setting affects our online environment but also to a growing extent our physical environment, since we increasingly get our information and make our purchases online, and since our household and wearable devices are increasingly connected to the internet.

There are many countermeasures that could be taken to mitigate harm and reduce manipulation. Carissa Véliz (2020) has a long list of interesting proposals when it comes to protecting privacy. In terms of public policy, efforts could be directed at limiting the potential for effective manipulation, for example by banning personalized advertising and trade in personal data (Véliz 2020, 119–30). Personalized advertising is a plausible target for a ban because it could be argued that it is inherently manipulative and morally problematic, and we could do fine without it. We could also ban the use of some surveillance technologies, like facial recognition (Véliz 2020, 154). Online default-setting, however, is not something we can do without. We could try to ban the most manipulative defaults, in the sense of the most covert and most non-rationally influential, but this seems very difficult, since defaults are pervasive, often imperceptible and always have non-rational influence.

I propose that one plausible policy response to harmful and manipulative default-setting is to impose limits on what sort of outcomes users may be defaulted into. Just as there are limits to what products may be at all legally marketed, there could be limits to what defaults products may have. These limits could apply to pure online services as well as to goods purchases online, and really to all products and services. We are used to the idea that vehicles and machinery may not be too dangerous to use; that pharmaceutical drugs must be verifiably effective; that loans may not have excessive interest rates; that consumer products may not be excessively priced: that online content may not include hate speech or child sexual abuse material. These are some options that should not be accessible at all, defaults or not. In addition, however, consumer products should not have default-settings that are too poor for most users in most circumstances, even if these options should be accessible. While it should of course be legal to share one's entire photo collection openly online, it would be outrageous if such sharing was the default outcome of using some social media platform. This would be outrageous even if information about the default rule was clearly and visibly presented and it was easy to opt out. Defaults that are too poor should not be legal. General quality limits for defaults would protect us from options that are very poor for most of us most of the time, while leaving it open to opt into those options when beneficial. As a complement, it would also be sensible to regulate opt-out costs. Just as default outcomes should not be too poor, opting out of them should not be too costly.

To impose legal limits on defaults is to implement a two-tier system: there would be two quality thresholds that a product would have to pass to be legal. First, the product would have to be of sufficient quality generally, which includes obstacles to outright dangerous or otherwise very poor options. Second, the products' default-settings and the associated opt-out costs would have to be of sufficient quality. One can certainly argue about where lines should be drawn. This goes for general quality as well as for defaults. Options that are harmful to most users on most occasions may be beneficial to some people in some circumstances and so should perhaps be legal. For example, it is rarely in anyone's interest to accept a loan with a 60% interest rate, which is the highest legally permitted interest rate in Canada. However, it may be rational to do so if one needs the money short-term to avoid losing one's home or to pursue some very lucrative trade opportunity. Usury laws should strike a balance between protecting the many people for whom it is not rational to accept extreme interest rates and benefiting those few people for whom it is on occasion rational. So too must regulation of defaults strike a balance between competing considerations.

From any user's perspective, it is best to be defaulted into an optimal option. However, it is not reasonable to expect that others will arrange one's choice situations in such a way or even try to do so. In some cases, corporate interests align with the interests of suppliers and customers. More rarely do they align with the interests of their product or resource, such as social media users. Whether we endorse a stakeholder or shareholder perspective on firms, market actors form their strategies and produce their products quite freely. This is why market actors need to be regulated to protect social interest.

6 Some principles for minimal defaults

Limits for defaults must to some extent be set piecemeally. What interest rates may be offered as the default option for different kinds of loans will have to be considered in the context of loans and considerations relating to usury. Similarly, defaults for online services must be considered in the context of the online environment. However, I believe there are some general principles that can guide policy development. Clearly, the principle that default-setting should optimize outcomes for choosers would be too demanding. On the other hand, the principle that default-setting must not be too harmful is too vague to be useful. Suitable guiding principles should be both morally plausible and practically action-guiding. In the following, I will propose four principles in three subsections. I will discuss the first principle in some detail and the others more briefly.

6.1 Non-consumption

Cass Sunstein is a strong proponent of benevolent default rules. In his book, *Choosing Not to Choose* (2015), he considers defaulting people into purchases of household items like books, sneakers, and toilet paper, in the sense that these goods would arrive at your door without you having ordered them. Sunstein calls this "predictive shopping" and seems to view it quite favorably, as it would spare people the effort of choosing. Predictive shopping is so far quite unusual and often illegal. In our online environment, however, we regularly face advanced predictive and personalized advertising and purchase options so salient and easily realizable that purchasing may sometimes be the default. In addition, we have products that, after an initial purchase, defaults us outright into further purchases or other consumption. One way to counter the overall pressure and inducement to consume is to regulate defaults. I propose, *pace* Sunstein, that *non-consumption* is a strong candidate for a regulative principle for defaults, online as well as offline. Non-consumption prevents any form of *automatic consumption*, that is, consumption that you are defaulted into, whether or not it includes purchases.

Non-consumption is a plausible regulative principle for several reasons. First, non-consumption is our historically established and perhaps biologically suitable baseline. If we allow automatic consumption, then as our world becomes increasingly digital, we may be defaulted into massive consumption. We are not cognitively equipped for a world where we shape our lives by opting out of otherwise automatic consumption, nor brought up to confront it. Most parents still encourage their children to make active choices, to actively pursue their ambitions and develop their talents, rather than bombard them with items and fill their schedules, waiting for them to become autonomous by breaking away from a stream of defaults. Though we could decide to purposefully change the way we interact with the world from opting in to opting out, we arguably should not accept such a radical change to happen as a result of blind market forces.

Second, as Sunstein admits, sellers may not accurately predict what their potential customers want or need: "Requiring active choosing in ordinary markets minimizes the cost of error" (Sunstein 2015, 94). This goes not only for defaults into purchasing or using a product but also for additional consumption that follows after the initial purchase. For example, it is best to require an active choice for someone to buy a subscription to a streaming service, but it is also best to require an active choice for another movie to be streamed to them, to ensure that this is what they really want.

Third, active choice is conducive to autonomy or authenticity, in the sense of creating one's own life. My life is to a larger extent of my own making if I own and use the products I do because I deliberately choose them.[11] This may not be true for all products all of the time. As Sunstein has argued elsewhere, there may be reasons to facilitate everyday choices in order to leave more room for reflection about important ones and thereby promote autonomy (e.g., Sunstein 2014, 21). However, I propose that this is at most an argument for stimulating the development of high-quality subscription services for everyday products, services that consumers can actively opt into, not for defaulting us into consumption.

Fourth, automatic consumption undermines accountability, since no real choice is made in order for consumption to take place. The majority of purchases, as well as the consumption of online products such as streaming, contribute to environmental degradation. While each instance has very little and untraceable impact, the aggregate environmental harm of allowing automatic consumption is great. Moreover, the high level of consumption in rich countries and among rich people in poor countries does not, for the most part, make consumers better off. Even if we disregard coordination problems such as investments in positional goods, many individual purchases do not promote individual happiness, especially if we consider the labor required for obtaining the means of purchase. Arguably, only a very small portion of our consumption promotes such constituents of (or contributors to) human well-being as love, community, friendship, accomplishment, creative expression, enjoyment of beauty, or even (net) pleasure. It may of course be debated both what is the nature of human happiness and what particular consumption contributes to it. Because consumption is harmful to the environment, however, and because it is at least not obviously justified by its benefits, we should hold ourselves and others accountable for our purchases and our consumption.

By allowing consumption to be the default order, we allow ourselves to be shaped by it. As Sunstein notes, we often do not have a preference independently of the situation we find ourselves in, so that "the default rule may help to construct that preference" (Sunstein 2015, 38). Once I have a pair of sneakers delivered to my doorstep, or a movie running on my screen, the endowment effect will kick in and I will be less prone to return the shoes or turn off the movie than I would have been to abstain from bringing the shoes home or starting the movie. Furthermore, opting out requires effort, and so inertia will contribute to cementing the established state of things, preventing me from returning the sneakers or turning off the movie. By allowing automatic consumption, we allow consumption to be the privileged alternative, not only supported by consumerist social norms and the pervasive advertising that fuels them, but also by a number of other psychological biases. Therefore, we are better off with non-consumption defaults.

How might a principle of non-consumption be used in practice: does it give any action-guidance? Non-consumption clearly excludes predictive shopping. Arguably, it also excludes the marginally more modest practice of providing potential buyers with unsolicited goods or services, hoping that they will purchase them. Unsolicited goods are a default into consumption, since once a product has been delivered to me, the default is to make use of it. The non-consumption principle therefore underpins UK law on unsolicited goods, according to which such goods become the property of the recipient and it is a criminal offence to assert a right of payment for them. On the other hand, the principle arguably undermines New Zealand law, which imposes legal duties on recipients of unsolicited goods, including to keep these goods available for collection for ten days (New Zealand Legislation 2021). The default reaction to having a consumer good delivered to you in your home is arguably not to store it for collection.

In addition to one-off, one-good purchases, the principle of non-consumption must handle amalgam products and services. Goods can be amalgamated, or bundled, in two dimensions. First, a number of discrete items can be sold together, at the same time. This is what is traditionally called "product bundling", a well-known marketing tool in both offline and online contexts (e.g., Yang and Lai 2006). Second, the same or similar products can be sold repeatedly, over time, such as under a subscription. Both strategies are enhanced in digital environments. Product bundling is enhanced online because bundled offers can be made based on your tracked previous shopping behavior. Subscription services are enhanced online because it is easier to include a preselected checkbox for a subscription during an online purchase than to default a customer in a physical store into a subscription.

There are very many possibilities when it comes to specifying non-consumption along the two dimensions of additional items and repeated purchases over time. Defaulting consumers into automatic renewal seems clearly in violation of non-consumption. There is room for debate on this point, but I do not find it implausible to impose the strictest possible interpretation of the principle, such that some minimal purchase option should always be the default, while any additional items must be actively chosen. Compliance with such an interpretation of the principle might do much to remove excess consumption. I could even countenance a maximum duration for subscriptions that are actively chosen, perhaps of varying length depending on the product but with most subscriptions limited to one year. After all, the option to renew will no doubt be made very accessible, so the cost to users of having to renew annually should not be large. On the other hand, indefinite subscriptions can be expensive and are not rarely unwanted.

Consider now the defaults imposed by social media and streaming services such as Facebook and Netflix. By autoplay, Netflix defaults its customers into consumption of content, even if it does not involve purchase. The principle of non-consumption would arguably rule out this default. The obvious conventional minimal consumption option when watching movies or episodes is one movie or one episode. To add additional episodes is to bundle distinct goods.

It could be argued that non-consumption would exclude services like Netflix altogether, since their very nature is to allow unlimited consumption (for some time period) after one purchase. Such consumption seems far from the minimal consumption option. The same argument could then be made, and would for consistency have to be made, against all-you-can-eat buffets and many other buffet-style goods, where the minimal consumption option might be one dish, one portion, or similar. However, content access and buffet meals are arguably different in kind from a series of purchases of individual movies or dishes. The experience of having access to all that content, and the experience of getting to eat as much as inclined of all that food, has an important quality in its own right. This is why streaming services (like Netflix and Spotify) were such novelties when they appeared. Hence, non-consumption would not exclude buffet-style goods. Similarly, there is a special quality to having a subscription to, for example, a weekly magazine, knowing that you will receive it regularly without taking further action. Indeed, non-consumption should not preclude *opting into* a (time-limited) system of predictive shopping.

As for social media feeds, it is less obvious what should be the minimal consumption option, because the product or service is itself a novelty. One possibility, however, is to impose the default that only some limited amount of content can be viewed before an active choice is required to view more. A simple "load more" feature could appear after scrolling through a few updates (cf. Sharma and Murano 2020). This simple feature might potentially reduce addictive and mindless use.

6.2 *Privacy without cost*

One of Véliz's (2020) proposals to protect privacy is to stop default collection of personal data. The idea is that the default option on any product or website should collect and keep only such data as is necessary for its functionality.[12] This is a principle for regulating default-setting: *minimal data collection*. It should be emphasized that the necessity in question refers to the individual user's perspective. Necessary functionality should not be taken to include, as Véliz notes, funding the site by trading in collected data, or general product development. For example, a navigation app needs to collect location data in order to provide continuous directions. It does not, however, need to store that data once a trip is completed. Storing location data should be an opt-in possibility if it should be available at all.

Arguably, non-collection is in the best interest of most users in most contexts. However, there may be benefits to tracking, for the individual who can be targeted with more relevant advertising, and for society, which can harness the wealth of personal information for various purposes. Sunstein argues, for these sorts of reasons, that "privacy is smart for each but dumb for all" (Sunstein 2015, 30). On the other hand, Stuart Russel proposes that general insights about aggregate behavior can be reached based on encrypted data "without compromising privacy in any way" (Russell 2019, 71). Hence, minimal collection may mean a prohibition of all collection of personal data but only of the collection of *non-encrypted* personal data.

When personal data must be collected for some product to function properly, Véliz argues that this should happen only after a person "meaningfully and freely consents" (Véliz 2020, 133). In that spirit, let me propose a complementary principle: *minimal opt-out costs*. In analogue to minimal data collection, this principle prohibits the unnecessary or artificial imposition of opt-out costs. It prohibits, in other words, any opt-out cost that is not necessary for functionality.[13] A website that collects personal data without providing an easy and effective "reject all" option would have to explain, ultimately in court, why such an option would harm functionality. Minimal opt-out costs may potentially apply to all default-setting, not only when it concerns privacy. In particular, whenever we deem that some treatment of a user is so potentially problematic that active consent should be required, it makes sense to also require that withholding consent should not be unnecessarily costly or cumbersome.

Sometimes, there is a trade-off between functionality and minimal collection and possibly also minimal opt-out costs, rather than a clear necessity for collection or for substantial opt-out costs. For example, it may be that the more extensively a person's behavior is tracked, the better recommendations can be issued for her. In such cases, meaningful consent should be required, and collection and storage should be kept to the minimum required for providing good service to the individual user. In addition, any blocking of content and any interruption to ask for consent reduces functionality. Hence, unless this reduction is outweighed by increased functionality for the individual user, requests for data collection must be minimally intrusive.[14]

6.3 *Truth*

After many years of criticism and calls to action, some social media corporations have recently started to manage their content for accuracy or truth. In mid 2019, Twitter began to flag content from influential politicians' accounts that violates their rules (Twitter 2019). These rules do not focus on the quality of information, however, but rather on various possible harms of information. In February 2020, Twitter introduced new rules specifically against synthetic or manipulated media likely to cause harm (Twitter 2020a), and in May 2020, rules specifically regarding COVID-19 against "misleading information" and "disputed claims" (Twitter 2020b). This trend may indicate a growing concern with truth. The storming of the Capitol in January 2021 led to various responses from social media corporations against President Donald Trump, including Twitter permanently suspending his account and Facebook and Google's service YouTube suspending his accounts temporarily. The immediate cause was Trump's incitement to violence, though this incitement rested on false claims about election fraud and would not have been as effective without them.

It has of course always been a problem that people are misled by false information. We cannot eradicate lying if we want to respect freedom of expression. However, this freedom is not unlimited. Among the few limits on free speech implied by Article 17 of the European Convention of Human Rights, one concerns "casting doubt on clearly established historical facts" (Council of Europe 2018). Similarly, prohibitions on defamation in different jurisdictions typically prevent statements that are harmful *and false*.

In democratic countries, limits to freedom of expression are few and specific. They mostly concern public communication (though forgery and fraud are illegal also in private contexts). However, it is arguably reasonable to demand more from information that people are defaulted into receiving than from information they actively seek out. While it is unfortunate that people actively seek out websites that reinforce prejudice and misunderstanding, it is arguably worse if people are served such information without even seeking it out, simply because they use a media service that presents content to them based on their previous behavior. Russell (2019, 197) proposes that there should be a right to "mental security – the right to live in a largely true information environment". We might add that there should at least be a right to do so unless one has actively opted out of such an environment. We might consider, therefore, a principle of *truth*, that is, a requirement that information provided by default is true or at least not demonstrably false or against expert consensus. It may be that some existing products, such as social media platforms, cannot function without frequent violation of such a principle. If so, we might accept that those services must be altered or developed in order to remain legal.

7 Conclusion

Markets are and should be regulated by principles that protect the social good. There are some absolute requirements on consumer products: they may not be too poor or too harmful and marketing of them must be truthful. Individual customers cannot opt into buying a very dangerous product or being targeted by untruthful advertising.

Many products come with default settings. Online, we are often defaulted into consumption, into giving up our privacy, and into receiving false information. This is often harmful and manipulative. One potential countermeasure is to set minimal quality limits for defaults, as a complement to the more established limits for what options are at all acceptable.

When developing minimal quality limits for defaults, we may be guided by general principles. I have proposed four such principles. I presented the principle of non-consumption in some detail. I considered more briefly the principles of minimal data collection, minimal opt-out cost, and truth. If my ideas are on the right track, more discussion would be useful in order to develop and specify these principles further, as well as to introduce and develop others, including more specific principles for various product types and contexts.[15]

Notes

1. See, for example, Ofcom 2018, section on "Internet & online content" for some UK data.
2. See Hummel and Maedche 2019 for a recent systematic overview over empirical nudging studies, which concludes that default setting is the most effective type of nudge.

3. I defend the ethical importance of the subjective perspective on defaults in my manuscript "The Ethics of Default-setting". Daniel Kahneman assumes such a perspective in discussing regret in his 2011, ch. 32.
4. Some websites are not as benign as to offer users a clear choice but rather hide their privacy settings behind several difficult-to-navigate pages, sometimes in different languages. In some cases, there is no visible option to be found other than to "accept" data collection, while there may or may not be a link to lengthy information about data collection, in which information about how to opt out may or may not be buried.
5. This is one instance of the so called privacy paradox – the fact that internet users report valuing privacy yet behave as if they did not. See Barth and Jong (2017) for a systematic review of studies that attempt to explain this divergence.
6. On some ideas behind designing for engagement (or addiction), see Eyal and Hoover (2013); for a more scholarly and critical analysis of the ideas behind and the consequences of the design of social media and in particularly Facebook, see Vaidhyanathan (2018); for Facebook co-founder Sean Parker's direct statement that Facebook was designed to maximize use, see Allen (2017).
7. For the health effects of sedentary lifestyles, see Ford and Caspersen (2012); for the connection between sedentary lifestyles and television watching, see Garcia et al. (2019).
8. For example, Goodin (1980), Coons and Weber (2014); for online manipulation in particular, Susser, Roessler, and Nissenbaum (2019). For critical overviews over the relationship between manipulation and deception, see Cohen (2018) and Klenk (2021).
9. On Whitfield's (2020) recent account, obscuring (or rendering "deniable") one's intentions is a core feature of manipulating.
10. Scholars who have emphasized this aspect of manipulation include Noggle (1996); Baron (2003); Wilkinson (2013); and Hanna (2015).
11. For development of this ideal, see, for example, Raz (1986, ch. 4). In his discussion of predictive shopping in *Choosing not to Choose*, Sunstein mentions a possible "standpoint of autonomy" but quickly dismisses it (Sunstein 2015, 176). In this context, Sunstein assumes that the only point of choosing for oneself is that one can predict better than others what will promote one's well-being, though earlier in the book, he mentions in passing that "authenticity" may have non-instrumental value.
12. Björn Lundgren (2020) has argued, very similarly, that "nothing should be shared beyond what is necessary to make the website function properly".
13. In my manuscript "The Ethics of Default-Setting", I argue that this requirement may be implied by a more general moral requirement to minimize harm from defaults, as well as from a distinct moral requirement that defaults minimize any loss of liberty.
14. Part of the technical solution might be that browsers store privacy preferences and inform websites accordingly, as proposed by Lundgren (2020).
15. Material that later made its way into this chapter was first presented at the 2018 Mancept Workshops in Political Theory panel on Paternalism, Nudging and the Digital Sphere. The chapter has benefited from discussion at that panel as well as at the later workshop series organized for this present volume. Comments on an in-between rendition of the material from an anonymous reviewer for *Moral Philosophy and Politics* proved helpful. Lars Lindblom, Björn Lundgren, and Michael Klenk provided very helpful written comments on later drafts. My work on this chapter is part of the research project "AI, Democracy and Self-determination", funded by the Marianne and Marcus Wallenberg Foundation (grant number 2018.0116).

8 References

Allen, Mike. 2017. "Sean Parker Unloads on Facebook: 'God Only Knows What It's Doing to Our Children's Brains'." *Axios*, December 15. Accessed August 23, 2021. www.axios.com/sean-parker-unloads-on-facebook-god-only-knows-what-its-doing-to-our-childrens-brains-1513306792-f855e7b4-4e99-4d60-8d51-2775559c2671.html.

Andersson, Hilary. 2018. "Social Media Apps are 'Deliberately' Addictive to Users." *BBC News*, April 7. Accessed August 23, 2021. www.bbc.com/news/technology-44640959.

Baron, Marcia. 2003. "Manipulativeness." *Proceedings and Addresses of the American Philosophical Association* 77 (2): 37. doi:10.2307/3219740.

Barth, Susanne, and Menno D. de Jong. 2017. "The Privacy Paradox: Investigating Discrepancies Between Expressed Privacy Concerns and Actual Online Behavior. A Systematic Literature Review." *Telematics and Informatics* 34 (7): 1038–58.

Bovens, Luc. 2009. "The Ethics of Nudge." In *Preference Change: Approaches from Philosophy, Economics and Psychology*, edited by Till Grüne-Yanoff and Sven O. Hansson, 207–19. Dordrecht: Springer.

Choi, J. J., D. Laibson, B. C. Madrian, and A. Metrick. 2003. "Optimal Defaults." *American Economic Review* 93: 180–85.

Cohen, Shlomo. 2018. "Manipulation and Deception." *Australasian Journal of Philosophy* 96 (3): 483–97. doi:10.1080/00048402.2017.1386692.

Coons, Christian, and Michael Weber. 2014. "Manipulation: Investigating the Core Concept and its Moral Status." In *Manipulation: Theory and Practice*, edited by Christian Coons and Michael Weber, 1–16. Oxford: Oxford University Press.

Council of Europe. 2018. "Hate Speech, Apology of Violence, Promoting Negationism and Condoning Terrorism. The Limits to the Freedom of Expression." Thematic Factsheet.

Engelen, Bart. 2019. "Nudging and Rationality: What is There to Worry?" *Rationality and Society* 31 (2): 204–32. doi:10.1177/1043463119846743.

European Commission. 2016. Regulation (EU) 2016/679 of the European Parliament and of the Council of 27 April 2016 on the protection of natural persons with regard to the processing of personal data and on the free movement of such data, and repealing Directive 95/46/EC (General Data Protection Regulation) [2016] OJ L119/1 ('GDPR'). GDPR. April 27.

Eyal, Nir, and Ryan Hoover. 2013. *Hooked: How to Build Habit-Forming Products*. London: Penguin Books.

Ford, Earl S., and Carl J. Caspersen. 2012. "Sedentary Behaviour and Cardiovascular Disease: A Review of Prospective Studies." *International Journal of Epidemiology* 41 (5): 1338–53. doi:10.1093/ije/dys078.

Foroughi, B., M. Iranmanesh, D. Nikbin, and S. S. Hyun. 2019. "Are Depression and Social Anxiety the Missing Link between Facebook Addiction and Life Satisfaction? The Interactive Effect of Needs and Self-regulation." *Telematics and Informatics* 43: 101247.

Frischmann, Brett M., and Evan Selinger. 2018. *Re-Engineering Humanity*. Cambridge: Cambridge University Press.

Garcia, Jeanette M., Andrea T. Duran, Joseph E. Schwartz, John N. Booth, Steven P. Hooker, Joshua Z. Willey, Ying K. Cheung et al. 2019. "Types of Sedentary Behavior and Risk of Cardiovascular Events and Mortality in Blacks: The Jackson Heart

Study." *Journal of the American Heart Association* 8 (13): e010406. doi:10.1161/JAHA.118.010406.
Goodin, Robert E. 1980. *Manipulatory Politics.* New Haven, CT: Yale University Press.
Grüne-Yanoff, Till. 2012. "Old Wine in New Casks: Libertarian Paternalism Still Violates Liberal Principles." *Social Choice and Welfare* 38 (4): 635–45.
Hanna, Jason. 2015. "Libertarian Paternalism, Manipulation, and the Shaping of Preferences." *Social Theory and Practice* 41 (4): 618–43.
Hummel, D., and A. Maedche. 2019. "How Effective is Nudging? A Quantitative Review on the Effect Sizes and Limits of Empirical Nudging Studies." *Journal of Behavioral and Experimental Economics* 80: 47–58.
Kahneman, Daniel. 2011. *Thinking, Fast and Slow.* New York, NY: Farrar Straus Giroux.
Klenk, Michael. 2021. "Manipulation (Online): Sometimes Hidden, Always Careless." *Review of Social Economy.* 80: 1, 85–105. doi:10.1080/00346764.2021.1894350.
Lazer, David M. J., Matthew A. Baum, Yochai Benkler, Adam J. Berinsky, Kelly M. Greenhill, Filippo Menczer, Miriam J. Metzger et al. 2018. "The Science of Fake News." *Science* 359 (6380): 1094–96. doi:10.1126/science.aao2998.
Lundgren, B. 2020. "How Software Developers Can Fix Part of GDPR's Problem of Click-through Consents." *AI & Society* 35: 759–60.
Maitland, Ian. 1997. "Virtuous Markets: The Market as School of the Virtues." *Business Ethics Quarterly* 7 (1): 17–32.
New Zealand Legislation. 2021. Fair Trading Act 1986 No 121 (as at 14 March 2021). www.legislation.govt.nz/act/public/1986/0121/latest/whole.htm.
Noggle, Robert. 1996. "Manipulative Actions: A Conceptual and Moral Analysis." *American Philosophical Quarterly* 33 (1): 43–55.
Ofcom. 2018. "The Communications Market 2018: Interactive Report."
Pariser, Eli. 2011. *The Filter Bubble: What the Internet Is Hiding from You.* New York, NY: Penguin Books.
Raz, Joseph. 1986. *The Morality of Freedom.* Oxford: Oxford University Press.
Russell, Stuart J. 2019. *Human Compatible: AI and the Problem of Control.* London: Penguin Books.
Scanlon, Thomas M. 1998. *What We Owe to Each Other.* Cambridge, MA: Harvard University Press.
Schmidt, Andreas T. 2019. "Getting Real on Rationality: Behavioral Science, Nudging, and Public Policy." *Ethics* 129 (4): 511–43.
Sunstein, Cass R. 2014. *Why Nudge? The Politics of Libertarian Paternalism.* New Haven, CT: Yale University Press.
Sunstein, Cass R. 2015. *Choosing Not to Choose: Understanding the Value of Choice.* New York, NY: Oxford University Press.
Susser, Daniel, Beate Roessler, and Helen Nissenbaum. 2019. "Online Manipulation: Hidden Influences in A Digital World." *Georgetown Law Technology Review* 4 (1): 1–45. Accessed February 27, 2020.
Twitter. 2019. "Defining Public Interest on Twitter." https://blog.twitter.com/en_us/topics/company/2019/publicinterest.html.
Twitter. 2020a. "Building Rules in Public: Our Approach to Synthetic & Manipulated Media." https://blog.twitter.com/en_us/topics/company/2020/new-approach-to-synthetic-and-manipulated-media.html.

Twitter. 2020b. "COVID-19 Misleading Information Policy." https://help.twitter.com/en/rules-and-policies/medical-misinformation-policy.

Vaidhyanathan, Siva. 2018. *Antisocial Media: How Facebook Disconnects US and Undermines Democracy*. New York, NY: Oxford University Press.

van Dijk, Jan A. 2020. *The Network Society*. London: SAGE Publications.

Véliz, Carissa. 2020. *Privacy is Power: Why and How You Should Take Back Control of Your Data*. London: Transworld Digital.

Whitfield, G. 2020. "On the Concept of Political Manipulation." *European Journal of Political Theory*, 1–25.

Wilkinson, T. M. 2013. "Nudging and Manipulation." *Political Studies* 61 (2): 341–55. doi:10.1111/j.1467-9248.2012.00974.x.

Yang, T.-C., and H. Lai. 2006. "Comparison of Product Bundling Strategies on Different Online Shopping Behaviors." *Electronic Commerce Research and Applications* 5: 295–304.

Zuboff, Shoshana. 2019. *The Age of Surveillance Capitalism: The Fight for a Human Future at the New Frontier of Power*. New York, NY: PublicAffairs.

20 Manipulation, Real-Time Profiling, and their Wrongs[1]

Jiahong Chen and Lucas Miotto

1 Introduction

Claims denouncing manipulation in the online sphere are pervasive and familiar (Susser, Roessler, and Nissenbaum 2019; Bradshaw and Howard 2018). Most writers, in academic circles and beyond, seem to agree that wrongful forms of online manipulation occur because of – or are at least facilitated by – new digital technologies. Despite such agreement, little has been done to pin down the exact wrongs that come from online manipulation. Like other contributions to this volume, this chapter aims to add to this discussion. We argue that paying closer attention to what we call "real-time profiling" allows us to identify the wrong-making features of an important subset of online manipulative practices; practices which are taken to be "manipulation's future" (Spencer 2020).

The chapter is divided into four sections. The first is where we characterise real-time profiling in socio-technical terms. In the second, we show that real-time profiling is analogous to some forms of interpersonal manipulation and that, for that reason, there is a presumption in favour of seeing real-time profiling as an example of manipulation. The third is where we propose our account of what makes real-time profiling wrong. Contrary to some extant accounts of the wrongness of manipulation (both in the online and offline spheres), our proposed account does not link the wrongness of real-time profiling to covertness, deception, harm, autonomy, or to bypassing individual's rational capacity. As we will argue, real-time profiling is wrong both because it involves what we call "psychological hijacking" and because it works as a gateway to further wrongs. In the final section we explore some implications of our account for the legal regulation of online manipulative technologies. We argue that existing legal frameworks are not fine-grained enough to deal with the wrongs associated with real-time profiling and related forms of online manipulation.

2 The Rise of Real-Time Profiling

Profiling is anything but new in the online world. As an important part of the digital advertising ecosystem, profiling has been a technique widely

DOI: 10.4324/9781003205425-24

employed by different categories of actors across the online marketing sector (Chen 2021, ch. 1). Traditional profiling systems are said to be "interest-based" since they are designed to make inferences about what might interest the user and what demographics the user is likely to fit in. The operation of such systems depends on the ability to observe users' online traces over time, so as to gain an increasingly precise understanding of their preferences. The predictions may change as more data accumulates, but generally, traditional interest-based profiling is mainly about grouping users according to stable and longitudinal features.

In recent years, and largely due to an array of socio-economic factors,[2] the capabilities of profiling techniques have developed beyond some of the technical constraints as seen in earlier days of online advertising. Profiling broke through the boundaries of browser-based tracking, as well as accuracy and speed constraints and culminated to what we call "real-time profiling". The practice of real-time profiling, as we describe it, has two main steps:

1. A private or public agent collects information about an individual's *present status*. This can cover an array of aspects: from the individual's current health status to how hungry, stressed, or annoyed she is. Once the information is evaluated, a profile of the individual's present status is built.
2. The private or public agent then attempts to influence the individual's actions, choices, or preferences in the immediate future based on the constructed profile.

As we can infer from these steps, real-time profiling differs from traditional interest-based profiling in that it is designed to track certain transient and dynamic characteristics of a user and to adjust interactive strategies in real time. Real-time profiling may or may not rely on the long-term construction of a user's profile; the goal is to work out the instant status "on the spot" rather than a relatively static aspect about the user. To illustrate the phenomenon, consider the following case involving Uber (Lindsay 2019; Mahwadi 2018):

> (Uber) It has been revealed that vehicle-for-hire app Uber could implement a technology that enables it to assess users' level of inebriation and battery level. The technology could be used to get users who are in a more vulnerable position to pay more for their standard rides or to push these users to choose a premium ride.

Similar examples of real-time profiling abound. We know of gambling firms that can influence users based on their location (e.g., at sports events, Rudgard 2018), or of smart devices that can influence individuals on the basis of their current level of stress or heart rate (Brown 2018; Shapiro 2016; Alvarez 2017; Charara 2020), or even of eye tracking technology that can be used to influence individuals based on what they are paying attention to at a particular moment (Metz 2016; Valliappan et al. 2020). In principle, as a set of targeting strategies, real-time profiling can be implemented on a variety of technical settings, theoretically including any human–machine interactions. These may include, for example, how online services present information, how smart devices change the ambience of a space, how robots adjust actions towards humans, or even how the urban infrastructures configure resources (e.g., "smart cities" or "smart transport" initiatives). More importantly, different strategies can be combined to make the assessment more accurate and to find the "optimal" way to interact with and influence the profiled individual (Sax 2021).

Each real-time profiling strategy may exhibit its own challenges, which requires specific discussion. Be this as it may, a general inquiry raised by the practice of real-time profiling concerns its moral status. Real-time profiling seems to be the sort of practice that calls for justification. And the reason for it is that, at face value, it resembles instances of wrongful manipulation in relevant respects. In what follows we briefly elaborate real-time profiling's resemblance to manipulation.

3 Is Real-Time Profiling Manipulative?

We need not assume that all instances of real-time profiling are instances of wrongful manipulation.[3] Our goal is simply to show that some cases of real-time profiling and typical cases of wrongful manipulation are analogous in relevant respects. To begin with, consider the following case of interpersonal manipulation:

> (Phil) Phil and Claire, a married couple, earn about the same salary. Phil plans to get Claire to pay for a much larger share of their household's expenses so that he can save up for a trip with his friends. He knows that Claire is much more receptive to his proposals when she is relaxed and after enjoying a good meal. Phil makes a plan: he gives Claire a spa-day voucher and spends the day cooking Claire's favourite meal while she is away. When back, Claire finds Phil at the dinner table, meal ready. After a pleasant dinner, Phil suggests that she pay a larger fraction of the household's expenses. As predicted by Phil, Claire accepts.

Phil got his own way with Claire not by persuading or reasoning with her. Nor did he get his own way by coercing, pressuring, blackmailing, deceiving, or lying to her. His act is – intuitively at least – manipulative. Some features are key to that assessment. First, Phil *attempted to influence* Claire's decision. Second, the *primary motive* for Phil's behaviour was the benefit he would get from influencing Claire (i.e., saving money). Finally, and more importantly, Phil's attempt to influence was specially tailored to take advantage of an aspect of Claire's *deliberative capacity*, namely the fact that she would be more receptive to Phil's proposals at a particular time. We find variations of these features in some definitions of manipulation.[4] However, here we take them not as necessary or sufficient features of manipulation but merely as features which are often salient in core cases of wrongful manipulation; features that would give us a *presumption* that a case is an example of manipulation in the absence of stronger countervailing considerations.

Now, when we compare Phil's influence with an example of real-time profiling, we can see that they share the aforementioned features, which gives us *some* reason to presume that at least some instances of real-time profiling are examples of manipulation. To make this point clearer, consider a hypothetical case of real-time profiling:

> (MoodX) A social media company, MoodX, develops an algorithm that predicts its users' current mood with high accuracy. With the help of the algorithm, MoodX advertises products tailored to users' current mood. Sales of advertised products skyrocket as a result.[5]

MoodX clearly attempted to influence users with its algorithm and advertisement strategy. And we can see that the way MoodX chose to attempt influencing users was part of MoodX's *unilateral* plan: a plan primarily motivated by its benefit to MoodX.[6] We can also see that MoodX's influence was tailored to take advantage of a particular aspect of users' deliberative capacity,[7] namely the fact that their moods affect their buying choices. In virtue of exhibiting these features, we can say that MoodX – and other examples of real-time profiling – can be presumed to be an example of manipulation in the absence of countervailing considerations.

While Phil's and MoodX's influences can be seen as (or presumed to be) examples of manipulation, it is worth drawing attention to one way in which these examples may deviate from some typical cases of manipulation (e.g., doing small favours for others to feel obliged, placing more expensive products at eye-level). Both Phil and MoodX exploited their target's *transient and dynamic features*. The content, message, action, or conditions set out by Phil and MoodX are not simply tailored to their targets on a personal level but more importantly, to a precise point of time when the internal or environmental circumstances have changed such that the impact of their move is maximised. The kind of manipulative influence that we are focusing on should thus be seen as a distinct subset of manipulative influence: one where the manipulator is sensitive not just to *who* to target but also to *when* and *where* to target someone.

Now we may wonder whether, despite their similarities, cases of real-time profiling would be in some sense distinct from interpersonal manipulative influences like Phil's. We believe that there is no difference in *kind*. The obvious difference, when there is one, has to do with the intensity and the scope of the real-time profiler's influence. Real-time profiling happens in the online environment and the profiler is typically either a corporation or a public agent who has information and technological resources that enable constant observation of the target's online and offline activities. In a sense, we can say that the profiler is *always around* (at least when the target is next to the right gadgets) and usually has information about the target which is unfeasible to obtain via everyday interpersonal interactions (in fact, profilers might be more insightful of targets' current status than the targets themselves). The profiler, therefore, typically does not face temporal, spatial, and access restrictions that interpersonal manipulators (like Phil) do. As such, their scope for interference with the target is much wider.

As mentioned earlier, some writers have suggested that this form of real-time interference is "manipulation's future" in the online domain and is bound to become more common (Spencer 2020, n2). Being the future of online manipulation or not, this form of manipulation raises ethical concerns and calls for justification. It is thus worth examining what makes real-time profiling (and related forms of manipulative practices) wrong when they are wrong. That is the task we take up next.

4 Why Is Real-Time Profiling Wrong?

As we have seen, there is a presumption in favour of seeing some cases of real-time profiling as instances of manipulation. Whether *all* instances of real-time profiling fall under a properly regimented concept of wrongful manipulation is not something that interests us. Instead, we are interested in what makes some cases[8] of real-time profiling pro tanto wrong when they are wrong.[9] In this section, we analyse the features of real-time profiling that make it wrong.

Let us return to MoodX. At face value, MoodX did something *pro tanto* wrongful. A few elements in this case can help explain why. From the description, and from what we have said about it in the previous section, we can infer that MoodX had a unilateral plan to profit from the sale of products and took steps towards making it successful. What seems to make MoodX's action wrong, however, is not simply that it had a unilateral plan and acted on it, but the *way* in which MoodX implemented its plan matters.

Recall that in the previous section we said that MoodX's influence was tailored to take advantage of a particular aspect of users' deliberative capacity. This can be fleshed out in more specific terms. What seems to be the case is that MoodX *hijacked* users' psychology; it worked out a way whereby users' own psychological states – that is, their moods – served MoodX's unilateral plan. And the act of hijacking someone's psychology, we submit, is an essential part of the explanation of what makes real-time profiling wrong. To explain why, we must be more precise about what *psychological hijacking*, as we call it, involves.

As we understand it, psychological hijacking is a means by which one attempts to implement one's unilateral plan. Hence, for it to take place, the hijacker must at least:

(a) Have a unilateral plan *P*.
(b) Intend that *P* is successful.[10]

(c) Believe that an action or a series of related actions, ϕ, is a means to the success of P.
(d) Perform ϕ.

Conditions (a)–(d) suggest that psychological hijacking can take place even when the hijacker's plan is unsuccessful. However, even though psychological hijacking does not depend on the success of the hijacker's *plan*, it can only be said that an individual has psychologically hijacked another if the action (or series of actions) performed by the hijacker – that is, ϕ – succeeds in generating a particular effect. Namely, by ϕ-ing the hijacker must:

(e) Make some of the target's psychological states[11] *subservient* to the hijacker's intention that P succeeds.

Two qualifications about (e) are in order. The first concerns the hijacker's intention. As per condition (b), the hijacker must intend to see his unilateral plan through. But one may think that the hijacker must also intend the specific effects mentioned in (e). Such a requirement, we submit, would make our account unnecessarily under-inclusive. Most evidently, it would rule out the possibility of one being engaged in psychological hijacking (and in real-time profiling) without realising it. For example, it is possible (though perhaps unlikely) that MoodX's executive board were unaware of how the new algorithm worked and decided to implement it solely based on the incomplete information that its implementation would maximise sales. Insofar as we can reduce MoodX's decisions to the decisions of its executive board, we could say that MoodX did not intend that users' moods become subservient to their plan to increase sales. But it would still be correct to say that users were psychologically hijacked (and we could even imagine MoodX's executive board making a public apology highlighting the fact that they would not have implemented the algorithm had they known how it worked).[12]

The second qualification concerns "subservient". By saying that the hijacker makes the target's psychological states subservient to the hijacker's intentions, we mean that the hijacker's influence establishes a hierarchy between the hijacker and the target. The hierarchy in question can be construed as a hierarchy between the target's psychological states and the hijacker's intentions. The hijacker behaves as if the target's psychological states (including the target's plans, intentions, and preferences) are *less valuable* than his own intentions. No, or little, regard is given to the target's standing to demand that her own psychological states are not placed at the service of the hijacker. The target's psychological states are treated as mere means to the success of the hijacker's unilateral plan.

The notion of subservience is, therefore, key to understanding why psychological hijacking features in the explanation of real-time profiling wrongness. Of course, similar forms of subservience are sometimes justified. For example, a social media company that intends to prevent users from engaging in self-harmful behaviour could use information about users' current moods or stress levels to induce them to seek professional help. In a way, we could say that the company also created a hierarchical and instrumental relationship whereby users' intentions, desires, or preferences were treated as less important than the company's intentions. But, contrary to MoodX's case, here the company's influence seems morally acceptable. And it seems so because the company's plan took the *interests* of individuals into account. But notice that *even* in this case, the company would not get off the hook with ease. Their interference, even if in the name of users' interests, would still stand in need of justification.[13] And that is so precisely because there seems to be something (pro tanto) wrong with creating hierarchical and instrumental relationships.[14]

Note that the wrongness of subservience is closely tied to its need and to the availability of alternative options. When an individual has other reasonable means to achieve their intended plan, doing so while making someone subservient is particularly condemnable. For example, MoodX had alternative ways to profit and to advertise its products. Sure, perhaps the alternatives would have been less effective, but choosing effectiveness over treating their users in a non-hierarchical and instrumental way would itself be a form of being reckless about morality (Chen 2021, 72–73).

What is striking about psychological hijacking – in the context of the cases of real-time profiling that we have been considering – is that there are often alternative ways to get targets to adopt the profiler's plan that show some regard for the target's consent, standing, or interests. But despite there being alternative ways to influence targets, profilers still choose to engage in a form of influence that gives rise to subservience. That is one of the reasons why this form of online manipulation often strikes many as deeply problematic.

Now, because psychological hijacking makes some aspect of the target's psychology subservient to the hijacker's intentions, it might be thought that psychological hijacking is a form of *domination*. Whether the hijacker–target relationship amounts to a relation of domination is debatable. One reason against seeing it as such concerns the scope of the subjection which is constitutive of relations of domination. In typical relations of domination (e.g., the slave–slaveowner relation), the dominated's "normative reasons to do what the [dominator] proposes constitutively track considerations that are dependent on the power-facts" (Vrousalis 2019, 8). Contrarily, because the hijacker influences the target by meddling with pre-existing reasons or other psychological features, we cannot say that the target's reasons (motivation, disposition, etc.) that are subservient to the hijacker's plan *constitutively* track considerations which are dependent on power facts.

Another reason against seeing the hijacker–target relation as one of domination concerns the transactional and transient – as opposed to structural and persistent – nature of their interaction. At least in the set of cases of psychological hijacking that interests us (i.e., cases of real-time profiling), the hijacker's influence over the target's psychology is episodic and dependent on the hijacker's *actual* exercise of his power over the target. By contrast, paradigmatic cases of domination are cases where we find an institutionally stabilised and enduring power relation whereby the mere subjection to the dominator's power, and not its exercise, is what calls for justification. As Dorothea Gädeke puts it, "conceptualizing both opportunistic and robust capacities to interfere as forms of domination risks losing sight of what is distinctive of non-domination as opposed to non- interference" and "[risks] misconstruing domination as an anomaly perpetrated by individual wrongdoers instead of as a feature that pervades society" (Gädeke 2020, 199). This is, of course, not to say that psychological hijacking can never constitute a relation of domination in the online domain.[15] The point is that we need not see psychological hijacking as necessarily constituting relations of domination to explain what is wrong with it.

A further clarification concerns the equation of psychological hijacking with bypassing rational capacity. The two forms of influence should not be conflated. In fact, sometimes psychological hijacking occurs only if the hijacked properly exercises their rational capacities. Consider another example of real-time profiling:

> (Election) To promote chaos and polarisation during elections, a search engine changes its algorithm to condition the information that users are exposed to according to their real-time online behaviour. Robust evidence favouring one's preferred candidate is presented at a calculated time when the user is believed to be less emotional and more likely to take evidence-based decisions. Polarisation rises as a result.

In (Election), the search engine did not bypass individuals' rational capacities – at least not in the sense of suppressing individuals' rational deliberation or disengaging individuals' "system 2", to borrow Kahneman's terminology (Kahneman 2011).[16] Raising one's confidence in a proposition based on stronger evidence is simply what should be expected from individuals who properly exercise their rational capacities.[17] The search engine's act, therefore, did not bypass users' rational capacities. In fact, in this case the exercise of their rational capacity was necessary for the search engine to achieve its plan. Be this as it may, the search engine is still engaged in psychological hijacking and its action still seems wrong for the reasons we have discussed earlier.

The idea that psychological hijacking creates a hierarchical and instrumental relation where there was none and where there need not be one is an important part of the explanation of what makes real-time profile wrong – and it might even help us explain why unsuccessful or benign instances of real-time profiling still raise moral concerns.

With all that said, it would be mistaken to assume that psychological hijacking alone provides the full explanation for what makes real-time profiling wrong. Another element should be included. To find it, let us once again return to (MoodX). A further fact that we can infer from this case is that the implementation of the algorithm *transformed* user's moods into vulnerabilities. While vulnerabilities are often associated with intrinsic characteristics, such as age, gender, or disability status, this is not the case in real-time profiling.

In cases of real-time profiling, like in (MoodX), the transient and dynamic status of profiled individuals gives the profiler a unique opportunity to exercise influence that is specific to the context (the current circumstances that the profiled subject is undergoing) and the relation (the personal, commercial, or political relationship between the profiler and the profiled subject). We can say, therefore, that the profiler's increased knowledge and his influence in the online environment work together as an *enabling* condition: they remove an obstacle for individuals to be wronged in different ways. For example, by figuring out how to influence individuals on the basis of their moods and by making this influence possible, MoodX *is now able to* get individuals to do more than buying products. In principle MoodX could rely on its ability to influence users on the basis of their moods to exploit, abuse, harm, or discriminate them. Just as it is said that gateway drugs prime or prepare someone's organism for heavier substances by removing some natural inhibitors, we can say that real-time profiling is a *gateway wrong*: By transforming some psychological features into vulnerabilities, the profiler removes obstacles and creates opportunities for individuals to be wronged in different ways. We take it that removing obstacles and creating opportunities for individuals to be wronged in different ways without a strong justification for doing so is itself pro tanto wrong. After all, this amounts to subjecting individuals to unnecessary risks.

Now, one may object by submitting that something similar often happens in interpersonal or online interactions without giving rise to moral concerns. For example, by befriending someone we might remove some obstacles to wrong the person. We might, for example, make the person more vulnerable to emotional blackmail or to abuse of trust. So why do we not say that befriending someone is also a gateway wrong? The reason is that despite making each other more vulnerable to some wrongs, when we form genuine friendships, these concerns are mitigated by the fact that we treat each other as equals, non-instrumentally, and by the fact that making each other vulnerable is not constitutive of the relation but simply an inevitable by-product. Contrarily, in cases of real-time profiling, creating a vulnerability on the profiled subject is an integral part of the way in which the profiler chooses to exert its influence. That is why it is worth highlighting that real-time profiling – but not befriending – is a gateway wrong.

Notice, however, that the fact that real-time profiling is a gateway wrong should not be seen as a de facto harm-based explanation. Cases where the profiled subject fails to adopt the profiler's unilateral plan can help us clarify the point. Arguably, no harm occurred to an individual who resisted the temptation to make a bet at a sports event despite being induced to doing so by their phone's real-time profiling. But the attempted influence did work as a gateway wrong. It removed an obstacle and created an opportunity for the individual to use his money against his own interests and subjected him to the risk of having his information about being at a sports event used for other detrimental purposes.

The claim that the profiler's interference works as a gateway wrong because it enables further wrongs should also not be conflated with a claim about the wrong-making features associated with the enabling conditions of real-time profiling. The occurrence of individual instances of real-time profiling typically depends on the satisfaction of a series of background conditions. In (MoodX), for example, we presupposed that the company had the relevant information about users' moods. But, in all likelihood, the company would not have been able to influence users had it not possessed such information. The possession of information then, in this case, works as an *enabling condition* for real-time profiling. Despite enabling conditions varying from case to case, we might say by way of generalisation that they are conditions the satisfaction of which places the hijacker in a position of power; in a position where he can, intentionally or not, interfere with the target's psychological state so as to cause the target to favour his plan.

Once we draw attention to the enabling conditions of real-time profiling, it is not difficult to see that an agent may wrong others by merely satisfying them. For example, it could be argued that MoodX has wronged users even before interfering with the online environment and users' moods. The mere acquisition of information about users' moods was (arguably) pro tanto wrongful because it gained access to intimate details about the users without a sound justification. Though such wrongs raise serious concerns, they should not be conflated with the wrongs *of* real-time profiling (one of them being that real-time profiling itself enables further wrongs). As such, they are less important for our purposes.

As per our account of real-time profiling, the profiler not only makes the profiled subject subservient, in some specific sense, to the profiler's unilateral plan but also does so whilst simultaneously enabling further wrongs. Therefore, when wrong, real-time profiling is wrong both because it involves psychological hijacking and because it works as a gateway wrong. These two aspects represent the key normative characteristics of real-time profiling but not exclusively to it, since we can observe the same characteristics in interpersonal counterparts of real-time profiling (e.g., Phil's case).

Having identified the wrong-making features of real-time profiling (and analogous interpersonal manipulative practices), we now move on to discuss what such normative reflections mean for regulatory initiatives.

5 Regulatory Implications

First, we must acknowledge that just because something is morally problematic it does not necessitate regulatory interventions. Other considerations, such as the scale of the impact, the costs of regulation, and the possibility of correction by less invasive mechanisms, may affect the policy outcome.

As much as we believe that at least the most blatant forms of real-time profiling should be regulated, the appropriate scope and venue of regulation will depend on further research. Nevertheless, we see the need to explain the regulatory implications of our theoretical findings, especially because ongoing public debates are taking place around the world about regulating online manipulative practices. Assuming that online manipulation is something that calls for legal regulation and that policymakers have good reasons to proceed with legal interventions, this section will briefly explore how our reflection on real-time profiling may help highlight the flaws in the current regulatory frameworks and perhaps more importantly, point towards a more promising direction. We have chosen the European Union (EU) regime for our analysis, but there is no obvious reason why the implications discussed in this section do not apply to other jurisdictions. We focus on two of the most relevant areas in the EU legal order, consumer protection and data protection law, before moving onto further comments on the latest developments in the proposed regulation on digital services and artificial intelligence (AI).

5.1 Consumer Protection Law

It is probably not difficult to think of the relevance of consumer protection law in addressing at least some of the challenges arising from real-time profiling. When it comes to *commercial* targeting (but not *political* targeting), individuals are usually protected as consumers. The EU's Unfair Commercial Practices Directive (UCPD),[18] for example, prohibits misleading, aggressive, and otherwise unfair commercial practices (European Commission 2005, art 5(3), para 28 Annex I).

Our theoretical discussions about real-time profiling could raise (and partly answer) the question as to whether the current consumer protection legal framework can fully address manipulative marketing practices.

First, can typical cases of real-time profiling be deemed as misleading in legal terms? Article 6(1) UCPD defines a misleading commercial practice as one that "contains false information and is therefore untruthful or . . . deceives or is likely to deceive the average consumer". We have already clarified that real-time profiling does not necessarily involve false, mis- or disinformation as such, and it can be simply presenting truthful information at an opportunistic time. As regards deception, we pointed out that core cases of real-time profiling are more likely to fall within the scope of non-deceptive manipulation. Whether a legal – as opposed to philosophical – concept of deception can capture this phenomenon would be a separate question, but in the current absence of legislative guidance or clear case law on this matter, it would probably at best be a stretch to consider real-time profiling as deceptive without a strong conceptual support.

Second, in terms of aggressive practices, the UCPD has a particular emphasis on "harassment, coercion, including the use of physical force, or undue influence" (European Commission 2005, art 8). Our earlier example of MoodX involves no harassment or coercion (although neither of those two terms are defined in the UCPD) but can nevertheless be seen as a form of manipulation. When it comes to undue influence, while conceptually it is debatable whether MoodX's practices are undue, the law has a relatively narrow definition of undue influence, namely "exploiting a position of power in relation to the consumer so as to apply pressure . . . in a way which significantly limits the consumer's ability to make an informed decision" (European Commission 2005, art 2(j)). Though we see that real-time profilers "exploit a position of power", the definition does not fit many cases of real-time profiling because real-time profilers, as a rule, do not pressurise users.

Third, and due to the unsatisfactory coverage of the legal definitions of "misleading" and "aggressive" practices, the next question would be whether real-time profiling falls within the more generic concept of unfair practices. Under the UCPD, an unfair practice is one that meets two criteria: (a) it breaches professional diligence; and (b) it distorts the economic behaviour of the average consumer. Real-time profiling presents a particularly interesting case to condition (b), because on the one hand, it clearly shows potentially distortive effect on the economic behaviour of the consumers, which rests at the heart of the very idea of psychological hijacking in a commercial context. On the other hand, however, condition (b) has a particular emphasis on the average consumer – whether with regard to the entire market or a targeted group – not an individual consumer, which can be problematic in the case of real-time profiling. With its hyper-personalisation nature, it is unclear how the average consumer standard may apply to individualised manipulation. Indeed, current online marketing practices have evolved from targeting a group audience to "an audience of one" (Summers, Smith, and Reczek 2016). Laux et al. have highlighted some of the similar challenges in the context of online behavioural advertising and call for a stricter average consumer test (Laux, Wachter, and Mittelstadt 2021). For consumer protection law to fully capture real-time profiling and similarly manipulative practices, it would either entail further legislative, judicial, or regulatory guidance to expand the legal concept of "misleading", "aggressive" or "unfair", or a new provision specifically covering manipulative practices.

5.2 Data Protection Law

To the extent that typical real-time profiling techniques involve the collection of personal data, data protection law may stand out as a promising regulatory forum in restricting the use of personal data and hence real-time profiling practices. The earlier discussions on the technical and moral nature of these practices, however, reveal some conceptual challenges in applying data protection law to real-time profiling.

First, on a technical level, it has been pointed out how real-time profiling can be particularly intrusive by identifying the exact moment where the targeted individual would be susceptible to the influence. Yet, this does not necessary involve "sensitive data" as defined by Article 9 of the General Data Protection Regulation (GDPR).[19] Under Article 9, sensitive data is defined as

> data revealing racial or ethnic origin, political opinions, religious or philosophical beliefs, or trade union membership, and the processing of genetic data, biometric data for the purpose of uniquely identifying a natural person, data concerning health or data concerning a natural person's sex life or sexual orientation. (European Commission 2016, art 9)

Real-time profiling does not necessarily involve any of such categories of data but can be equally revealing and exploitative. The challenge, as such, lies at the current data protection regime's inability to clearly capture manipulative practices that do not rely on data that is classically considered prone to discrimination or manipulation. It should be noted that other parts of the GDPR still apply to non-sensitive uses of personal data, but the level of protection would be significantly lower.

Second, on a moral level, and perhaps on a more optimistic note, our conceptualisation of real-time profiling elaborates why such practices are legally challengeable in the absence of clearly applicable rules or guidance. The disrupted power dynamics exhibited in the case of real-time profiling provides an articulation of how such practices can possibly be held unlawful. For example, as a matter of data protection principle, Article 5(1)(a) requires that personal data must be "processed lawfully, fairly and in a transparent manner", rendering any unfair uses of personal data illegal. While the law does not further clarify what amounts to "unfairness", our discussion on psychological hijacking and gateway wrongs presents a conceptual case against the acceptance of real-time profiling as a "fair use" of personal data. Another example is how the profiler's exploitation of its position of power, as fleshed out with the notion of *subjection*, would create a power imbalance. The legal consequence of the establishment of such an imbalance is that any consent given by the data subject would no longer be considered "freely given" (European Data Protection Board 2020, 7–9), rendering the uses of personal data reliant on such consent no longer lawful. Of course, specific provisions directly addressing real-time profiling would be the most effective way to regulate it, but before new rules are put in place, courts and regulators would have to rely on a theoretical explanation of the moral wrongness of real-time profiling.

5.3 The Digital Services Act and the Artificial Intelligence Act: An Opportunity?

Given the limitations of consumer protection and data protection law, arguably a more targeted regulatory approach is needed to effectively address the unique challenges of real-time profiling. There have been ongoing regulatory efforts initiated in the EU on manipulation. In December 2020, for example, the European Commission published the long-expected proposal for a Digital Services Act (DSA) (European Commission 2021). Two draft provisions might be of particular interest. The proposed Article 24 requires online platforms to disclose the factors that determine how adverts are targeted to internet users. The transparency requirement here may partly cover commercial real-time profiling through third-party platforms, but as discussed earlier, the fact that the manipulee is aware of the manipulator's intention does not fundamentally change the moral status of the action. The proposed Article 26(1), on the other hand, would impose a duty on key online platforms to monitor the spread of information with regard to public interest, which could cover manipulative political – but not necessarily commercial – real-time profiling.

More recently, in April 2021, the Commission tabled a proposal for the Artificial Intelligence Act (AIA). While not all real-time profiling techniques will involve what the AIA defines as AI, the relatively broad definition[20] would likely capture a large part of real-time profiling systems, especially the more sophisticated ones.

Article 5 of the draft AIA prohibits, among other things, two types of manipulative AI systems, one that "deploys subliminal techniques beyond a person's consciousness", the other that "exploits any of the vulnerabilities of a specific group of persons" (European Commission 2021, art 5). Both banned practices must however "materially distort a person's behaviour" and cause "physical or psychological harm" (European Commission 2021, art 5). As a preliminary assessment, it seems typical real-time profiling practices may count as a "subliminal technique" but would probably not involve vulnerabilities as currently limited to only "age, physical or mental disability". More importantly, while our analysis shows successful attempts of real-time profiling could create behavioural distortion, the "physical or psychological harm" bar is perhaps too high a legal test to cover the more subtle, yet wrongful, forms of real-time profiling.

The Commission is clearly mindful of the interplays between the AIA and other areas of law by stating "[o]ther manipulative or exploitative practices affecting adults that might be facilitated by AI systems could be covered by the existing data protection, consumer protection and digital service legislation" (European Commission 2021, 13), but our analysis has exposed some of the regulatory challenges in those areas. Building on our theoretical enquiry into the nature of real-time profiling, further legal research could – and should – be carried out to uncover how the regulatory regime could be better equipped to address novel forms of online manipulation.

6 Conclusion

Real-time profiling is already a part of our online environment. All suggests that it is here to stay. We have shown that some cases of real-time profiling closely resemble wrongful manipulative practices and, thus, raise similar ethical concerns. To highlight such concerns, we have provided an account of what makes real-time profiling wrong. Real-time profiling is wrong both because the profiler engages in what we have called "psychological hijacking" and because it is a gateway wrong. This diagnosis has led us to identify shortcomings that might help the potential regulation of real-time profiling and related online manipulative practices. Whether real-time profiling needs to be regulated and how to precisely go about it are questions that we cannot tackle in this chapter. But if, as some have envisaged, real-time profiling is the future of online manipulation, these questions cannot be ignored in further discussions.

Notes

1. *We thank Fleur Jongepier, Himani Bhakuni, Kalle Grill, Michael Klenk, Moti Gorin, and Pei-Hua Huang for their helpful written comments and suggestions on a previous draft. For discussion and feedback, we also thank the audiences of the Manipulation Online Workshop Series and the Maastricht Law and Tech Lab.
2. Among them, the increase in the use of smart devices, the increase in computing capacity which allowed for more sophisticated forms of data collection/analysis, the growth of digital services that monetise users' data, among others. For more examples, see Zuboff (2019).
3. We assume here that the category "manipulative but justified" is not an empty one. It is worth noting, however, that this is not uncontroversial, as some philosophers may adopt a thick conception of manipulation according to which manipulation would be wrongful by definition. Given that we focus on the wrongful instances of manipulation (and on real-time profiling), nothing in our argument would change if the thick conception of manipulation turns out to be correct. For an overview of thick and thin conceptions of manipulation, see Jongepier and Klenk, in this volume.
4. For example, Sunstein (2016, 82) defines manipulation as "an effort to influence people's choices . . . to the extent that it does not sufficiently engage or appeal to their capacity for reflection and deliberation." Along the same lines, others have highlighted the fact that manipulators influence behaviour by "adjusting [the manipulee's] psychological levers" (Noggle 1996, 44).
5. The scenario is fictional, but not fictitious. See Sam Levin (2017).
6. There are different ways in which a plan can be said to be unilateral: when (i) the *design* of one's plan is underpinned by an agenda set out without the manipulee's input, consent, or awareness; when (ii) the *implementation* of one's plan is not actually accepted or would not be accepted by the target in idealised conditions; or when (iii) the *primary* motivating reason for implementing the plan is its benefit to the planner. Though many instances of manipulative influences (and real-time profiling) are unilateral in all three senses, we circumscribe our use of "unilateral" to the third sense just specified. It is this sense that helps in explaining why manipulative influences in general, and real-time profiling in particular, seem morally suspicious even when it favours the manipulee's interests or well-being. We return to this in Section 3. Thanks to Kalle Grill for pressing us on this point.
7. Note here that we are not suggesting that MoodX's influence *bypasses* users' deliberative capacity. Some accounts of manipulation do require the manipulator to either bypass or disengage the manipulee's deliberative capacity. We explain why we think this is inadequate in Section 3.

8. The kind of cases that interest us are cases like (MoodX), (Uber), and other examples we cited in Section 1.
9. From this point onwards whenever we use "wrong" and related terms we mean "*pro tanto* wrong". For short, we also suppress the qualifier "when wrong".
10. Conditions (a) and (b) are stated separately to highlight the fact that one can *have* a plan (in the sense of having a layout of the steps that will lead to a certain end) but can still decide to commit to the plan or not.
11. We use "psychological states" broadly and include phenomena that might not be strictly or purely part of someone's psychology. For example, we would include moods, feelings, preferences, motives, reasons, dispositions, beliefs, and other propositional attitudes.
12. Our point here follows Marcia Baron's general claims about what she calls the "*Mens Rea* of manipulation" (Baron 2003, 2014).
13. It could, for example, be seen as wrongfully paternalistic. See Grill (2012).
14. On why purely instrumental (and hierarchical) relations such as the ones we have been considering are (at least) *pro tanto* wrongful, see Jongepier and Wieland, in this volume.
15. Also, we do not deny the possibility that large-scale and continued imposition of psychological hijacking may lead to the materialisation of domination in the long term. For accounts that associate forms of online manipulation with relations of domination, see Gorin and Capasso, both in this volume.
16. For a helpful discussion on whether manipulation necessarily involves bypassing, see Gorin (2014).
17. A detailed and recent argument along these lines can be found in Dorst (2020).
18. European Commission 2005.
19. European Commission 2016.
20. The draft AIA defines AI as, in short, software developed with machine learning, logic- and knowledge-based, and statistical approaches, Bayesian estimation, search and optimization methods. See art 3(1), Annex 1, ibid.

7 References

Alvarez, Sandra. 2017. "Mood Tracking & Emotional Advertising: What Does the Future Hold – NMPi." Accessed August 23, 2021. https://nmpidigital.com/us/mood-tracking-emotional-advertising-future-hold/.

Baron, Marcia. 2003. "Manipulativeness." *Proceedings and Addresses of the American Philosophical Association* 77 (2): 37. doi:10.2307/3219740.

Baron, Marcia. 2014. "The Mens Rea and Moral Status of Manipulation." In Coons and Weber 2014, 98–109.

Bradshaw, Samantha, and Philip N. Howard. 2018. "Challenging Truth and Trust: A Global Inventory of Organized Social Media Manipulation." https://demtech.oii.ox.ac.uk/research/posts/industrialized-disinformation/.

Brown, Bruce. 2018. "Fitbit Mental Health Physiology Monitoring App." Accessed August 23, 2021. https://healthtechinsider.com/2018/07/13/fitbit-mental-health-physiology-monitoring-app/.

Capasso, Marianna. 2022. "Manipulation as Digital Invasion: A Neo-republican Approach." In *The Philosophy of Online Manipulation*, edited by Jongepier, F. and Klenk, M., 180–198. New York, NY: Routledge.

Charara, Sophie. 2020. "A New Fitbit Claims to Track Your Stress Levels. Can it Really do it?" *WIRED UK*, August 28. Accessed August 23, 2021. www.wired.co.uk/article/fitbit-stress-tracking-eda.

Chen, Jiahong. 2021. *Regulating Online Behavioural Advertising through Data Protection Law*. Cheltenham: Edward Elgar Publishing.
Coons, Christian, and Michael Weber, eds. 2014. *Manipulation: Theory and Practice*. Oxford: Oxford University Press.
Dorst, Kevin. 2020. "Reasonably Polarized: Why Politics is More Rational Than You Think." Accessed August 23, 2021. www.kevindorst.com/stranger_apologies/rp.
European Commission. 2005. Directive 2005/29/EC of the European Parliament and of the Council of 11 May 2005 concerning unfair business-to-consumer commercial practices in the internal market and amending Council Directive 84/450/EEC, Directives 97/7/EC, 98/27/EC and 2002/65/EC of the European Parliament and of the Council and Regulation (EC) No 2006/2004 of the European Parliament and of the Council ('Unfair Commercial Practices Directive') (Text with EEA relevance). UCPD. May 11. https://eur-lex.europa.eu/eli/dir/2005/29/oj.
European Commission. 2016. Regulation (EU) 2016/679 of the European Parliament and of the Council of 27 April 2016 on the protection of natural persons with regard to the processing of personal data and on the free movement of such data, and repealing Directive 95/46/EC (General Data Protection Regulation) [2016] OJ L119/1 ('GDPR'). GDPR. April 27.
European Commission. 2021. Proposal for a Regulation of the European Parliament and of the Council laying down harmonised rules on Artificial Intelligence (ARTIFICIAL INTELLIGENCE ACT) and amending certain Union legislative acts. https://eur-lex.europa.eu/resource.html?uri=cellar:e0649735-a372-11eb-9585-01aa75ed71a1.0001.02/DOC_1&format=PDF.
European Data Protection Board. 2020. Guidelines 05/2020 on consent under Regulation 2016/679.
Gädeke, D. 2020. "Does a Mugger Dominate? Episodic Power and the Structural Dimension of Domination." *Journal of Political Philosophy* 28: 199–221.
Gorin, Moti. 2022. "Gamification, Manipulation, and Domination." In *The Philosophy of Online Manipulation*, edited by Jongepier, F. and Klenk, M., 199–215. New York, NY: Routledge.
Gorin, Moti. 2014. "Towards a Theory of Interpersonal Manipulation." In Coons and Weber 2014, 73–97.
Grill, Kalle. 2012. "Paternalism." In *Encyclopedia of Applied Ethics*, 2nd ed, edited by R. Chadwick, 359–69. London: Elsevier.
Jongepier, Fleur, and Michael Klenk. 2022a. "Online Manipulation: Charting the Field." In *The Philosophy of Online Manipulation*, edited by Jongepier, F. and Klenk, M., 15–48. New York, NY: Routledge.
Jongepier, Fleur, and Michael Klenk, eds. 2022b. *The Philosophy of Online Manipulation*. New York, NY: Routledge.
Jongepier, Fleur, and J. W. Wieland. 2022. "Microtargeting People as a Mere Means." In *The Philosophy of Online Manipulation*, edited by Jongepier, F. and Klenk, M., 156–179. New York, NY: Routledge.
Kahneman, Daniel. 2011. *Thinking, Fast and Slow*. New York, NY: Farrar Straus Giroux.
Laux, Johann, Sandra Wachter, and Brend Mittelstadt. 2021. "Neutralizing Online Behavioural Advertising: Algorithmic Targeting with Market Power as an Unfair Commercial Practice." *Common Market Law Reviews* 58 (3): 719.
Levin, Sam. 2017. "Facebook Told Advertisers it Can Identify Teens Feeling 'Insecure' and 'Worthless'." *The Guardian*, January 5. Accessed August 23, 2021. www.theguardian.com/technology/2017/may/01/facebook-advertising-data-insecure-teens.

Lindsay, Jessica. 2019. "Does Uber Charge More if Your Battery is Lower?" *Metro.co.uk*, September 27. Accessed August 23, 2021. https://metro.co.uk/2019/09/27/uber-charge-battery-lower-10778303/.

Mahwadi, Arwa. 2018. "Uber Developing Technology that Would Tell if You're Drunk." *The Guardian*, November 6. Accessed August 23, 2021. www.theguardian.com/technology/2018/jun/11/uber-drunk-technology-new-ai-feature-patent.

Metz, Rachel. 2016. "Control Your Smartphone with Your Eyes." *MIT Technology Review*, January 7. Accessed August 23, 2021. www.technologyreview.com/2016/07/01/159012/control-your-smartphone-with-your-eyes.

Noggle, Robert. 1996. "Manipulative Actions: A Conceptual and Moral Analysis." *American Philosophical Quarterly* 33 (1): 43–55.

Rudgard, Olivia. 2018. "Gambling Firms Could Use GPS to Tempt 'Vulnerable' Customers." *The Telegraph*. Accessed August 23, 2021. www.telegraph.co.uk/news/2018/06/25/gambling-firms-could-use-gps-tempt-vulnerable-customers/.

Sax, Marijn. 2021. "Optimization of What? For-profit Health Apps as Manipulative Digital Environments." *Ethics and Information Technology* 23 (3): 345–61.

Shapiro, Tom. 2016. "How Emotion-Detection Technology Will Change Marketing." October 17. Accessed August 23, 2021. https://blog.hubspot.com/marketing/emotion-detection-technology-marketing.

Spencer, Shaun B. 2020. "The Problem of Online Manipulation." *University of Illinois Law Review* 2020 (3): 959–1006. doi:10.2139/ssrn.3341653.

Summers, Christopher A., Robert W. Smith, and Rebecca W. Reczek. 2016. "An Audience of One: Behaviorally Targeted Ads as Implied Social Labels." *Journal of Consumer Research* 43 (1): 156.

Sunstein, Cass R. 2016. *The Ethics of Influence: Government in the Age of Behavioral Science*. Cambridge: Cambridge University Press.

Susser, Daniel, Beate Roessler, and Helen Nissenbaum. 2019. "Online Manipulation: Hidden Influences in A Digital World." *Georgetown Law Technology Review* 4 (1): 1–45. Accessed February 27, 2020.

Valliappan, Nachiappan, Na Dai, Ethan Steinberg, Junfeng He, Kantwon Rogers, Venky Ramachandran, Pingmei Xu et al. 2020. "Accelerating Eye Movement Research via Accurate and Affordable Smartphone Eye Tracking." *Nature Communications* 11 (1): 4553. doi:10.1038/s41467-020-18360-5.

Vrousalis, Nicholas. 2019. "How Exploiters Dominate." *Review of Social Economy* 1.

Zuboff, Shoshana. 2019. *The Age of Surveillance Capitalism: The Fight for a Human Future at the New Frontier of Power*. New York, NY: PublicAffairs.

Index

Printed in the United States
by Baker & Taylor Publisher Services